KV-045-209

8
Modern Mathematics for Schools

Second Edition

Scottish Mathematics Group

Chambers

Blackie & Son Limited
Bishopbriggs · Glasgow G64 2NZ
5 Fitzhardinge Street · London W1H 0DL

W. & R. Chambers Limited
11 Thistle Street · Edinburgh EH2 1DG

First published 1974

Designed by James W. Murray

International Standard Book Numbers
Pupils' Book
Blackie 0 216 89428 X
Chambers 0 550 75919 0
Teachers' Book
Blackie 0 216 89429 8
Chambers 0 550 75929 8

Printed in Great Britain by
Smith & Ritchie Ltd, Edinburgh
Set in 10pt Monophoto Times Roman

Members associated with this book

W. T. Blackburn
Dundee College of Education

W. Brodie
Trinity Academy

C. Clark
Formerly of Lenzie Academy

D. Donald
Formerly of Robert Gordon's College

R. A. Finlayson
Jordanhill College School

Elizabeth K. Henderson
Westbourne School for Girls

J. L. Hodge
Madras College

J. Hunter
University of Glasgow

R. McKendrick
Langside College

W. More
Formerly of High School of Dundee

Helen C. Murdoch
Hutchesons' Girls' Grammar School

A. G. Robertson
John Neilson High School

A. G. Sillitto
Formerly of Jordanhill College of Education

A. A. Sturrock
Grove Academy

Rev. J. Taylor
St. Aloysius' College

E. B. C. Thornton
Bishop Otter College

J. A. Walker
Dollar Academy

P. Whyte
Hutchesons' Boys' Grammar School

Book 1 of the original series *Modern Mathematics for Schools* was first published in July 1965. This revised series has been produced in order to take advantage of the experience gained in the classroom with the original textbooks and to reflect the changing mathematical needs in recent years, particularly as a result of the general move towards some form of comprehensive education.

Throughout the whole series, the text and exercises have been cut or augmented wherever this was considered to be necessary, and nearly every chapter has been completely rewritten. In order to cater more adequately for the wider range of pupils now taking certificate-oriented courses, the pace has been slowed down in the earlier books in particular, and parallel sets of A and B exercises have been introduced where appropriate. The A sets are easier than the B sets, and provide straightforward but comprehensive practice; the B sets have been designed for the more able pupils, and may be taken in addition to, or instead of, the A sets. Often from Book 4 onwards a basic exercise, which should be taken by all pupils, is followed by a harder one on the same work in order to give abler pupils an extra challenge, or further practice; in such a case the numbering is, for example, Exercise 2 followed by Exercise 2B. It is hoped that this arrangement, along with the *Graph Workbook for Modern Mathematics*, will allow considerable flexibility of use, so that while all the pupils in a class may be studying the same topic, each pupil may be working examples which are appropriate to his or her aptitude and ability.

Each chapter is backed up by a summary, and by revision exercises; in addition, cumulative revision exercises have been introduced at the end of alternate books. A new feature is the series of Computer Topics from Book 4 onwards. These form an elementary introduction to computer studies, and are primarily intended to give pupils some appreciation of the applications and influence of computers in modern society.

Books 1 to 7 provide a suitable course for many modern Ordinary Level and Ordinary Grade syllabuses in mathematics, including the University of London GCE Syllabus C, the Associated Examining Board Syllabus C, the Cambridge Local Syndicate Syllabus C, and the Scottish Certificate of Education. Books 8 and 9 complete the work for the Scottish Higher Grade Syllabus, and provide a good preparation for all Advanced Level and Sixth Year Syllabuses, both new and traditional.

Related to this revised series of textbooks are the *Modern Mathematics Newsletters* (No. 4, February 1974), the *Teacher's Editions* of the textbooks, the *Graph Workbook for Modern Mathematics*, the *Three-Figure Tables for Modern Mathematics*, and the booklets of *Progress Papers for Modern Mathematics*. These new Progress Papers consist of short, quickly marked objective tests closely connected with the textbooks. There is one booklet for each textbook, containing A and B tests on each chapter, so that teachers can readily assess their pupils' attainments, and pupils can be encouraged in their progress through the course.

The separate headings of Algebra, Geometry, Arithmetic, and later Trigonometry and Calculus, have been retained in order to allow teachers to develop the course in the way they consider best. Throughout, however, ideas, material and method are integrated *within* each branch of mathematics and *across* the branches; the opportunity to do this is indeed one of the more obvious reasons for teaching this kind of mathematics in the schools – for it is *mathematics* as a whole that is presented.

Pupils are encouraged to find out facts and discover results for themselves, to observe and study the themes and patterns that pervade mathematics today. As a course based on this series of books progresses, a certain amount of equipment will be helpful, particularly in the development of geometry. The use of calculating machines, slide rules, and computers is advocated where appropriate, but these instruments are not an essential feature of the work.

While fundamental principles are emphasised, and reasonable attention is paid to the matter of structure, the width of the course should be sufficient to provide a useful experience of mathematics for those pupils who do not pursue the study of the subject beyond school level. An effort has been made throughout to arouse the interest of all pupils and at the same time to keep in mind the needs of the future mathematician.

The introduction of mathematics in the Primary School and recent changes in courses at Colleges and Universities have been taken into account. In addition, the aims, methods, and writing of these books have been influenced by national and international discussions about the purpose and content of courses in mathematics, held under the auspices of the Organisation for Economic Co-operation and Development and other organisations.

The authors wish to express their gratitude to the many teachers who have offered suggestions and criticisms concerning the original series of textbooks; they are confident that as a result of these contacts the new series will be more useful than it would otherwise have been.

Algebra

Geometry

Trigonometry

Calculus

Notation

Sets of numbers

Different countries and different authors
give different notations and definitions
for the various sets of numbers.
In this series the following are used:

E The universal set

$\emptyset$ The empty set

N The set of natural numbers $\{1, 2, 3, \ldots\}$

W The set of whole numbers $\{0, 1, 2, 3, \ldots\}$

Z The set of integers $\{\ldots, -2, -1, 0, 1, 2, \ldots\}$

Q The set of rational numbers

R The set of real numbers

The set of prime numbers $\{2, 3, 5, 7, 11, \ldots\}$

Algebra

Note to the Teacher on Chapter 1

The solution of systems of equations of the first degree in two variables was first studied in Book 4, Chapter 2, the methods of solution considered being (*a*) graphical, (*b*) elimination of one of the variables and (*c*) substitution. The restriction of our attention to systems of linear equations in two variables enabled us to translate the problem of solution into geometrical language by virtue of the correspondence between a line and its representation as a first-degree equation in x and y, so that statements about solutions became statements about lines and their intersection. This geometrical picture helps to give insight into the two important principles of elimination and substitution, which enable us to replace a system of equations by successive equivalent systems until we end up with a system whose solution set is obvious. These two principles are perfectly general. The important idea of linear combination, which is used in many branches of mathematics, justifies the elimination method of solving systems of linear equations.

Section 1 revises the methods of solving systems of linear equations in two variables, and in particular the method of elimination of one of the variables. This is followed by a short exercise to prepare the way for *Section* 2 in which the method of elimination is extended to systems of linear equations in three variables. Ample practice in solving such systems is provided by Exercise 2, which also contains several forward-looking examples in both algebra and geometry.

Since the pupil is not sufficiently advanced in three-dimensional coordinate geometry at this stage, it is not possible to set up the basic geometric-algebraic correspondence for the discussion of solution sets in *Section* 2. Nevertheless, it may be useful to the teacher if this aspect of the problem is developed here. By way of example, consider the case of the equation $x-2y+z = 6$.

Take some ordered triple which satisfies the equation, say $(1, -1, 3)$ and regard this as giving the coordinates of a point Q. Now let (x, y, z) be the coordinates of a point P and consider the two vectors

$\boldsymbol{t} = \begin{pmatrix} x-1 \\ y+1 \\ z-3 \end{pmatrix}$, represented by $\overrightarrow{\mathrm{QP}}$, and $\boldsymbol{n} = \begin{pmatrix} 1 \\ -2 \\ 1 \end{pmatrix}$; and also their scalar product $\boldsymbol{n}.\boldsymbol{t}$. The reason for the choice of $\boldsymbol{n}$ will soon appear.

Since $\boldsymbol{n} \neq \boldsymbol{0}$ the open sentence $\boldsymbol{n}.\boldsymbol{t} = 0$ is equivalent to the open sentence

$$\boldsymbol{t} = \boldsymbol{0} \textit{ or } \boldsymbol{t} \text{ is perpendicular to } \boldsymbol{n},$$

and this is equivalent to

$$\text{P is at Q } \textit{or } \overrightarrow{\text{QP}} \text{ is perpendicular to } \boldsymbol{n}.$$

But $\boldsymbol{n}.\boldsymbol{t} = 0$ is equivalent to

$$1(x-1)-2(y+1)+1(z-3) = 0$$

i.e.
$$x-2y+z = 6$$

Thus the two sets

$A = \{\text{P: P is at Q } \textit{or } \overrightarrow{\text{QP}} \text{ is perpendicular to } \boldsymbol{n}\}$

and

$B = \{\text{P}(x,y,z)\text{: } x-2y+z = 6\}$

are equal sets.

Now A is the plane through Q perpendicular to the direction of $\boldsymbol{n}$, and B is the set of points in space whose coordinates satisfy the equation $x-2y+z = 6$: these are therefore the same (see Figure 1).

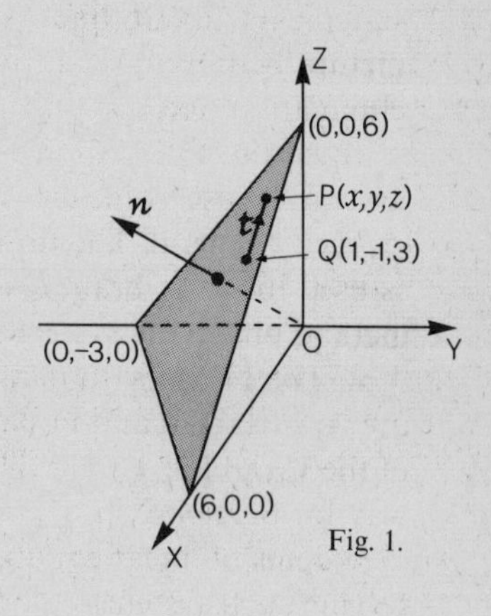

Fig. 1.

Using the usual terminology for the equation corresponding to a locus, the equation $x-2y+z = 6$ can be called the equation of the plane A.

Notes. (i) Notice the connection between the components of $\boldsymbol{n}$ and the cofficients of x, y, z in the equation of the plane. Compare with the explanation in Book 7 of the angle between two planes.

(ii) In terms of this illustration, the process used in the Worked Example of Section 2 can be thought of as finding new planes through the point common to the three given planes.

(iii) Do three planes always meet in a point? It is interesting to consider in how many different ways three planes can fail to meet in one point.

For a system of linear equations in three variables, there are four types of solutions corresponding to the following four geometrical cases of intersection:

(1) a single point—an ordered triple
(2) a line—an infinite set of ordered triples corresponding to points on the line
(3) a plane—an infinite set of ordered triples corresponding to points on the plane
(4) no common point of intersection—the empty set.

For the reason stated, work is confined to those systems in which the solution is an ordered triple of numbers. There are no examples of systems which are inconsistent (the solution set is ϕ) or dependent (the solution set is the same as one of the equations of the system).

In *Section* 3, systems of equations in two variables in which one equation is linear and the other quadratic are solved by the substitution method which is justified in conjunction with a geometrical argument. It is important to notice that when one element of an ordered pair of the solution set has been found, the second element of the ordered pair must be obtained by substitution of the first element in the *linear* equation. This eliminates the possibility of *extraneous roots* which may be met if substitution is made in the second-degree equation. To illustrate the procedure to be avoided consider the system of equations $x^2+y^2 = 25$, $y = x-1$ solved in Example 1, page 7.

$x^2+y^2 = 25$ and $y = x-1 \Rightarrow x^2+(x-1)^2 = 25 \Leftrightarrow x = -3$ or $x = 4$. Replacing x by -3 in equation $x^2+y^2 = 25$, we obtain $y = \pm 4$. Repeating for $x = 4$ gives $y = \pm 3$. We now have ordered pairs $(-3, 4)$, $(-3, -4)$, $(4, 3)$ and $(4, -3)$. Of these, only $(-3, -4)$ and $(4, 3)$ belong to the solution set since neither $(-3, 4)$ and $(4, -3)$ satisfy the linear equation $y = x-1$. The latter pair are 'extraneous'. In Exercise 3, the systems to be solved are such that the solution set contains either one or two pairs of real numbers. Systems involving the empty set, or a set whose graph is a line, will be postponed until the theory of quadratic equations has been discussed.

1 Systems of Equations in Two and Three Variables

It is not uncommon to have to deal with three (or more) equations in three (or more) variables in applied mathematics. For example, in connection with work on an electrical circuit it may be necessary to solve the system of equations:

$$\begin{aligned} 2{\cdot}1x - 0{\cdot}1y - z &= 0{\cdot}9 \\ -0{\cdot}1x + 2{\cdot}1y - z &= 3{\cdot}1 \\ -x + 1{\cdot}2y + z &= 2{\cdot}4 \end{aligned}$$

In practice, the coefficients of x, y and z usually have a larger number of significant figures than is shown above, and calculating machines or computers may be used to find the solution set of the system of equations.

Various methods of solution are available, and we shall base ours on the techniques which we used in Book 4 to solve systems of equations in two variables.

Throughout this chapter we assume that the variables are on the set of real numbers.

1 *Systems of linear equations in two variables (revision)*

Consider the system $\left.\begin{aligned} 3x+4y-7 &= 0 \\ 4x-3y-1 &= 0 \end{aligned}\right\}$ ... (1)

which has solution set $\{(1, 1)\}$.

The method of Book 4 depends on the fact that, for all p and q, the line $p(3x+4y-7)+q(4x-3y-1) = 0$ passes through the point of intersection of the graphs of the two equations in (1). This is shown in Figure 1.

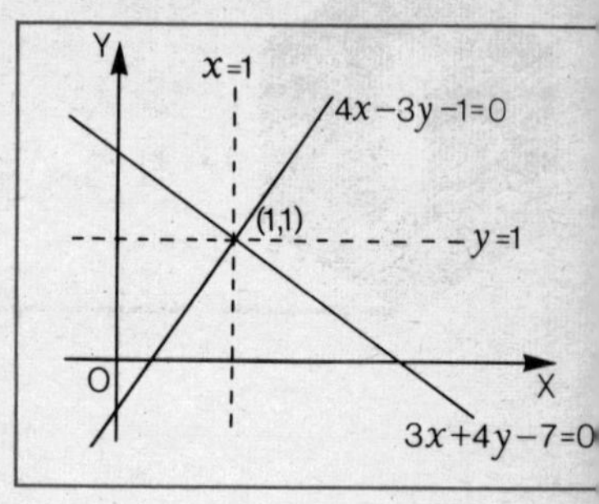

Thus for all p and q the system

$$\left.\begin{aligned} 3x+4y-7 &= 0 \\ 4x-3y-1 &= 0 \\ p(3x+4y-7)+q(4x-3y-1) &= 0 \end{aligned}\right\} \ldots(2)$$

has the same solution set as (1). Systems (1) and (2) are equivalent. By choosing suitable p and q, equations can be formed in which only the variable x or the variable y occurs, i.e. equations from which y or x has been eliminated.

Thus, $p = 3$ and $q = 4$ gives $25x-25 = 0$
$p = 4$ and $q = -3$ gives $25y-25 = 0$,
and so systems (1) and (2) are equivalent to the simpler system

$$\left.\begin{aligned} 25x-25 &= 0 \\ 25y-25 &= 0 \end{aligned}\right\}$$

which has solution set $\{(1, 1)\}$.

Notice that the left-hand side of each of the new equations is a *linear combination* of the left-hand sides of the original equations.

We can write system (1) in the form $\left.\begin{aligned} 3x+4y &= 7 \\ 4x-3y &= 1 \end{aligned}\right\}$.

and the third equation as $p(3x+4y)+q(4x-3y) = 7p+q$.

Taking $p = 3$ and $q = 4$, we obtain $25x = 25$, i.e. $x = 1$.

We now have $\left.\begin{aligned} 3x+4y &= 7 \\ x &= 1 \end{aligned}\right\}$ which has solution set $\{(1, 1)\}$.

The working may be set down conveniently as follows.

Example. Solve $\left.\begin{aligned} 3x+4y &= 7 \\ 4x-3y &= 1 \end{aligned}\right\}$

$$\begin{array}{llcr} 3x+4y = 7 & \times 3 & \Leftrightarrow & 9x+12y = 21 \\ 4x-3y = 1 & \times 4 & \Leftrightarrow & 16x-12y = 4 \\ & & \textit{Add} & 25x \quad = 25 \\ & & \Leftrightarrow & x \quad = 1 \end{array}$$

Substituting $x = 1$ in the first equation $3+4y = 7$

$$\Leftrightarrow \quad 4y = 4$$

$$\Leftrightarrow \quad y = 1$$

The solution set is $\{(1, 1)\}$.

Exercise 1

Solve the following systems of equations.

1 $\left.\begin{aligned} 2x+3y &= 2 \\ x-y &= 1 \end{aligned}\right\}$ **2** $\left.\begin{aligned} 7x+4y &= 1 \\ 5x+2y &= -1 \end{aligned}\right\}$ *3* $\left.\begin{aligned} 2x-3y &= 0 \\ 6x+6y &= 5 \end{aligned}\right\}$

4 $\left.\begin{aligned} 5x+y &= 5 \\ 17x-y &= -5 \end{aligned}\right\}$ **5** $\left.\begin{aligned} 2x+3y &= 8 \\ 3x-3y &= 12 \end{aligned}\right\}$ **6** $\left.\begin{aligned} 3x-y &= 1 \\ x-2y &= -8 \end{aligned}\right\}$

7 $\left.\begin{aligned} x+y+1 &= 0 \\ 2x-y-5 &= 0 \end{aligned}\right\}$ *8* $\left.\begin{aligned} x-y+2 &= 0 \\ 2x+y-2 &= 0 \end{aligned}\right\}$ **9** $\left.\begin{aligned} 4x-5y &= 22 \\ 7x+3y &= 15 \end{aligned}\right\}$

10 $\left.\begin{aligned} 3x-5y &= 2 \\ 7x+3y &= 12 \end{aligned}\right\}$ *11* $\left.\begin{aligned} 2x-5y &= 1 \\ 4x-3y &= 9 \end{aligned}\right\}$ *12* $\left.\begin{aligned} 11x+3y+7 &= 0 \\ 2x+5y-21 &= 0 \end{aligned}\right\}$

---

2 *Systems of linear equations in three variables*

We now use the same methods to solve systems of linear equations in three variables. We first eliminate one of the variables, and then solve the resulting system of two equations in two variables as in Section 1.

Example. Solve $\left.\begin{aligned} x-2y+z &= 6 \\ 3x+y-2z &= 4 \\ 7x-6y-z &= 10 \end{aligned}\right\}$

We decide to start by eliminating z.

$$\begin{aligned} x-2y+z &= 6 \quad \times 2 \quad \Leftrightarrow \quad 2x-4y+2z = 12 \\ 3x+y-2z &= 4 \quad \times 1 \quad \Leftrightarrow \quad 3x+y-2z = 4 \\ &\qquad \textit{Add} \quad 5x-3y = 16 \quad \ldots (1) \end{aligned}$$

Also,

$$\begin{aligned} x-2y+z &= 6 \\ 7x-6y-z &= 10 \\ \textit{Add} \quad 8x-8y &= 16 \quad \Leftrightarrow \quad x-y = 2 \quad \ldots (2) \end{aligned}$$

We now decide to eliminate y from (1) and (2).

$$\begin{array}{llll} 5x-3y=16 & \times 1 & \Leftrightarrow & 5x-3y=16 \\ x-\ y=\ 2 & \times 3 & \Leftrightarrow & 3x-3y=\ 6 \\ \hline & & \textit{Subtract} & 2x \qquad = 10 \\ & & \Leftrightarrow & x \qquad = \ 5 \end{array}$$

Substituting $x = 5$ in (2), $5-y=2 \Leftrightarrow y=3$.
Substituting $x = 5$ and $y = 3$ in the first given equation;

$$5-6+z=6$$
$$\Leftrightarrow \qquad z=7$$

The solution set is $\{(5, 3, 7)\}$.

Exercise 2

Solve the systems of equations in questions ***1*** to ***18***.

1 $\begin{array}{l} x+y+z=4 \\ 2x+2y-z=5 \\ x-y=1 \end{array}$ ***2*** $\begin{array}{l} 2x+y+z=9 \\ x+2y-z=6 \\ 3x-y+z=8 \end{array}$ ***3*** $\begin{array}{l} x+y+z=3 \\ 3x-y+2z=4 \\ x+y-z=1 \end{array}$

4 $\begin{array}{l} 3x-y+4z=8 \\ 5x+y+2z=12 \\ 2x+2y+3z=14 \end{array}$ ***5*** $\begin{array}{l} x+y+z=2 \\ 3x-y+2z=4 \\ 2x+3y+z=7 \end{array}$ ***6*** $\begin{array}{l} x+2y+3z=-1 \\ -x+y+3z=4 \\ 2x-y-z=-4 \end{array}$

7 $\begin{array}{l} 2x+y-z=9 \\ x+2y+z=6 \\ 3x-y+2z=17 \end{array}$ ***8*** $\begin{array}{l} 3u+2v-4w=10 \\ u+v+2w=3 \\ 2u+v-3w=7 \end{array}$ ***9*** $\begin{array}{l} x+y+2z=1 \\ 4x+2y+z=4 \\ 9x+3y+z=17 \end{array}$

10 $\begin{array}{l} u-2v+w=0 \\ 3u+4v+2w=17 \\ 4u-3v+2w=4 \end{array}$ ***11*** $\begin{array}{l} a+b+2c=3 \\ 4a+2b+c=13 \\ 2a+b-2c=9 \end{array}$ ***12*** $\begin{array}{l} p+q+r=-1 \\ p+2q+3r=-2 \\ 3p-2q-r=2 \end{array}$

13 $\begin{array}{l} 3x+2y-z=11 \\ x-2y+2z=-9 \\ 2x-y-3z=5 \end{array}$ ***14*** $\begin{array}{l} x-3y+7z=-16 \\ x+2y-z=-2 \\ x-8y+4z=3 \end{array}$ ***15*** $\begin{array}{l} 5u-v+2w=25 \\ 3u+2v-3w=16 \\ 2u-v+w=9 \end{array}$

16 $\begin{array}{l} a+b+2c=3 \\ 4a+2b+c=13 \\ 2a+b-2c=9 \end{array}$ ***17*** $\begin{array}{l} 4x-3y+2z=4 \\ 5x+z=2 \\ z=-3 \end{array}$ ***18*** $\begin{array}{l} 3x+2y-z=-1 \\ 2x+5y=16 \\ z=3 \end{array}$

19 The parabola $y = ax^2+bx+c$ passes through the points (1, 2), (2, 4), (3, 8). Find a, b, c, and write down the equation of the parabola.

20 Repeat question ***19*** for the points (1, 1), (−1, −5), (3, 23).

21 The circle $x^2+y^2+ax+by+c=0$ passes through the points (2, 1), (1, 2), (1, 0). Find a, b, c, and write down the equation of the circle.

22 Repeat question ***21*** for the points (5, 5), (2, 6), (7, 1).

23 The quadratic expression ax^2+bx+c has the value −1 when $x = 1$, 4 when $x = 2$, and 17 when $x = 3$. Find a, b and c. Find a simple formula for the nth term of a sequence which starts −1, 4, 17, ...

24 The quadratic expression px^2+qx+r has the value 1 when $x = 0$, 6 when $x = 1$, and 2 when $x = -1$. Find p, q, r. Write down a simple formula for the nth term of a sequence which starts 6, 17, 34, ...

25 Solve the system of equations at the beginning of the chapter:

$$\begin{aligned} 2{\cdot}1x-0{\cdot}1y-z &= 0{\cdot}9 \\ -0{\cdot}1x+2{\cdot}1y-z &= 3{\cdot}1 \\ -x+1{\cdot}2y+z &= 2{\cdot}4 \end{aligned}$$

26 Solve the system
$$\begin{aligned} 0{\cdot}5x+0{\cdot}3y+0{\cdot}2z &= 46 \\ 0{\cdot}2x-0{\cdot}5y+0{\cdot}4z &= 0 \\ 0{\cdot}1x+0{\cdot}8y-0{\cdot}6z &= 26 \end{aligned}$$

---

3 *Systems of equations in two variables, one linear and one quadratic*

Example 1. Solve the system of equations:

$$x^2+y^2 = 25 \quad \ldots (1)$$
$$y = x-1 \quad \ldots (2)$$

Using the method of substitution, (1) and (2) imply that

$$\begin{aligned} & x^2+(x-1)^2 = 25 \\ \Leftrightarrow \quad & x^2+x^2-2x+1-25 = 0 \\ \Leftrightarrow \quad & 2x^2-2x-24 = 0 \\ \Leftrightarrow \quad & x^2-x-12 = 0 \\ \Leftrightarrow \quad & (x+3)(x-4) = 0 \\ \Leftrightarrow \quad & x = -3 \quad or \quad x = 4 \end{aligned}$$

The graphs of (1) and (2) are shown in Figure 2. The points of intersection A and B correspond to the elements of the solution set of the given system of equations.

To find the coordinates of A and B, we must solve the systems:

$\left.\begin{array}{l} x = -3 \\ y = x-1 \end{array}\right\}$, obtaining $y = -4$ and solution $(-3, -4)$;

$\left.\begin{array}{l} x = 4 \\ y = x-1 \end{array}\right\}$, obtaining $y = 3$ and solution $(4, 3)$.

Hence the solution set of (1) and (2) is $\{(-3, -4), (4, 3)\}$.

Note. Can we be sure that *both* ordered pairs above do belong to the solution set? Strictly speaking we should check that each pair satisfies (1) and (2). However, the geometrical argument indicates that in fact both pairs do check. This always happens, although we shall not prove it.

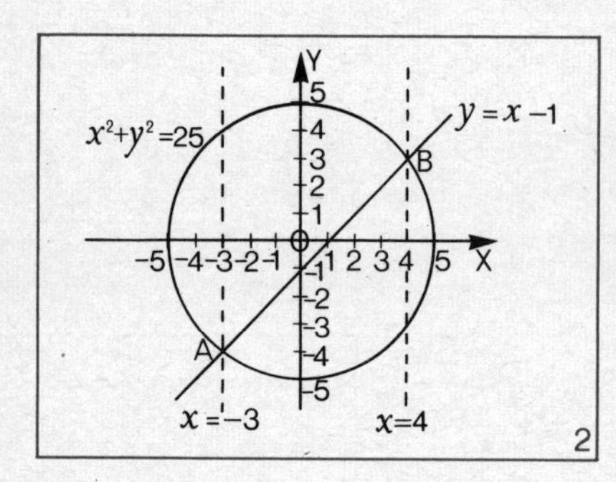

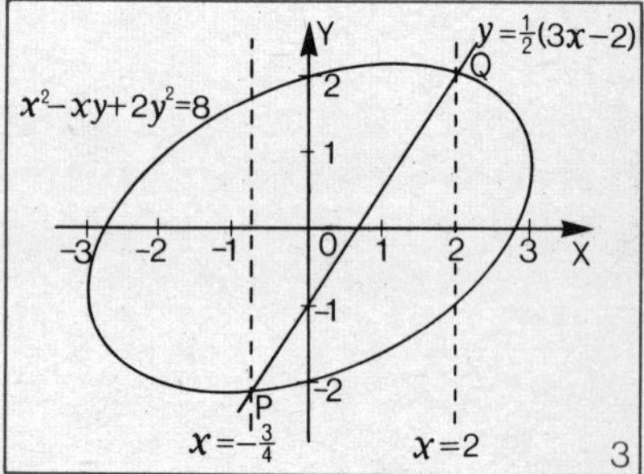

Example 2. Solve the system of equations:

$$x^2 - xy + 2y^2 = 8 \qquad \ldots (1)$$

$$3x - 2y = 2 \qquad \ldots (2)$$

Rewriting (2) in the equivalent form $y = \frac{1}{2}(3x-2)$, ... (3)

(1) and (3) imply that

$$\begin{aligned} & x^2 - \tfrac{1}{2}x(3x-2) + \tfrac{2}{4}(3x-2)^2 = 8 \\ \Leftrightarrow \quad & 2x^2 - x(3x-2) + (3x-2)^2 = 16 \\ \Leftrightarrow \quad & 2x^2 - 3x^2 + 2x + 9x^2 - 12x + 4 = 16 \\ \Leftrightarrow \quad & 8x^2 - 10x - 12 = 0 \\ \Leftrightarrow \quad & 4x^2 - 5x - 6 = 0 \\ \Leftrightarrow \quad & (4x+3)(x-2) = 0 \\ \Leftrightarrow \quad & x = -\tfrac{3}{4} \quad \text{or} \quad x = 2 \end{aligned}$$

The graphs of (1) and (2) are shown in Figure 3. The points of intersection P and Q correspond to the elements of the solution set of the given system.

Solving the systems $\left.\begin{array}{l} x = -\frac{3}{4} \\ y = \frac{1}{2}(3x-2) \end{array}\right\}$ *and* $\left.\begin{array}{l} x = 2 \\ y = \frac{1}{2}(3x-2) \end{array}\right\}$,
we obtain $y = \frac{1}{2}(-\frac{9}{4}-2) = -\frac{17}{8}$ and $y = \frac{1}{2}(6-2) = 2$ respectively. So P is the point $(-\frac{3}{4}, -2\frac{1}{8})$ and Q is the point $(2, 2)$. Hence the required solutions are $(-\frac{3}{4}, -2\frac{1}{8})$ and $(2, 2)$.

The method used in Examples 1 and 2 may be applied to some systems of two equations in two variables of which both equations are quadratic (see Exercise 3, question **27**).

Exercise 3

In questions *1* to *6*, find the solution set of each of the systems of equations and illustrate by sketch graphs.

1 $y = x^2$, $y = x+6$

2 $y = x^2$, $y = 3-2x$

3 $y = 8x-x^2$, $y = 2x$

4 $y = x-1$, $y = x^2-6x+5$

5 $x-y = 0$, $x^2+y^2 = 18$

6 $x^2+y^2 = 20$, $2y-x = 0$

Solve the following systems of equations:

7 $x^2 = 4y$, $y = 9$

8 $x^2-y^2 = 9$, $x = 5$

9 $y = x^2$, $y = 8-2x$

10 $y = x^2-2x+5$, $y = 4x$

11 $y = \dfrac{8}{x}$, $y = 7+x$

12 $xy = 4$, $y = 2x+2$

13 $x^2+y^2 = 25$, $y = x+1$

14 $x^2+4y^2 = 4$, $x = 2-2y$

15 $y^2 = 4x$, $2x+y = 4$

16 $4y^2-3x^2 = 1$, $x-2y = 1$

17 $x^2+xy+y^2 = 7$, $2x+y = 1$

18 $x^2+5x+y = 4$, $x+y = 8$

19 $x^2-xy-y^2 = -11$, $2x+y = 1$

20 $x^2+y^2+4x+6y-40 = 0$, $x-y = 10$

21 $3x^2-2y^2+5 = 0$, $3x-2y = 1$

22 $2x^2+3y^2+x = 13$, $2x+3y-7 = 0$

23 $2x^2-3xy-2y^2 = 12$
$2x-3y = 4$

24 $5x^2-5xy+2y^2 = 12$
$3x-2y+2 = 0$

25 $2x+5y-1 = 0$
$x^2+5xy-4y^2 = -10$

26 $3x+1 = 2y$
$9x^2+6xy-4y^2 = 1$

27*a* Solve the system of equations $y = 8x-x^2$ and $3y = x^2$, and illustrate the solution by a sketch Cartesian graph.

b Show in the sketch the solution set of the system of inequations $y \leqslant 8x-x^2$ and $3y \geqslant x^2$ by means of shading.

Summary

1 *Systems of linear equations in 2 and 3 variables* can be solved by a process of elimination and substitution.

Example. Solve $\left.\begin{aligned} x+2y+3z &= 0 \\ 3x+y+2z &= 2 \\ -2x+3y-4z &= 5 \end{aligned}\right\} x, y, z \in R.$

$$\begin{array}{rcccr} x+2y+3z=0 & \times 3 & \Leftrightarrow & 3x+6y+9z=0 & \\ 3x+y+2z=2 & \times 1 & \Leftrightarrow & 3x+y+2z=2 & \\ & & \textit{Subtract} & 5y+7z=-2 & \ldots(1) \end{array}$$

$$\begin{array}{rcccr} x+2y+3z=0 & \times 2 & \Leftrightarrow & 2x+4y+6z=0 \\ -2x+3y-4z=5 & \times 1 & \Leftrightarrow & -2x+3y-4z=5 \\ & & \textit{Add} & 7y+2z=5 \end{array}$$

$$\begin{array}{rcccr} 5y+7z=-2 & \times 7 & \Leftrightarrow & 35y+49z=-14 \\ 7y+2z=5 & \times 5 & \Leftrightarrow & 35y+10z=25 \\ & & \textit{Subtract} & 39z=-39 \\ & & \Leftrightarrow & z=-1 \end{array}$$

Substituting $z=-1$ in (1), we obtain $y=1$.
Substituting $z=-1$ and $y=1$ in $x+2y+3z=0$, $x=1$.
The solution set is $\{(1, 1, -1)\}$.

2 *Systems of equations in 2 variables, one of the equations being quadratic*, can be solved by the method of substitution.

Example. Solve $\left.\begin{aligned} x^2+y^2 &= 5 \\ 2x+3y &= 7 \end{aligned}\right\} x, y \in R.$

$$2x+3y=7 \quad \Leftrightarrow \quad x=\tfrac{1}{2}(7-3y)$$

Substituting in $x^2+y^2=5$, $\frac{1}{4}(7-3y)^2+y^2=5$

$$\begin{aligned} \Leftrightarrow \quad (49-42y+9y^2)+4y^2 &= 20 \\ \Leftrightarrow \quad 13y^2-42y+29 &= 0 \\ \Leftrightarrow \quad (y-1)(13y-29) &= 0 \\ \Leftrightarrow \quad y=1 \text{ or } y &= 2\tfrac{3}{13} \end{aligned}$$

Substituting in $x=\frac{1}{2}(7-3y)$, we obtain $x=2$ or $x=\frac{2}{13}$.
The solution set is $\{(2, 1), (\frac{2}{13}, 2\frac{3}{13})\}$.

2 Sequences and Series

1 Sequences

In earlier books you met many sequences of numbers, usually with instructions to suggest the next few terms.

By way of revision, suggest a possible next two terms for each of the following sequences, and describe a rule which could be used to form the sequence in each case:

(i) 1, 3, 5, …
(ii) 500, 400, 320, 256, …
(iii) 1, 1, 2, 3, 5, …
(iv) 1, 2, 6, 24, 120, …
(v) 2, 3, 5, 7, 11, 13, 17, …
(vi) 3, 9, 19, 34, 55, …

Such sequences often arise in everyday life. You might come across part of the sequence (i) when searching for a house number 18; you would guess that the house you were looking for was on the other side of the road. Sequence (ii) might give the value in £ sterling of a motor car that is year by year depreciating at 20% per annum. Sequence (vi) gives the odds against picking at random the runners who will finish in the first three in a race where there are 4, 5, 6, 7, 8, … runners.

Sequences occur frequently in mathematics. Essentially the elements or terms of a sequence are the values of some *function* u whose domain is the *set of natural numbers*. In sequence (i), the third term is 5, so the function u for this sequence has value 5 at 3, i.e. $u(3) = 5$.

For every sequence we have a mapping u from the set $N = \{1, 2, 3, \dots\}$ to the set of elements of the sequence. A common way to define this mapping is by a *formula*. For the function u associated with sequence (i), one possible formula is $u(n) = 2n-1$. This formula gives the nth term of the sequence. The formula is usually written $u_n = 2n-1$, where $n \in \{1, 2, 3, \dots\}$.

Note to the Teacher on Chapter 2

The pupils have already met sequences and series in earlier parts of the course, but the approach was quite informal and intuitive. The subject is now discussed in more precise terms.

The fundamental idea introduced in *Section* 1 is that a sequence is a mapping from the set of positive integers to the set of real numbers. If the associated function is u then the nth term is $u(n)$, more usually written u_n. The revision examples at the beginning include the most common of the Fibonacci sequences, the sequence of factorials, and the primes. The last of the examples has formula $u_n = {}^{n+3}C_3 - 1$, the difference between two terms being a triangular number; this example points towards Pascal's triangle, which contains many interesting sequences.

In Exercise 1, questions ***8*** and ***9*** make it clear that a sequence is *not* specified uniquely by giving a few terms; to define a sequence it is necessary to give a formula defining u_n for all $n \in N$.

Closely related to the topic of *sequences* is that of *series*. Thus we have the sequence 1, 3, 5, 7, and the series $1+3+5+7$ whose sum is 16.

Section 2 covers arithmetic sequences and series concisely, and places some emphasis on the formulae for the nth term of an arithmetic sequence and the sum of n terms of an arithmetic series. In the related Exercises, as in most of the Exercises in this Chapter, a sufficient number of numerical questions are given to provide practice with the various formulae; in addition, there is an element of purposeful manipulation, where appropriate, and suitable opportunities are taken to widen the scope of the topic to include related problems and investigations.

In *Section* 3, the formulae for the nth term of a geometric sequence and the sum of n terms of a geometric series are studied and applied. Manipulations of indices and surds are included in Exercises 4 and 6; Exercise 5 explores some interesting and useful practical applications of geometric sequences, and indicates their widespread nature. It may be noted that compound interest appears in Exercise 5, questions ***1*** and ***2*** (where the general formula is suggested) and in Exercise 6, questions ***17–19*** and ***20*** (in connection with life assurance). The ideas involved are not easy, but are worthy of study.

Section 4 approaches the sum to infinity of a geometric series through numerical examples and related illustrations. The two questions to be considered are: (i) Does the series have a sum to infinity? (ii) If so, what is the sum to infinity? A common misconception by the pupil is to suppose that when we say 'The sum to infinity of the infinite series $\frac{1}{2}+\frac{1}{4}+\frac{1}{8}+\ldots$ is 1' we mean that by adding an unending string of terms the result is 1. Instead, the pupil should understand that the sum of the first n terms can be made as close as we wish to 1 by taking n sufficiently large. It follows that care should be taken in introducing the S_∞ notation and the formula $S_\infty = \dfrac{a}{1-r}$. Pupils will also meet the idea of a limit in calculus, both in differentiation and integration. The concept will gradually become clearer and more meaningful as experience and maturity increase.

The related Topics to Explore on the Fibonacci sequence, hire purchase and logarithms provide useful background knowledge for pupils (see page 78).

If $u_n = 2n-1$, then $u_1 = (2\times 1)-1 = 1$
$u_2 = (2\times 2)-1 = 3$
$u_3 = (2\times 3)-1 = 5$, and so on.

Hence we obtain the sequence 1, 3, 5, ...

Figure 1 shows part of the graph of the function u for each of the sequences (i), (ii) and (iii).

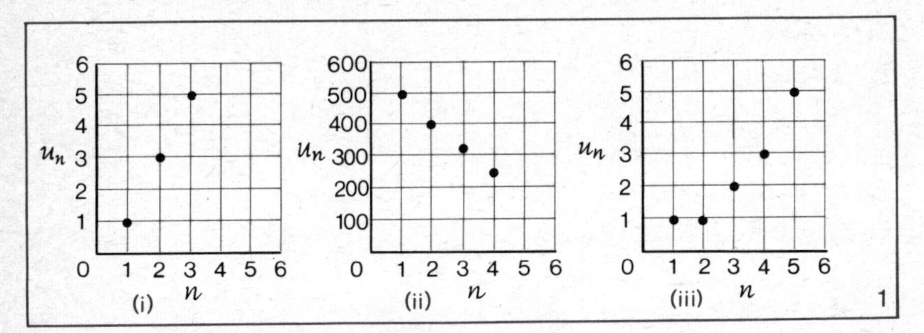

Example. Find a formula for the nth term of a sequence whose first four terms are: 1, 4, 7, 10, ... (ii) 3, 9, 27, 81, ... (iii) $-2, 2, -2, 2, \ldots$

(i) Pairs of terms differ by 3: $u_n = 3n-2$
(ii) Powers of 3: $u_n = 3^n$
(iii) $(-1)^1 = -1, (-1)^2 = 1$, etc.; $u_n = 2\times(-1)^n$.

Exercise 1

1 Find the first four terms of the sequences given by the formulae:

a $u_n = 2n$ ***b*** $u_n = n^2$ ***c*** $u_n = n-1$

d $u_n = 2n+1$ ***e*** $u_n = n^2-1$ ***f*** $u_n = n(n+1)$

g $u_n = 3^n$ ***h*** $u_n = (-1)^n$ ***i*** $u_n = \dfrac{n}{n+1}$

2 Write down two more terms in each of the following, and then find a simple formula for the nth term of each sequence:

a 1, 2, 3, 4, ... ***b*** 2, 4, 6, 8, ... ***c*** 3, 6, 9, 12, ...

d 2, 3, 4, 5, ... ***e*** 1, 4, 9, 16, ... ***f*** 1, 8, 27, 64, ...

g 2, 4, 8, 16, ... ***h*** 1, 2, 4, 8, ... ***i*** 1, 3, 5, 7, ...

3 Find a simple formula for the nth term of sequences starting:

a $\frac{1}{2}, \frac{1}{4}, \frac{1}{8}, \frac{1}{16}, \ldots$ ***b*** $1, \frac{1}{2}, \frac{1}{4}, \frac{1}{8}, \ldots$

c $1, \frac{1}{2}, \frac{1}{3}, \frac{1}{4}, \frac{1}{5}, \ldots$ ***d*** $\frac{1}{1 \times 2}, \frac{1}{2 \times 3}, \frac{1}{3 \times 4}, \ldots$

e $\frac{1}{3}, \frac{1}{6}, \frac{1}{9}, \frac{1}{12}, \ldots$ ***f*** $\frac{1}{3}, \frac{1}{9}, \frac{1}{27}, \frac{1}{81}, \ldots$

4 Find a simple formula for the nth term of each of the following:

a $-1, 1, -1, 1, \ldots$ ***b*** $1, -1, 1, -1, \ldots$

c $1, -2, 4, -8, \ldots$ ***d*** $1, -\frac{1}{3}, \frac{1}{9}, -\frac{1}{27}, \ldots$

5 ***a*** Give the first four terms of the sequence defined by $u_n = 3n+1$.
b Which term of the sequence is 196?

6 ***a*** Give the first four terms of the sequence defined by $u_n = 2n^2-1$.
b Which term of the sequence is 161?

7 ***a*** Give the first four terms of the sequence defined by $u_n = n^2+n$.
b Which term of the sequence is 132?

8 Find the first four terms of the sequences given by the formulae:

a $u_n = 2^{n-1}$ ***b*** $u_n = \frac{1}{2}(n^2-n+2)$

What do you notice?

9 ***a*** Give a simple formula for the nth term of the sequence 1, 3, 5, ...
b Find the first three terms of the sequences given by:

(*1*) $u_n = n^3-6n^2+13n-7$ (*2*) $u_n = 2n^3-12n^2+24n-13$

c Comment on the results for ***a*** and ***b***.

---

2 *Arithmetic sequences and series*

Arithmetic sequences

1, 3, 5, 7, ...; 2, 6, 10, 14, ...; 100, 90, 80, 70, ... are examples of arithmetic (pronounced arithme'tic) sequences.

$u_1, u_2, u_3, \ldots, u_n$ is an arithmetic sequence provided that:

$$u_2-u_1 = u_3-u_2 = \ldots = u_n-u_{n-1} = \text{constant.}$$

This constant is called the *common difference*, and is denoted by d.

For 1, 3, 5, 7, ... the common difference is $3-1 = 5-3 = \dots = 2$.

For 100, 90, 80, ... the common difference is $90-100 = 80-90 = \dots = -10$.

Formula for nth term

Denoting the first term u_1 by a, we have:

$$u_2-u_1 = d \quad\Rightarrow\quad u_2 = u_1+d = a+d$$
$$u_3-u_2 = d \quad\Rightarrow\quad u_3 = u_2+d = (a+d)+d = a+2d$$
$$u_4-u_3 = d \quad\Rightarrow\quad u_4 = u_3+d = (a+2d)+d = a+3d, \text{ and so on.}$$

This gives the standard arithmetic sequence

$$a, a+d, a+2d, a+3d, \dots, a+(n-1)d.$$

Notice that the nth term is $u_n = a+(n-1)d$.

Example. Find the 100th term of the arithmetic sequence 2, 5, 8, 11, ...

Here $a = 2, d = u_2-u_1 = 5-2 = 3, n = 100$

$$u_n = a+(n-1)d$$
$$\Rightarrow u_{100} = 2+(100-1)3 = 2+(99\times 3) = 299.$$

Exercise 2

1 Find the common difference in each of these arithmetic sequences:

a 2, 4, 6, 8, ... ***b*** 3, 7, 11, 15, ... ***c*** 1, 6, 11, 16, ...

d 10, 9, 8, 7, ... ***e*** 70, 53, 36, 19, ... ***f*** 1, 1·5, 2, 2·5, ...

g $\frac{3}{4}, \frac{1}{4}, -\frac{1}{4}, -\frac{3}{4}, \dots$ ***h*** 2·6, 4, 5·4, ... ***i*** 40, 20, 0, -20, ...

2 Find the first four terms of the arithmetic sequences with given term and common difference (d):

a $u_1 = 5, d = 3$ ***b*** $u_1 = 0, d = -2$ ***c*** $u_1 = 10, d = 10$

d $u_3 = 15, d = 2$ ***e*** $u_5 = 20, d = -1$ ***f*** $u_4 = 33, d = 5$

3 Find the required term in each of these arithmetic sequences:

a 2, 4, 6, 8, ...; 100th term ***b*** 1 4, 7, 10, ...; 16th term *c* 3, 5, 7, 9, ...; 20th term *d* 14, 10, 6, 2, ...; 15th term

e −5, −1, 3, 7, ...; 12th term *f* 10, 5, 0, −5, ...; 21st term

4 Find a formula for the nth term of each of these sequences:

a 1, 3, 5, 7, ... *b* 4, 7, 10, 13, ... *c* 1, 5, 9, 13, ...

d 5, 7, 9, 11, ... *e* 2, 5, 8, 11, ... *f* −5, −4, −3, −2, ...

Form a system of equations for each of the following arithmetic sequences, and hence find the first term and the common difference in each case:

5 The tenth term is 41 and the fifth term is 21.

6 The eighth term is −18 and the third term is 12.

7 The fourth term is −9 and the fifteenth term is −31.

The sum of n terms of an arithmetic series

The story is told of the great mathematician Karl Friedrich Gauss (1777–1855) that in his early days in primary school the teacher told the class to add up one hundred large numbers that were successive terms in an arithmetic sequence, hoping that this would keep them quiet for some time. Gauss gave the answer in a few seconds. Here we apply a method similar to his to find the sum of the first 100 natural numbers:

$$\begin{aligned} S_{100} &= 1+2+\dots\dots+99+100 \\ \Rightarrow \quad S_{100} &= 100+99+\dots\dots+2+1 \\ \Rightarrow \quad 2S_{100} &= 101+101+\dots\dots+101+101 = 100\times 101 \\ \Rightarrow \quad S_{100} &= 5050 \end{aligned}$$

$1+2+\dots+100$ is an example of an *arithmetic series.* The sum of this series is 5050.

We can find a formula for the sum of the standard arithmetic series

$$a+(a+d)+(a+2d)+\dots\dots+[a+(n-1)d]$$

in the same way.

Let the last term be l. Then the second last term is $l-d$, the third last term is $l-2d$, and so on. Hence

$$S_n = a+(a+d)+(a+2d)+\dots\dots+(l-2d)+(l-d)+l \quad \dots (1)$$

Reversing the order of terms,

$$S_n = l+(l-d)+(l-2d)+\dots\dots+(a+2d)+(a+d)+a \quad \dots (2)$$

Adding (1) and (2),

$$2S_n = (a+l)+(a+l)+(a+l)+\dots\dots+(a+l)+(a+l)+(a+l)$$
$$= n(a+l)$$
$$\Rightarrow \quad S_n = \tfrac{1}{2}n(a+l), \text{ i.e. } n \times (\text{average of first and last terms})$$
$$= \tfrac{1}{2}n[a+a+(n-1)d], \text{ since } l = a+(n-1)d$$
$$\text{i.e. } S_n = \tfrac{1}{2}n[2a+(n-1)d].$$

Example 1. Find the sum of the first 25 terms of the arithmetic series $44+40+36+\dots$

Here $a = 44, d = 40-44 = -4, n = 25$

$$S_n = \tfrac{1}{2}n[2a+(n-1)d]$$
$$\Rightarrow S_{25} = \tfrac{1}{2}\times 25[88+24(-4)] = \tfrac{1}{2}\times 25 \times -8 = -100.$$

Example 2. Find the sum of all the natural numbers between 1 and 100 which are divisible by 3.

Here $a = 3$, $d = 3$ and $l = 99$; we must first find n.

$$l = a+(n-1)d$$
$$\Rightarrow 99 = 3+(n-1)3$$
$$\Rightarrow \quad n = 33.$$

$$S_n = \tfrac{1}{2}n(a+l)$$
$$\Rightarrow S_{33} = \tfrac{1}{2}\times 33(3+99) = 1683$$

Exercise 3

Find the sum of each of these arithmetic series:

1 $1+3+5+\dots$ to 20 terms

2 $2+3+4+\dots$ to 40 terms

3 $80+70+60+\dots$ to 12 terms

4 $-4-5-6-\dots$ to 18 terms

5 $3{\cdot}5+3{\cdot}7+3{\cdot}9+\dots$ to 16 terms

6 $1+2+3+\dots$ to 100 terms

7 $1+2+3+\dots+40$

8 $2+4+6+\dots+100$

9 $1+3+5+\dots+21$

10 $3+8+13+\dots+98$

11 Calculate the sum of all the two-digit natural numbers which are divisible by 3.

12 Find n if: **a** $1+2+3+\dots+n = 55$ **b** $1+2+3+\dots+n = 120$.

13 How many terms of the series $5+7+9+\dots$ give a sum of 192?

14a How many terms of the series $24+20+16+\dots$ give a sum of 72? Interpret your answer.

b How many terms give a sum of 0?

15 The sixth term of an arithmetic sequence is 22 and the tenth term is 34. Find the sum of the first 16 terms of the corresponding series.

16 For a certain arithmetic series, $S_n = \frac{1}{2}n(11-n)$. Calculate S_1, S_2, S_3 and S_4. Hence find the first four terms of the corresponding sequence, and a formula for the nth term.

17 Repeat question ***16*** for the series with $S_n = \frac{1}{2}n(3n-17)$.

18a Show that $1+2+3+\dots+n = \frac{1}{2}n(n+1)$.

b The ancient Greeks were very interested in number patterns. The first five triangular numbers are 1, 3, 6, 10, 15. What are the next three numbers of the sequence? Is it an arithmetic sequence?

c Use the result in ***a*** to find the twentieth term of the sequence.

19 Two men start work at a salary of £1600 per annum. One is to have an increase of £100 each year, the other an increase of £230 every two years. Write down for each a series to show the total earnings in the first ten years, and sum these series.

20 Draw two bar graphs superimposed on the same horizontal time scale to show the individual terms of the two series in question ***19***. Does this suggest anything about the relative merits of the two schemes?

---

3 Geometric sequences and series

Geometric sequences

$1, 2, 4, 8, \dots;\ 27, -9, 3, -1, \dots;\ -1, 1, -1, 1, \dots$ are examples of geometric sequences.

$u_1, u_2, u_3, \dots, u_n$ is a geometric sequence provided that:

$$\frac{u_2}{u_1} = \frac{u_3}{u_2} = \dots = \frac{u_n}{u_{n-1}} = \text{constant}.$$

This constant is called the *common ratio*, and is denoted by r.

For 1, 2, 4, 8, ... the common ratio is $\frac{2}{1} = \frac{4}{2} = \frac{8}{4} = \ldots = 2$.
For 27, -9, 3, ... the common ratio is $\frac{-9}{27} = \frac{3}{-9} = \ldots = -\frac{1}{3}$.

Formula for nth term

Denoting the first term u_1 by a, we have:

$$\frac{u_2}{u_1} = r \quad \Rightarrow \quad u_2 = u_1 r = ar$$

$$\frac{u_3}{u_2} = r \quad \Rightarrow \quad u_3 = u_2 r = ar^2$$

$$\frac{u_4}{u_3} = r \quad \Rightarrow \quad u_4 = u_3 r = ar^3, \text{ and so on.}$$

This gives the standard geometric sequence

$$a, ar, ar^2, ar^3, \ldots, ar^{n-1}$$

Notice that the nth term is $u_n = ar^{n-1}$.

Example. In a geometric sequence, $u_1 = 64$ and $u_4 = 1$. Find r, and state the first five terms of the sequence.

Here $a = 64$, and $u_n = ar^{n-1}$

$$\Rightarrow u_4 = 64r^3 = 1$$

$$\Rightarrow \quad r^3 = \tfrac{1}{64}$$

$$\Rightarrow \quad r = \tfrac{1}{4}$$

The first five terms are 64, 16, 4, 1, $\frac{1}{4}$.

Exercise 4

1 Find the common ratio for each of these geometric sequences:

a 1, 3, 9, 27, ... *b* 12, 6, 3, $1\frac{1}{2}$, ... *c* 1, -2, 4, -8, ...

d 18, 54, 162, ... *e* $2\frac{1}{4}$, $1\frac{1}{2}$, 1, ... *f* 7, 0·7, 0·07, ...

g 1, -1, 1, -1, ... *h* $\frac{1}{4}, \frac{1}{8}, \frac{1}{16}, \ldots$ *i* $\sqrt{2}, \sqrt{6}, 3\sqrt{2}, \ldots$

2 Write down the first four terms of the geometric sequences defined by:

a $u_n = 3^{n-1}$ *b* $u_n = 3(-2)^{n-1}$ *c* $u_n = 6(-\frac{1}{2})^{n-1}$

3 Find the required terms in the following geometric sequences. (Do not simplify ***c*** and ***d***):

a 1, 2, 4, ...; u_5 ***b*** 2, 6, 18, ...; u_6
c 1, 1·2, 1·44, ...; u_8 ***d*** 100, −110, 121, ...; u_{21}

4 Write down a formula for the *n*th term of geometric sequences beginning:

a 1, 2, 4, ... ***b*** $\frac{1}{2}, \frac{1}{4}, \frac{1}{8}, \ldots$ ***c*** 2, −6, 18, ...
d 3, 6, 12, ... ***e*** 4, 2, 1, ... ***f*** 9, 3, 1, ...

Find the common ratio (r) and the fifth term of sequences in which:

5 $a = 6, u_3 = 24$ *6* $a = 50, u_4 = 400$ 7 $a = 36, u_2 = -12$

8 The legendary wise man, as a reward for solving some problem, was offered whatever he chose to name. His request, which the King thought very modest, was 'One grain of rice on the first square of a chessboard, two on the second, four on the third, and so on'. Calculate the number of grains of rice needed for the last square.

9 A piece of newspaper 0·005 cm thick is torn in two and the pieces placed on top of one another. These pieces are then torn in two and the four pieces placed on top of one another.

a Write down the first five terms of the sequence which gives the height of the pile of paper in hundredths of a centimetre after 1, 2, 3, ... tears.
b Calculate the height after 8 tears. Is a ninth tear practicable?
c Supposing the process could be continued, guess the height in kilometres after 50 tears, and check by calculation with logarithms.

Exercise 5 (some practical applications)

Compound interest
If you buy government stock, or debentures, you will be paid interest at regular intervals, but your capital value of stock will not change (apart from variations in market value). If you buy savings certificates, however, the interest is added to the capital, and the interest for the next year is calculated for the *new* capital. This is *compound interest*. It is the usual form of interest for financial calculations.

1 Suppose your capital is 100 units at the beginning of the year and the rate of interest is 5% per annum.

a Give the interest on the first year and the capital at the beginning of the second year.

b Show that the capital at the end of the second year can be expressed in the form $100(1{\cdot}05)^2$.

c Write down expressions for the capital at the end of the third, fourth and nth years.

2 Repeat question *1* for a rate of interest of r% per annum.

Growth

3 Suppose that a plant increases its height by 5% each month. If it is one metre high to begin with, use logarithms or slide rule to calculate its height after 0, 1, 2, 3, 4, 5 months, and exhibit the results in the form of a geometric sequence. Note that sometimes, as here, it is convenient to use u_0 to denote the first element of a sequence. Roughly how high would the plant be after 10 months? Show the sequence on a graph.

Note. In this sequence $u_n = (1{\cdot}05)u_{n-1}$. 1·05 is called the *growth factor* for the sequence.

Decay

4 A neutron can transform spontaneously into a proton and an electron, and does this in such a way that if we have a number of neutrons, about 5% of them will have changed by the end of one minute.

Use logarithms or slide rule to calculate how many neutrons will be left at the end of 0, 1, 2, 3, 4 minutes if we start with 1 000 000. Display these numbers as a sequence, and also on a graph.

Note. In this sequence $u_n = (0{\cdot}95)u_{n-1}$. 0·95 is called the *decay factor* for the sequence. The neutrons are in fact said to 'decay'.

Depreciation

5 What is the decay factor for the value of a car which depreciates by 20% per annum? Give the value at the end of 0, 1, 2, 3, 4 years for a car that costs £600: **a** as a sequence **b** on a graph.

6 Each year the value of a certain machine depreciates by 15% of its value at the beginning of the year. Show that after n years the value of a machine which cost £1000 is $£1000 \times (0{\cdot}85)^n$, and estimate its value after four years.

The sum of n terms of a geometric series

We can find a formula for the sum of the standard geometric series

$$a+ar+ar^2+\ldots\ldots+ar^{n-1}$$

as follows.

$$\begin{aligned} S_n &= a+ar+ar^2+\ldots\ldots+ar^{n-1} \\ \Rightarrow \quad rS_n &= \qquad ar+ar^2+\ldots\ldots+ar^{n-1}+ar^n \\ \textit{Subtract.}\ (1-r)S_n &= a-ar^n \\ &= a(1-r^n) \\ \Rightarrow \quad S_n &= \frac{a(1-r^n)}{1-r},\ r \neq 1 \\ &= \frac{a(r^n-1)}{r-1},\ \text{which is useful when } r > 1. \end{aligned}$$

Example 1. Find the sum of seven terms of the geometric series $4+2+1+0{\cdot}5+\ldots$

Here $a = 4$ and $r = \frac{2}{4} = \frac{1}{2}$.

$S_n = \dfrac{a(1-r^n)}{1-r} \Rightarrow S_7 = \dfrac{4(1-(\frac{1}{2})^7)}{1-\frac{1}{2}} = 8\left(1-\dfrac{1}{2^7}\right) = 7{\cdot}94$, to 3 significant figures.

Example 2. £1000 is invested at the beginning of each of four successive years at 8% per annum compound interest. Calculate the total amount due at the end of the four-year period.

The 'growth factor' is 1·08, so the first £1000 grows to $£1000 \times 1{\cdot}08^4$, the second to $£1000 \times 1{\cdot}08^3$, the third to $£1000 \times 1{\cdot}08^2$, the fourth to $£1000 \times 1{\cdot}08$.

$$\begin{aligned} \text{Amount} &= £1000(1{\cdot}08+1{\cdot}08^2+1{\cdot}08^3+1{\cdot}08^4) \\ &= £1000 \times 1{\cdot}08(1+1{\cdot}08+1{\cdot}08^2+1{\cdot}08^3) \\ &= £1080 \times \frac{1{\cdot}08^4-1}{1{\cdot}08-1} \\ &= £1080 \times \frac{0{\cdot}36}{0{\cdot}08} = £4860 \end{aligned}$$

nos	*logs*
$1{\cdot}08^4$	0·033
	4
1·36	0·132

Exercise 6

Use the formula to find the sum of each of the following geometric series, and simplify the answers as far as possible:

1 $1+2+4+\dots$ to 8 terms

2 $2+6+18+\dots$ to 6 terms

3 $1+\frac{1}{2}+\frac{1}{4}+\dots$ to 6 terms

4 $1+\frac{1}{3}+\frac{1}{9}+\dots$ to 5 terms

5 $2-4+8-\dots$ to 5 terms

6 $2-6+18-\dots$ to 5 terms

7 $1+x+x^2+\dots$ to n terms

8 $1-y+y^2-\dots$ to n terms

9 Show that $1+\sqrt{2}+2+2\sqrt{2}+\dots$ to 12 terms $=63(\sqrt{2}+1)$

10 Investigate the sum of n terms of:

a $1+1+1+1+\dots$ *b* $1-1+1-1+\dots$

11 Find n if:

a $3+3^2+3^3+\dots+3^n=120$ *b* $2+2^2+2^3+\dots+2^n=510$

12 After hitting the ground a ball bounces to a height of 3 m, the next time to a height of 1·5 m, then 0·75 m, and so on. How far has it travelled in its first six flights? (A sketch showing the first few bounces may help.)

13 Repeat question ***12*** for a first bounce of 30 cm, a second of 10 cm, a third of $3\frac{1}{3}$ cm, and so on.

14 Show that $\sqrt{5}$, $5+\sqrt{5}$ and $10+6\sqrt{5}$ are consecutive terms of a geometric series with common ratio $1+\sqrt{5}$. If these are the first three terms, find an expression for the sum of n terms.

15 Express the formula for the sum of n terms of $1+2+4+\dots$ in its simplest form. Hence find the smallest value of n such that $S_n > 60$.

16 Find the sum of the series $1+1{\cdot}1+1{\cdot}1^2+\dots+1{\cdot}1^{10}$.

Compound interest with annual investments

17 £100 is invested at the beginning of each of five successive years at 4% per annum compound interest. Find the total amount after five years.

18 £1000 is invested at the beginning of each of eight successive years at 7% per annum compound interest. Find the total amount after eight years.

19 In a Savings Scheme you invest £20 per *month* for a period of five years. Find the total sum paid in, and the amount due at the end of five years if interest is paid at the rate of 1% per month, compound.

Life insurance

20 If you pay £100 every year to an Assurance Company, and they let it grow at 5% compound interest, how much should you receive on your 60th birthday? Evidently this depends on when you start paying and also other factors connected with taxation and with the fact that you are also *insuring* your life. Ignoring the other factors and supposing you pay the first £100 on your 20th birthday, then this instalment will have 40 years to grow, and so grows to £100 $(1{\cdot}05)^{40}$. The next instalment grows to £100 $(1{\cdot}05)^{39}$, and the last to £100 (1·05). How much should you receive at age 60?

4 *The sum to infinity of a geometric series*

(i) For the geometric series $\frac{1}{2}+\frac{1}{4}+\frac{1}{8}+\ldots+\frac{1}{2^n}$, in which the common ratio is $\frac{1}{2}$,

$$S_n = \frac{\frac{1}{2}[1-(\frac{1}{2})^n]}{1-\frac{1}{2}} = 1-\frac{1}{2^n}.$$

By taking *n sufficiently large*, S_n can be made *as close as we wish* to 1 but no n exists such that $S_n = 1$.

For example, if we wish S_n to be within 0·01 of 1, then

$$\frac{1}{2^n} < 0{\cdot}01 \Leftrightarrow \frac{1}{2^n} < \frac{1}{100} \Leftrightarrow 2^n > 100 \Leftrightarrow n \geqslant 7.$$

Similarly, for S_n to be within 0·000 001 of 1, $n \geqslant 20$.

We say that S_n tends to 1 as n tends to infinity, and we write

$$S_n \to 1 \text{ as } n \to \infty, \quad \text{or} \quad \lim_{n\to\infty} S_n = 1,$$

where 'lim' is an abbreviation for the word 'limit'.

The sequence of sums S_1, S_2, S_3, ... is illustrated in Figure 2(i), which was constructed from the following table:

n	1	2	3	4	5	6	7
S_n	0·500	0·750	0·875	0·938	0·969	0·984	0·992

It is clear that the points get closer and closer to the line where $S_n = 1$.

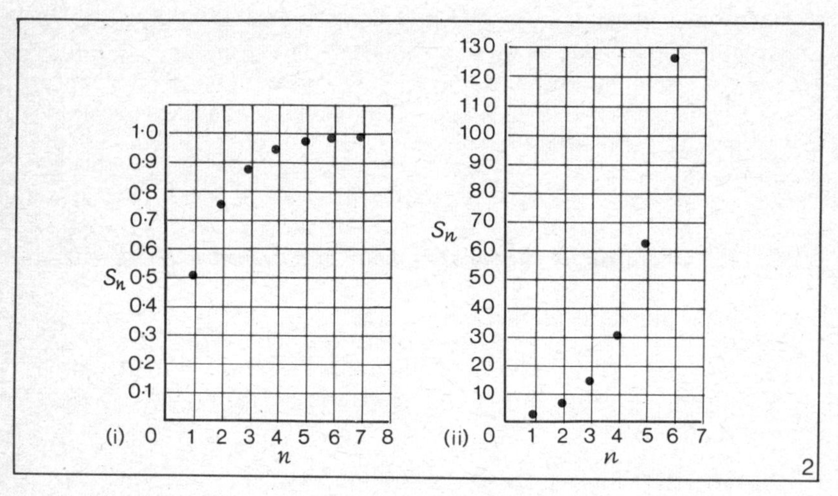

(ii) For the geometric series $2+4+8+\ldots+2^n$, in which the common ratio is 2,

$$S_n = \frac{2(2^n-1)}{2-1} = 2(2^n-1).$$

Clearly S_n can be made as large as we wish, i.e. it increases without limit, as n increases indefinitely. This is illustrated by the sequence of sums $S_1, S_2, S_3, \ldots$ in Figure 2(ii) which was constructed from the following table.

n	1	2	3	4	5	6
S_n	2	6	14	30	62	126

For $n \geqslant 19$, $S_n > 1\,000\,000$.

Sum to infinity

When the sequence of sums $S_1, S_2, S_3, \ldots$ does have a limit, S_n is as close as we wish to this limit for sufficiently large n. The limit is called the *sum to infinity* of the series and a special notation is used. Thus for Example 1 we write:

$$\tfrac{1}{2}+\tfrac{1}{4}+\tfrac{1}{8}+\ldots = 1$$

Notice that this does *not* say that if the series were to be continued indefinitely (as indicated by the three dots) then its sum would be 1. This is not true, since after n terms it would still differ from 1 by

$\frac{1}{2^n}$. What it does say is that the left side can be made *as close as we wish* to 1 by taking a sufficiently large number of terms. In other words, the left side as written means *the limit* as n tends to infinity *of the sum of n terms of the series.* The conventional way of saying this briefly is 'the sum to infinity of the series is 1'. This is sometimes expressed in the notation:

$$S_\infty = 1.$$

Formula for the sum to infinity of a geometric series

For the standard geometric series, $S_n = \dfrac{a(1-r^n)}{1-r}$.

If $-1 < r < 1$, r^n can be made as close as we wish to zero by taking n sufficiently large, i.e. r^n tends to zero as n tends to infinity, or $\lim_{n\to\infty} r^n = 0$.

Therefore for $-1 < r < 1$, $a+ar+ar^2+\ldots$

$$= \frac{a(1-0)}{1-r} = \frac{a}{1-r}, \quad \text{i.e.} \quad S_\infty = \frac{a}{1-r}.$$

If $r \geqslant 1$, or $r \leqslant -1$, the sum does not have a limit, i.e. there is no 'sum to infinity'. See the Worked Example on page 25.

Example. A geometric series is defined by $u_n = \dfrac{1}{3^n}$. Find S_n and the smallest value of n for which the sum of n terms and the sum to infinity differ by less than $\frac{1}{100}$.

$u_1 = \frac{1}{3}$ and $u_2 = \frac{1}{9}$, so $a = \frac{1}{3}$ and $r = \dfrac{u_2}{u_1} = \dfrac{\frac{1}{9}}{\frac{1}{3}} = \dfrac{1}{3}$.

$$S_n = \frac{a(1-r^n)}{1-r} = \frac{\frac{1}{3}(1-(\frac{1}{3})^n)}{1-\frac{1}{3}} = \tfrac{1}{2}\left(1-\frac{1}{3^n}\right)$$

$$S_\infty = \frac{a}{1-r} = \frac{\frac{1}{3}}{1-\frac{1}{3}} = \tfrac{1}{2}$$

$$S_\infty - S_n = \tfrac{1}{2} - \tfrac{1}{2}\left(1-\frac{1}{3^n}\right) = \frac{1}{2\times 3^n}$$

$\dfrac{1}{2\times 3^n} < \dfrac{1}{100} \quad \Leftrightarrow \quad \dfrac{1}{3^n} < \dfrac{1}{50} \quad \Leftrightarrow \quad 3^n > 50 \quad \Rightarrow \quad n > 3$, so the smallest value of n is 4.

Exercise 7

Find the common ratio for each of the following geometric series, and hence state whether a sum to infinity exists; find this sum where it exists.

1 $1+\frac{1}{3}+\frac{1}{9}+\ldots$ ***2*** $1+2+4+\ldots$ ***3*** $4+1+\frac{1}{4}+\ldots$

4 $8+4+2+\ldots$ ***5*** $1-\frac{1}{2}+\frac{1}{4}-\ldots$ ***6*** $9{\cdot}6+7{\cdot}2+5{\cdot}4+\ldots$

7 $0{\cdot}1+0{\cdot}05+0{\cdot}025+\ldots$ ***8*** $1-5+25-\ldots$ ***9*** $2+\frac{4}{3}+\frac{8}{9}+\ldots$

10 $1+1+1+\ldots$ ***11*** $1-1+1-\ldots$ ***12*** $10-9+8{\cdot}1-\ldots$

Find the sum to infinity of each of these geometric series:

13 $\frac{3}{10}+\frac{3}{100}+\frac{3}{1000}+\ldots$ ***14*** $\frac{9}{100}+\frac{9}{10\,000}+\frac{9}{1\,000\,000}+\ldots$

15a Write down S_1, S_2, S_3, S_4 and S_5 for the geometric series $5+2{\cdot}5+1{\cdot}25+\ldots$.

b Estimate, and then calculate, the sum to infinity of the series.

16 The first term of a geometric series is 2, and the sum to infinity is 4. Find the common ratio.

17 The common ratio of a geometric series is $-\frac{2}{3}$, and the sum to infinity is -12. Find the first term.

18 Is there an infinite geometric series with first term 6 and sum to infinity $\frac{2}{3}$? Give a reason for your answer.

19 The first term of a geometric series is 100, and the third term is 1. Find the sum to infinity of each of the two possible series.

20 The nth term of a geometric series is $\dfrac{1}{4^n}$. Find the first and second terms, the common ratio and the sum to infinity.

21 Repeat question ***20*** for the geometric series whose nth term is $\dfrac{1}{2^{2n-1}}$.

22a Show that $5x^{4/3}$, $10x^{-1/3}$ and $20x^{-2}$ could be the first three terms of a geometric series.

b If $x = 8$, show that the sum to infinity exists, and find it.

23 A ball rebounds from the ground to a height of 72 cm, and each time it rebounds to two thirds of the height from which it falls. Estimate the total distance the ball travels before it comes to rest. Does it ever come to rest?

24 Repeat question **23** for an initial bounce of 2 metres and successive bounces each 0·6 times the height of the preceding one.

25 Find the smallest value of n for which the sum of n terms and the sum to infinity of the series $1+\frac{1}{5}+\frac{1}{25}+\ldots$ differ by less than $\frac{1}{1000}$.

26 Find the smallest value of n for which the sum of n terms and the sum to infinity of the series $\frac{1}{4}+\frac{1}{8}+\frac{1}{16}+\ldots$ differ by less than 0·01.

27 Achilles is overtaking the tortoise at a metre a second. One second before he would catch up he reaches a point one metre behind the tortoise, which is then at a point A. One hundredth of a second before he catches up he reaches A but the tortoise has then moved on to B. 10^{-4} second before he catches up he reaches B but the tortoise is at C by that time. When he reaches C, the tortoise is at D. 'And so he never catches the tortoise.' Discuss.

Summary

1 *Sequences.* The terms of a sequence may be defined by a *formula.*

e.g. $u_n = 2n-1$ gives the sequence 1, 3, 5, ...

2 In an *arithmetic sequence*, $u_2 - u_1 = u_3 - u_2 = \ldots = u_n - u_{n-1}$; that is, it has a *common difference.*

The nth term of the arithmetic sequence

$$a, a+d, a+2d, \ldots \text{ is } u_n = a+(n-1)d$$

3 *Arithmetic series.*
The standard series is $a+(a+d)+(a+2d)+\ldots+[a+(n-1)d]$.

The sum of n terms, $S_n = \frac{1}{2}n[2a+(n-1)d] = \frac{1}{2}n(a+l)$, where l is the last term.

4 In a *geometric sequence,* $\dfrac{u_2}{u_1} = \dfrac{u_3}{u_2} = \ldots = \dfrac{u_n}{u_{n-1}}$;

that is, it has a *common ratio.*

The nth term of the geometric sequence

$$a, ar, ar^2, \ldots \text{ is } u_n = ar^{n-1}.$$

5 *Geometric series.*
The standard series is $a+ar+ar^2+\ldots+ar^{n-1}$.

The sum of n terms, $S_n = \dfrac{a(1-r^n)}{1-r}$, or $\dfrac{a(r^n-1)}{r-1}$ $(r \neq 1)$.

The sum to infinity, $S_\infty = \dfrac{a}{1-r}$, provided that $-1 < r < 1$.

3 Matrices 2

1 Description of a matrix

A *matrix* (plural: matrices) is a rectangular array of numbers arranged in rows and columns, the array being enclosed in round (or square) brackets.

Example 1. Specimen matrices:

(i) $\begin{pmatrix} x \\ y \end{pmatrix}$ 2 rows, 1 column

(ii) $\begin{pmatrix} 3 & 1 \\ 0 & 5 \end{pmatrix}$ 2 rows, 2 columns

(iii) $\begin{pmatrix} 6 & 8 & 10 \\ 3 & 4 & 5 \end{pmatrix}$ 2 rows, 3 columns

(iv) $\begin{pmatrix} 4 & -2 & 5 \end{pmatrix}$ 1 row, 3 columns

Each number in the array is called an entry or an element of the matrix and is identified by first stating the row and then the column in which it appears. In Example 1(iii), 8 is the entry in the first row and second column.

Information is often presented in matrix form in everyday life and in mathematics and science.

Example 2. Bus fares, in pence.

	Number of stages		
	1–3	4–6	over 6
Adult	6	8	10
Juvenile	3	4	5

This information may be conveyed in a concise form by the matrix

$$\begin{pmatrix} 6 & 8 & 10 \\ 3 & 4 & 5 \end{pmatrix}.$$

Example 3. A mapping $(x, y) \rightarrow (x', y')$ is defined by the equations

$$\left.\begin{aligned} x' &= 3x + y \\ y' &= \quad 3y \end{aligned}\right\} \quad \text{i.e.} \quad \begin{aligned} x' &= 3x + 1y \\ y' &= 0x + 3y. \end{aligned}$$

Note to the Teacher on Chapter 3

An introduction to matrices was given in Book 5, Algebra, Chapter 2. The present chapter revises and extends the ideas discussed in the earlier book so that the pupil will be able to sense the development of an algebra of matrices. For this reason emphasis is placed on square matrices, and their properties under addition and multiplication. The vector space structure is brought out by showing that addition is associative and commutative, there is a zero element, and every matrix has a negative; that scalar multiplication for matrices obeys two distributive laws and an associative law $r(sA) = (rs)A$, and that there is a unity element for scalar multiplication. Furthermore, together with matrix addition and multiplication, the system of 2×2 matrices has the structure of a ring. These properties are made explicit in the exercises; nevertheless, the treatment is kept informal.

Section 1 revises the definition of a matrix, and presents examples of information which can be set out in matrix form. The important subscript notation for describing elements of a matrix is introduced in *Section* 2 to give pupils practice in identifying elements, rows and columns. The a_{ij} notation is indispensible in Sixth Form work in proving theorems. The transpose of a matrix is also defined and used in simple cases throughout the chapter.

Now that the pupils have some appreciation of what a matrix is, the operations of addition and subtraction are discussed and developed in *Sections* 3 and 4. It is important to stress that for addition and subtraction matrices must have the same order. A comparison with the behaviour of real numbers under addition and subtraction is worth while at this stage. Teachers who wish to introduce the group concept could well begin here. It is also worth stressing that for the set of matrices of the same order, subtraction is always possible since every matrix in the set has an additive inverse. When solving a matrix equation such as $X + A = B$, it is recommended that expressions such as ‘transpose’ or ‘take over to the other side’ be avoided since they lessen understanding of the process. It is preferable to say ‘adding the inverse (or negative) of A to each side’.

Section 5 is concerned with the multiplication of a matrix by a real number. To avoid possible confusion between this kind of multiplication and multiplication of matrices, the term *scalar multi-*

plication is used for this operation. The class should appreciate that the product of a matrix and a real number is a matrix of the same order as the original matrix. The fundamental properties of scalar multiplication are investigated in Exercise 5, and are summarised in colour for ease of reference immediately after question *4*.

The rule for multiplication of matrices may seem strange at first, and to justify the 'row into column' process, an appeal is made to linear transformations. The method is consolidated by carefully graded questions which proceed step by step from the multiplication of an $m \times p$ matrix by a $p \times 1$ matrix to an $m \times p$ matrix by a $p \times n$ matrix. In this Section, attention is given to the operational aspect rather than the applied aspect so that the pupil is not concerned with too many ideas at once. **In Book 9, matrices will be used extensively in the study of transformations and their composition.** Meanwhile, the pupil will discover that matrix multiplication is not in general commutative, but is associative and distributive with respect to matrix addition. Furthermore, the product $AB = \boldsymbol{O}$ does not necessarily mean that $A = \boldsymbol{O}$ or $B = \boldsymbol{O}$, and for square matrices there is a unity element I such that $AI = IA = A$, which has an important part to play in the next two Sections of the chapter.

Since multiplication of matrices has some of the properties of multiplication of real numbers, it is natural to ask the question: 'Given a square matrix A, does there exist a matrix B such that $AB = BA = I$?' The investigation of this question for 2×2 matrices is the subject of *Sections* 7 and 8. The investigation begins by exploring the possibility of finding inverses for (i) matrices whose determinants are 1, and (ii) matrices whose determinants are not 1. This investigation leads to the consideration of singular matrices for which no inverse exists. The following theorem arises from this study:

$$A \text{ is a non-singular matrix} \quad \Leftrightarrow \quad \det A \neq 0.$$

The pupil should carefully memorise the 'cross-multiplication' rule for finding det A. At this stage, it may be instructive to discuss the similarities between real numbers and matrices, illustrated in the table opposite.

It is equally valuable to discuss some of the differences that exist, such as:

(i) Multiplication of real numbers is commutative, whereas matrix multiplication is not.

(ii) The product of two non-zero real numbers is not zero but the product of two non-zero matrices may be the zero matrix,

e.g. $$A = \begin{pmatrix} 2 & 1 \\ 2 & 1 \end{pmatrix},\ B = \begin{pmatrix} -1 & 3 \\ 2 & -6 \end{pmatrix} \Rightarrow AB = \boldsymbol{O}$$

but $A \neq \boldsymbol{O}$ and $B \neq \boldsymbol{O}$.

(iii) For real numbers, $ab = ac$ with $a \neq 0$ implies $b = c$; but for matrices, $AB = AC$ with $A \neq \boldsymbol{O}$ does *not* imply $B = C$,

e.g. $$A = (2 \quad 3),\ B = \begin{pmatrix} -1 & 3 \\ 6 & 0 \end{pmatrix},\quad C = \begin{pmatrix} 8 & 0 \\ 0 & 2 \end{pmatrix}$$

give $AB = AC = (16 \quad 6)$, but $B \neq C$.

Operation	*Real Numbers*	*Matrices*	*Remarks*
Addition	$a+b = b+a$	$A+B = B+A$	Addition is commutative.
	$(a+b)+c = a+(b+c)$	$(A+B)+C = A+(B+C)$	Addition is associative.
	$a+0 = 0+a = a$	$A+\boldsymbol{O} = \boldsymbol{O}+A = A$	Identity for addition.
	$a+(-a) = (-a)+a = 0$	$A+(-A) = (-A)+A = \boldsymbol{O}$	Inverse elements for addition.
Multiplication	$(ab)c = a(bc)$	$(AB)C = A(BC)$	Multiplication is associative.
	$a \times 1 = 1 \times a = a$	$AI = IA = A$	Identity for multiplication.
	$a(b+c) = ab+ac$	$A(B+C) = AB+AC$	Multiplication is distributive over addition.
	$aa^{-1} = a^{-1}a = 1$	$AA^{-1} = A^{-1}A = I$	Inverse elements for multiplication.
	$a \times 0 = 0 \times a = 0$	$A\boldsymbol{O} = \boldsymbol{O}A = \boldsymbol{O}$	Property of zero in multiplication.

Near the end of Exercise 8, a basic method of finding the inverse of a matrix is suggested. This second method is instructive for the pupil at this stage.

Section 9 completes the chapter by using matrices and their inverses to solve systems of linear equations and also equations of the type $AX = B$, in which X is a 2×2 matrix. The techniques used are very important, and will be highlighted in the Geometry of Book 9.

Note. For the historical background to matrices, reference may be made to *Men of Mathematics*, Chapter 21, E.T. Bell (Penguin Books), and for a more advanced treatment suitable for Sixth Forms, *Algebra and Number Systems*, J. Hunter and others (Blackie/Chambers) is recommended.

The matrix of the coefficients of x and y, $\begin{pmatrix} 3 & 1 \\ 0 & 3 \end{pmatrix}$, may be used to specify the mapping.

Note that a matrix is not a number but an array of numbers which may be meaningful in a given context as in Examples 2 and 3.

Exercise 1

1 Answer questions ***a–e*** for the matrix $\begin{pmatrix} 1 & 2 & 3 & 4 \\ 5 & 6 & 7 & 8 \\ 9 & 10 & 11 & 12 \end{pmatrix}$.

a State: (*1*) the number of rows (*2*) the number of columns.
b List the elements in the second row.
c List the elements in the third column.
d Write down the entry in: (*1*) the first row and first column (*2*) the third row and third column (*3*) the second row and fourth column.
e State the rows and columns which describe the positions of these entries:

(*1*) 4 (*2*) 9 (*3*) 6 (*4*) 11 (*5*) 2 (*6*) 5

2 For each of the following matrices, state the number of rows and columns, and the entry in the *first row* and *second column.*

a $\begin{pmatrix} 2 & -4 \\ 5 & 7 \end{pmatrix}$ *b* $\begin{pmatrix} a & b & c \\ p & q & r \end{pmatrix}$ *c* $(9 \quad 0 \quad -3)$

d $\begin{pmatrix} 1 & 3 \\ 5 & 7 \\ 0 & 9 \end{pmatrix}$ *e* $\begin{pmatrix} 3 & -1 & 5 \\ 0 & 2 & 1 \\ 1 & 4 & -8 \end{pmatrix}$ *f* $\begin{pmatrix} -1 & 2 & -3 & 4 \\ 5 & 0 & 1 & -9 \end{pmatrix}$

3 In each of the following systems of equations write down the matrix of coefficients of the variables x and y.

a $3x+2y = 5$
$4x+5y = 2$

b $2x-y = 4$
$x+2y = 3$

c $2x+y = 4$
$3y = 2$

4 Write down examples of matrices with numerical elements arranged in:

a 2 rows and 2 columns *b* 3 rows and 2 columns

c 1 row and 4 columns *d* 5 rows and 1 column

5 A linear mapping $(x, y) \rightarrow (x', y')$, defined by the equations

$$\left.\begin{aligned} x' &= ax+by \\ y' &= cx+dy \end{aligned}\right\},\ \text{can be expressed in matrix form } \begin{pmatrix} x' \\ y' \end{pmatrix} = \begin{pmatrix} a & b \\ c & d \end{pmatrix}\begin{pmatrix} x \\ y \end{pmatrix},$$

where a, b, c, d are constants. Express the following in matrix form:

a $x' = 3x+5y$, $y' = 2x+4y$ ***b*** $x' = 2x+3y$, $y' = 3x-2y$ ***c*** $x' = 3x-4y$, $y' = -2x+y$

d $x' = x\cos\alpha+y\sin\alpha$, $y' = -x\sin\alpha+y\cos\alpha$ ***e*** $x' = x$, $y' = 2x+y$ ***f*** $x' = y$, $y' = x$

6 Figure 1 shows a *network* which consists of *vertices* A, B, C, D, E and *edges* AB, BC, BD, BE, CE, DE. The table describes the network by giving the number of edges joining the various pairs of vertices. Copy and complete the table, and then write the entries in matrix form.

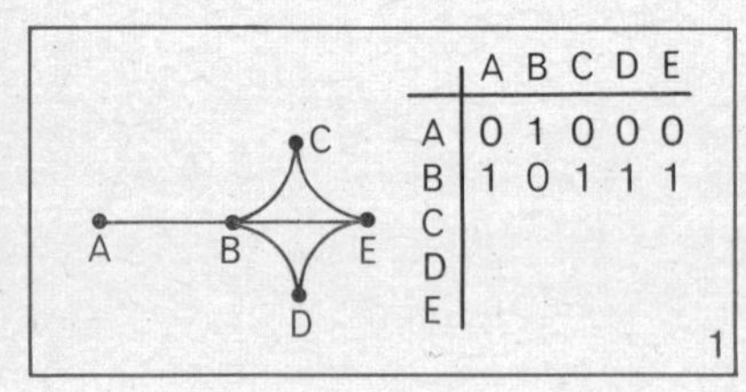

	A	B	C	D	E
A	0	1	0	0	0
B	1	0	1	1	1
C					
D					
E					

1

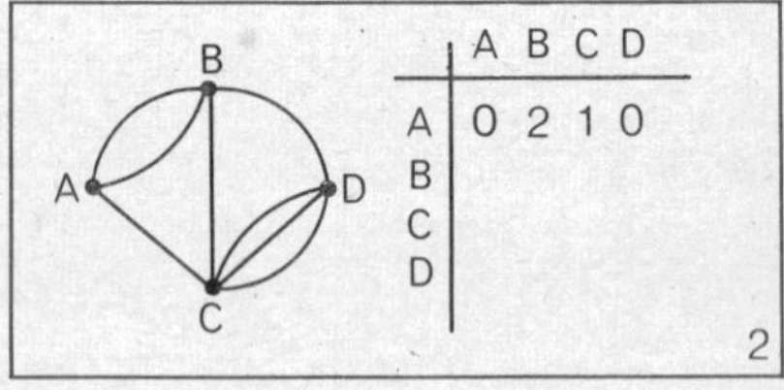

	A	B	C	D
A	0	2	1	0
B				
C				
D				

2

7 Repeat question **6** for the network and table in Figure 2.

8 Write down the addition and multiplication tables for numbers in binary form, then show the content of each table as a matrix.

9 Find examples of information presented in matrix form in newspapers or magazines.

2 *The order of a matrix; equal matrices*

A matrix is often denoted by a capital letter for easy reference. For example,

$$A = \begin{pmatrix} 4 & 6 & 8 \\ 2 & 3 & 4 \end{pmatrix} \qquad B = \begin{pmatrix} 3 & 4 \\ 2 & -6 \end{pmatrix}$$

The order of a matrix is given by stating the number of rows, followed by the number of columns. A has 2 rows and 3 columns and is said to be of order 2×3 (read '2 by 3'). When the number of rows in a matrix is the same as the number of columns, the matrix is called a *square matrix*; B is a *square matrix of order* 2.

Exercise 2

1 State the order of each of the following matrices:

a $\begin{pmatrix} 3 & 1 & 4 & 2 \\ 5 & 4 & 0 & 7 \end{pmatrix}$ ***b*** $\begin{pmatrix} 2 & -1 \\ 4 & 8 \\ 1 & -2 \end{pmatrix}$ ***c*** $\begin{pmatrix} a & h & g \\ h & b & f \\ g & f & c \end{pmatrix}$

d $\begin{pmatrix} 1 & 2 \\ 3 & 4 \\ 5 & 6 \\ 7 & 8 \end{pmatrix}$ ***e*** $(-1 \quad 0 \quad 1)$ ***f*** $\begin{pmatrix} u \\ v \\ w \end{pmatrix}$

Note. ***e*** is a *row matrix*, and ***f*** is a *column matrix*.

2 Write down the total number of entries in each matrix in question ***1***. Do you see a quick way to find the answers?

3 How many entries are there in:

a a 3×3 matrix ***b*** a 4×3 matrix ***c*** a 1×1 matrix

d a $1 \times n$ matrix ***e*** an $m \times n$ matrix ***f*** an $n \times n$ matrix?

4 Write down an example of a:

a 2×3 matrix ***b*** 3×3 matrix ***c*** 1×1 matrix

d 4×1 matrix ***e*** 1×4 matrix

Two matrices A and B are said to be *equal* when
(i) they are of the same order, and
(ii) their corresponding entries are equal.

For example, if $A = \begin{pmatrix} 2 & 3 \\ 1 & 5 \end{pmatrix}$, $B = \begin{pmatrix} 2 & 3 \\ 1 & 5 \end{pmatrix}$ and $C = \begin{pmatrix} 2 & 3 & 0 \\ 1 & 5 & 0 \end{pmatrix}$, then $A = B$ but $A \neq C$.

5 List any equalities for pairs of the following matrices.

$A = (1 \quad 2 \quad 3)$ $\qquad B = (3 \quad 2 \quad 1)$ $\qquad C = (1 \quad 2 \quad 3)$

$D = \begin{pmatrix} 2 \\ -1 \end{pmatrix}$ $\qquad E = \begin{pmatrix} 1 \\ 2 \end{pmatrix}$ $\qquad F = \begin{pmatrix} 2 \\ 1 \end{pmatrix}$

$G = \begin{pmatrix} 2 \\ 1 \end{pmatrix}$ $\qquad H = \begin{pmatrix} 1 & 2 \\ 3 & 4 \end{pmatrix}$ $\qquad J = \begin{pmatrix} -1 & -2 \\ -3 & -4 \end{pmatrix}$

$K = \begin{pmatrix} 1 & 3 \\ 2 & 4 \end{pmatrix}$ $\qquad L = \begin{pmatrix} 1 & 2 \\ 3 & 4 \end{pmatrix}$ $\qquad M = (1 \quad 2)$

6 What is the order of each matrix in question **5**?

7 Find x and y in each of the following:

a $(3x \quad -y) = (12 \quad 3)$ $\qquad$ ***b*** $\begin{pmatrix} x+3 \\ 4-y \end{pmatrix} = \begin{pmatrix} 7 \\ 5 \end{pmatrix}$

c $\begin{pmatrix} x+2y \\ 2x-y \end{pmatrix} = \begin{pmatrix} 9 \\ 8 \end{pmatrix}$ $\qquad$ ***d*** $\begin{pmatrix} x^2 & y^2 \\ y^3 & x^3 \end{pmatrix} = \begin{pmatrix} 4 & 9 \\ -27 & 8 \end{pmatrix}$

8 The entry in the ith row and jth column of a matrix A is often written a_{ij}.

For the matrix $A = \begin{pmatrix} 3 & 4 & 7 & 9 \\ 5 & 1 & 6 & 2 \\ 3 & 8 & 0 & 5 \end{pmatrix}$, state the entry:

a a_{13} $\qquad$ ***b*** a_{31} $\qquad$ ***c*** a_{24} $\qquad$ ***d*** a_{22} $\qquad$ ***e*** a_{32}

9 For the matrix A in question **8**, find:

a i, j for which $a_{ij} = 0$ $\qquad$ ***b*** i, j for which $a_{ij} = 5$.

10 A general 2×3 matrix B can be written $\begin{pmatrix} b_{11} & b_{12} & b_{13} \\ b_{21} & b_{22} & b_{23} \end{pmatrix}$.

In a similar way, write out: ***a*** a 3×3 matrix C ***b*** a 3×4 matrix D.

From a given matrix A, a new matrix can be formed by writing row 1 as column 1, row 2 as column 2, and so on. This new matrix is called the *transpose* of A, and is denoted by A' (read as 'A transpose'). For example, if $A = \begin{pmatrix} 1 & 2 \\ 3 & 4 \\ 5 & 6 \end{pmatrix}$, then $A' = \begin{pmatrix} 1 & 3 & 5 \\ 2 & 4 & 6 \end{pmatrix}$.

11 Write down the transpose of each matrix in question *1*, and state the order of each transpose.

12 $P = \begin{pmatrix} x & 9 \\ -3 & y \end{pmatrix}$ and $Q = \begin{pmatrix} 5 & -3 \\ 9 & -4 \end{pmatrix}$. Find x and y, given that $P' = Q$.

3 *Addition of matrices*

Now that we have described matrices, we shall define operations of addition and multiplication in such a way that an algebra of matrices can be constructed. In this Section, addition and its properties will be considered.

If A and B are matrices of the *same order*, the sum of A and B, denoted by $A+B$, is the matrix obtained by adding to each entry of A the corresponding entry of B.

Example 1. $\begin{pmatrix} 3 & 1 & 2 \\ 5 & 0 & 7 \end{pmatrix} + \begin{pmatrix} 4 & 2 & -1 \\ -3 & 8 & 2 \end{pmatrix} = \begin{pmatrix} 7 & 3 & 1 \\ 2 & 8 & 9 \end{pmatrix}$

Example 2. $\begin{pmatrix} a & b \\ c & d \end{pmatrix} + \begin{pmatrix} p & q \\ r & s \end{pmatrix} = \begin{pmatrix} a+p & b+q \\ c+r & d+s \end{pmatrix}$

Two important facts follow from the definition:

(1) The matrix $A+B$ will be of the same order as each of A and B.

(2) It is not possible to add two matrices of different orders.

Exercise 3

Find the sums of the matrices in questions *1* to *16*:

1 $\begin{pmatrix} 3 \\ 1 \end{pmatrix} + \begin{pmatrix} 4 \\ 5 \end{pmatrix}$ *2* $\begin{pmatrix} 2 \\ 4 \end{pmatrix} + \begin{pmatrix} -1 \\ 3 \end{pmatrix}$ *3* $\begin{pmatrix} 2a \\ b \end{pmatrix} + \begin{pmatrix} 7a \\ -3b \end{pmatrix}$

4 $\begin{pmatrix} m \\ n \end{pmatrix} + \begin{pmatrix} 1 \\ 2 \end{pmatrix}$ *5* $\begin{pmatrix} p \\ q \end{pmatrix} + \begin{pmatrix} r \\ s \end{pmatrix}$ *6* $\begin{pmatrix} 2u \\ -3v \end{pmatrix} + \begin{pmatrix} -2u \\ 3v \end{pmatrix}$

7 $(2 \quad 5)+(1 \quad 4)$ *8* $(2 \quad -3)+(-5 \quad 8)$

9 $(3i \quad 4j \quad -k)+(4i \quad -2j \quad 5k)$ *10* $\begin{pmatrix} 1 & 0 \\ 0 & 1 \end{pmatrix} + \begin{pmatrix} 2 & 3 \\ 4 & 5 \end{pmatrix}$

11 $\begin{pmatrix} 4 & 7 \\ 2 & 3 \end{pmatrix} + \begin{pmatrix} 8 & 0 \\ 3 & 9 \end{pmatrix}$ *12* $\begin{pmatrix} 1 & 2 \\ 3 & 4 \end{pmatrix} + \begin{pmatrix} -3 & 1 \\ 1 & -5 \end{pmatrix}$

13 $\begin{pmatrix} 4 & -1 \\ 3 & 2 \end{pmatrix} + \begin{pmatrix} 2 & 1 \\ -3 & 0 \end{pmatrix}$ *14* $\begin{pmatrix} 2 & 3 & 1 \\ 5 & 1 & 0 \end{pmatrix} + \begin{pmatrix} 1 & -2 & 3 \\ -7 & 1 & -4 \end{pmatrix}$

15 $\begin{pmatrix} 2a & b \\ 3a & -b \end{pmatrix} + \begin{pmatrix} a & 2b \\ -4a & b \end{pmatrix}$ *16* $\begin{pmatrix} x & -y \\ -x & 2y \end{pmatrix} + \begin{pmatrix} 2x & 3y \\ -x & 2y \end{pmatrix}$

17 $P = \begin{pmatrix} a & b \\ c & d \end{pmatrix}$ and $Q = \begin{pmatrix} f & g \\ h & k \end{pmatrix}$. Write down the matrices $P+Q$ and $Q+P$. What law for matrix addition does this result suggest?

18a $A = \begin{pmatrix} 3 & 2 \\ 1 & -1 \end{pmatrix}$, $B = \begin{pmatrix} 1 & 4 \\ 5 & 2 \end{pmatrix}$ and $C = \begin{pmatrix} -2 & 3 \\ 1 & -4 \end{pmatrix}$. Find the matrices:

(*1*) $A+B$ (*2*) $B+C$ (*3*) $(A+B)+C$ (*4*) $A+(B+C)$

b Is it true that $(A+B)+C = A+(B+C)$? What law for addition of matrices does this result suggest?

A *zero matrix*, denoted by $\boldsymbol{O}$, is one whose elements are all zero. $\begin{pmatrix} 0 & 0 & 0 \\ 0 & 0 & 0 \end{pmatrix}$ is the 2×3 zero matrix; $\begin{pmatrix} 0 & 0 \\ 0 & 0 \end{pmatrix}$ is the square zero matrix of order 2.

19 Given that $\boldsymbol{O} = \begin{pmatrix} 0 & 0 \\ 0 & 0 \end{pmatrix}$ and $A = \begin{pmatrix} a & b \\ c & d \end{pmatrix}$, verify that

$$\boldsymbol{O} + A = A + \boldsymbol{O} = A$$

The zero matrix is the *identity element* for addition of matrices.

20 Given that $A = \begin{pmatrix} 3 & -4 \\ -5 & 1 \end{pmatrix}$ and $B = \begin{pmatrix} -3 & 4 \\ 5 & -1 \end{pmatrix}$, find the matrices $A+B$ and $B+A$. Comment on your results.

Note. Each entry in B is the negative of the corresponding entry in A. For this reason, B is called the *negative* of A and is written $-A$.

If A is a matrix, then the *negative* of A, written $-A$, is the matrix in which each entry is the negative of the corresponding entry of A. Since $A+(-A) = (-A)+A = \boldsymbol{O}$, we call $-A$ the *additive inverse* of A, so that $-(-A) = A$. Clearly $-\boldsymbol{O} = \boldsymbol{O}$.

21 Write down the negative of each of the following matrices:

a $\begin{pmatrix} 3 \\ 2 \end{pmatrix}$ ***b*** $\begin{pmatrix} 5 \\ -3 \\ 4 \end{pmatrix}$ ***c*** $\begin{pmatrix} 4 & -7 \\ -5 & 8 \end{pmatrix}$ ***d*** $\begin{pmatrix} -3 & 1 & 0 \\ 4 & -2 & -1 \end{pmatrix}$

22 Given that X is a 2×3 matrix, solve the equation:

$$X + \begin{pmatrix} 2 & -1 & 4 \\ 7 & 6 & -3 \end{pmatrix} = \boldsymbol{O}$$

4 *Subtraction of matrices*

We have seen that if a and b are two real numbers, then $a-b = a+(-b)$. In the same way, since each matrix has a negative, we can write $A+(-B)$ as $A-B$, and talk about *subtracting* one matrix from another. Hence, we define subtraction as follows:

$$A-B = A+(-B).$$

i.e. to subtract B from A, *add* the negative of B to A.

Example 1. If $P = \begin{pmatrix} 3 & 2 \\ -1 & 4 \end{pmatrix}$ and $Q = \begin{pmatrix} 2 & 0 \\ -1 & 5 \end{pmatrix}$, then

$$\begin{aligned} P-Q &= \begin{pmatrix} 3 & 2 \\ -1 & 4 \end{pmatrix} - \begin{pmatrix} 2 & 0 \\ -1 & 5 \end{pmatrix} \\ &= \begin{pmatrix} 3 & 2 \\ -1 & 4 \end{pmatrix} + \begin{pmatrix} -2 & 0 \\ 1 & -5 \end{pmatrix} \\ &= \begin{pmatrix} 1 & 2 \\ 0 & -1 \end{pmatrix} \end{aligned}$$

If A, B and X are matrices of the same order, the equation

$$X+A = B$$

can always be solved for X by adding the negative of A to both sides.

Example 2. Solve the equation $X+\begin{pmatrix} 2 & -3 \\ -1 & 4 \end{pmatrix}=\begin{pmatrix} -1 & 5 \\ -3 & 9 \end{pmatrix}$, given that X is a 2×2 matrix.

$$X+\begin{pmatrix} 2 & -3 \\ -1 & 4 \end{pmatrix}=\begin{pmatrix} -1 & 5 \\ -3 & 9 \end{pmatrix}$$

$$\Leftrightarrow \quad X=\begin{pmatrix} -1 & 5 \\ -3 & 9 \end{pmatrix}+\begin{pmatrix} -2 & 3 \\ 1 & -4 \end{pmatrix}$$

$$\Leftrightarrow \quad X=\begin{pmatrix} -3 & 8 \\ -2 & 5 \end{pmatrix}$$

Exercise 4

1 Simplify each of the following:

a $\begin{pmatrix} 4 \\ 3 \end{pmatrix}-\begin{pmatrix} 1 \\ 2 \end{pmatrix}$ *b* $\begin{pmatrix} 3 \\ 4 \end{pmatrix}-\begin{pmatrix} -2 \\ 2 \end{pmatrix}$ *c* $\begin{pmatrix} -1 \\ 2 \end{pmatrix}-\begin{pmatrix} -1 \\ -2 \end{pmatrix}$

d $\begin{pmatrix} -a \\ b \end{pmatrix}-\begin{pmatrix} 2a \\ -3b \end{pmatrix}$ *e* $\begin{pmatrix} x+2 \\ 3-y \end{pmatrix}-\begin{pmatrix} 3 \\ 2+y \end{pmatrix}$ *f* $\begin{pmatrix} a \\ b \end{pmatrix}-\begin{pmatrix} c \\ d \end{pmatrix}$

2 Simplify:

a $\begin{pmatrix} 9 & 5 \\ 4 & 2 \end{pmatrix}-\begin{pmatrix} 7 & 1 \\ 3 & 0 \end{pmatrix}$ *b* $\begin{pmatrix} 2 & -1 \\ 3 & 7 \end{pmatrix}-\begin{pmatrix} 0 & -1 \\ -2 & 0 \end{pmatrix}$

c $\begin{pmatrix} 1 & 2 \\ -3 & 1 \end{pmatrix}-\begin{pmatrix} 4 & -1 \\ -2 & 3 \end{pmatrix}$ *d* $\begin{pmatrix} 3x & 2 \\ 4 & 5y \end{pmatrix}-\begin{pmatrix} x & -1 \\ 2 & -y \end{pmatrix}$

3 Given $A=\begin{pmatrix} 1 & 2 \\ 3 & 4 \end{pmatrix}$, $B=\begin{pmatrix} -2 & 3 \\ 0 & 1 \end{pmatrix}$ and $C=\begin{pmatrix} 5 & 2 \\ -1 & 0 \end{pmatrix}$, find in simplest form:

a $A+B$ **b** $A+C$ c $A+B+C$ *d* $A-B$
e $C-B$ *f* $C-A$ *g* $(A+C)+(A+B)$ *h* $(A+C)-(A+B)$

Which two of these matrices are equal?

4 Solve the equation $X+\begin{pmatrix} 2 \\ -1 \end{pmatrix}=\begin{pmatrix} -4 \\ 6 \end{pmatrix}$, for the 2×1 matrix X.

5 If $(x \;\; y \;\; z)-(3 \;\; 0 \;\; -5)=(-5 \;\; 3 \;\; 2)$, find x, y, z.

6 Solve each of the following equations for the 2×2 matrix X:

a $X+\begin{pmatrix}1 & 0\\0 & 1\end{pmatrix}=\begin{pmatrix}0 & 2\\2 & 1\end{pmatrix}$ *b* $\begin{pmatrix}2 & 3\\4 & 5\end{pmatrix}+X=\begin{pmatrix}4 & -1\\3 & 2\end{pmatrix}$

c $X-\begin{pmatrix}3 & 5\\-2 & 1\end{pmatrix}=\begin{pmatrix}4 & -7\\5 & 3\end{pmatrix}$ *d* $\begin{pmatrix}3 & -4\\2 & 7\end{pmatrix}-X=\begin{pmatrix}-1 & -5\\3 & -6\end{pmatrix}$

7 Find p, q, r, s in each of the following:

a $\begin{pmatrix}p & q\\r & s\end{pmatrix}-\begin{pmatrix}2 & 4\\1 & 0\end{pmatrix}=\begin{pmatrix}3 & 5\\-1 & 2\end{pmatrix}$ *b* $\begin{pmatrix}3 & -2\\5 & 0\end{pmatrix}-\begin{pmatrix}p & q\\r & s\end{pmatrix}=\begin{pmatrix}-2 & 5\\6 & -1\end{pmatrix}$

8 If $A=\begin{pmatrix}1 & 2\\3 & 4\\5 & 6\end{pmatrix}$, $B=\begin{pmatrix}3 & -2\\1 & 5\\4 & 6\end{pmatrix}$, and $C=\begin{pmatrix}4 & 2\\1 & 0\\-3 & 5\end{pmatrix}$, simplify:

a $A+B$ *b* $B-C$ *c* $(A+B)-C$ *d* $A+(B-C)$

5 *Multiplication of a matrix by a real number*

If k is a real number, and A is a matrix, then kA is the matrix obtained by multiplying each entry of A by k.

Note. In matrix algebra, a real number is often called a *scalar*. The above operation of multiplying k by A is called *scalar multiplication.*

Example. $A=\begin{pmatrix}2 & 1 & 3\\-1 & 0 & 4\end{pmatrix}$ and $B=\begin{pmatrix}2 & -1 & 0\\3 & 4 & 5\end{pmatrix}$. Find in simplest form the matrix $3A-2B$.

$$\begin{aligned}3A-2B &= 3\begin{pmatrix}2 & 1 & 3\\-1 & 0 & 4\end{pmatrix}-2\begin{pmatrix}2 & -1 & 0\\3 & 4 & 5\end{pmatrix}\\ &= \begin{pmatrix}6 & 3 & 9\\-3 & 0 & 12\end{pmatrix}-\begin{pmatrix}4 & -2 & 0\\6 & 8 & 10\end{pmatrix}\\ &= \begin{pmatrix}2 & 5 & 9\\-9 & -8 & 2\end{pmatrix}\end{aligned}$$

From the meaning of kA, it readily follows that (i) $0A=\boldsymbol{O}$ and (ii) $k\boldsymbol{O}=\boldsymbol{O}$, where $\boldsymbol{O}$ is a zero matrix of suitable order.

Exercise 5

1 Work out the following:

a $3\begin{pmatrix}2\\3\end{pmatrix}$ *b* $2\begin{pmatrix}3\\1\\0\end{pmatrix}$ *c* $5(3 \quad -1 \quad 2)$

d $2\begin{pmatrix}3 & 1\\2 & 4\end{pmatrix}$ *e* $3\begin{pmatrix}2 & -1\\0 & 1\end{pmatrix}$ *f* $-\frac{1}{2}\begin{pmatrix}6 & 4\\0 & -2\end{pmatrix}$

g $2\begin{pmatrix}3 & 1 & -2\\1 & 4 & 0\end{pmatrix}$ *h* $-3\begin{pmatrix}2 & -1 & 0\\5 & 0 & -3\end{pmatrix}$ *i* $5\begin{pmatrix}a & 2b & c\\2a & -b & -c\end{pmatrix}$

2 If $X = \begin{pmatrix}3 & 4\\-1 & -2\end{pmatrix}$, find the following matrices:

a $2X$ *b* $3X$ *c* $-5X$ *d* $-X$ *e* $(-1)X$

Which two matrices are equal? Verify that for each $m \times n$ matrix A, $-A = (-1)A$.

3 Using the results in question **2**, find:

a $2X+3X$, and hence show that $2X+3X = 5X$.
b $5X-3X$, and hence show that $5X-3X = 2X$.
c $2X+1X$, and hence show that $2X+X = 3X$.
d $3X-X$, and hencc show that $3X-X = 2X$.

4 $X = \begin{pmatrix}3 & 4 & 1\\2 & 0 & 3\end{pmatrix}$ and $Y = \begin{pmatrix}2 & -1 & 3\\4 & 5 & 0\end{pmatrix}$. Find in simplest form:

a $X+Y$ *b* $2(X+Y)$ *c* $2X$ *d* $2Y$ *e* $2X+2Y$
f $6X$ *g* $3(2X)$ *h* $8Y$ *i* $4(2Y)$

Questions **3** and **4** illustrate the properties of multiplication of a matrix by a scalar. If A and B are $m \times n$ matrices and $r, s \in R$,

(*1*) $(r+s)A = rA+sA$ (*2*) $r(A+B) = rA+rB$
(*3*) $r(sA) = (rs)A$ (*4*) $1A = A$ (*5*) $(-1)A = -A$

5 Copy and complete the entries in each of the following:

a $\begin{pmatrix}4 & 6\\8 & 2\end{pmatrix} = 2\begin{pmatrix}\cdot & \cdot\\\cdot & \cdot\end{pmatrix}$ *b* $\begin{pmatrix}6 & -3\\-9 & 0\end{pmatrix} = 3\begin{pmatrix}\cdot & \cdot\\\cdot & \cdot\end{pmatrix}$

c $\begin{pmatrix}-5 & 10 & 0\\15 & 0 & -10\end{pmatrix} = -5\begin{pmatrix}\cdot & \cdot & \cdot\\\cdot & \cdot & \cdot\end{pmatrix}$ *d* $\begin{pmatrix}4 & 6 & 3\\1 & 2 & 0\end{pmatrix} = \frac{1}{2}\begin{pmatrix}\cdot & \cdot & \cdot\\\cdot & \cdot & \cdot\end{pmatrix}$

6 $A = \begin{pmatrix} 2 & -3 \\ 4 & 1 \end{pmatrix}$ and $B = \begin{pmatrix} -4 & 1 \\ 3 & -2 \end{pmatrix}$. Simplify:

a $2A$ *b* $2B$ *c* $3A$ *d* $5A$
e $A+B$ *f* $2(A+B)$ *g* $A-B$ *h* $2(A-B)$
i $2A+2B$ *j* $2A-2B$ *k* $3A+2A$ *l* $5A-3A$

7 Simplify:

a $3\begin{pmatrix} 2 & 1 & -3 \\ 5 & 4 & 0 \end{pmatrix} + 2\begin{pmatrix} 1 & -2 & 0 \\ 3 & -5 & 2 \end{pmatrix}$ *b* $3\begin{pmatrix} 3 & -2 & 1 \\ 1 & 3 & 4 \end{pmatrix} - 2\begin{pmatrix} 1 & -3 & 2 \\ -2 & 3 & 4 \end{pmatrix}$

8 Solve each of the following equations for the 2×2 matrix X:

a $3X = \begin{pmatrix} 6 & -3 \\ 12 & 9 \end{pmatrix}$ *b* $2X + \begin{pmatrix} 3 & 1 \\ 4 & 2 \end{pmatrix} = \begin{pmatrix} 9 & 5 \\ 2 & 8 \end{pmatrix}$

c $4X - \begin{pmatrix} 3 & 1 \\ 4 & 7 \end{pmatrix} = \begin{pmatrix} 5 & 3 \\ 0 & 13 \end{pmatrix}$ *d* $\begin{pmatrix} 7 & 1 \\ -4 & 3 \end{pmatrix} - 3X = \begin{pmatrix} -5 & 10 \\ 8 & 9 \end{pmatrix}$

9 Find the matrix X in each of the following:

a $2\begin{pmatrix} 1 & -1 & 3 \\ 2 & -7 & 5 \end{pmatrix} + X = 3\begin{pmatrix} 1 & 2 & -4 \\ 3 & -5 & 1 \end{pmatrix}$

b $5\begin{pmatrix} 1 & 2 \\ 3 & 4 \end{pmatrix} - 3X = 4\begin{pmatrix} -4 & 7 \\ 3 & 8 \end{pmatrix}$

10 Given that $2\begin{pmatrix} p & q \\ r & s \end{pmatrix} + \begin{pmatrix} 7 & -2 \\ -4 & 5 \end{pmatrix} = \begin{pmatrix} 5 & 6 \\ 2 & 1 \end{pmatrix}$, find the entries p, q, r, s.

---

6 *Multiplication of matrices*

Although addition and subtraction of matrices are straightforward operations, multiplication of matrices is not so obvious. Historically, it arose from a study of systems of linear mappings (transformations) by Arthur Cayley about the middle of the 19th century.

(i) Multiplication of an $m \times p$ matrix by a $p \times 1$ matrix

Consider the two linear expressions $\left.\begin{matrix} ax+by \\ cx+dy \end{matrix}\right\}$ ... (1)

Write these in matrix form $\begin{pmatrix} a & b \\ c & d \end{pmatrix}\begin{pmatrix} x \\ y \end{pmatrix}$. ... (2)

We can think of (2) as the product of a 2×2 matrix and a 2×1 matrix.

To obtain the linear expressions in (1) from (2), *multiply each entry of a row in the first matrix by the corresponding entry of the column in the second matrix and then add the products* to give the 2×1 matrix:

$$\begin{pmatrix} ax+by \\ cx+dy \end{pmatrix} \qquad \dots (3)$$

The multiplication process described can be readily extended.

Example 1. Find the product $\begin{pmatrix} 3 & 2 & 1 \\ 5 & -3 & 7 \end{pmatrix}\begin{pmatrix} x \\ y \\ z \end{pmatrix}$.

$$\begin{pmatrix} 3 & 2 & 1 \\ 5 & -3 & 7 \end{pmatrix}\begin{pmatrix} x \\ y \\ z \end{pmatrix} = \begin{pmatrix} 3x+2y+\ z \\ 5x-3y+7z \end{pmatrix}$$

Example 2. Perform the matrix multiplication $(4 \quad 3 \quad 2)\begin{pmatrix} 3 \\ 1 \\ -5 \end{pmatrix}$.

$$(4 \quad 3 \quad 2)\begin{pmatrix} 3 \\ 1 \\ -5 \end{pmatrix} = (4 \times 3 + 3 \times 1 + 2 \times (-5))$$

$$= (12+3-10) = (5)$$

(*Questions* **1–3** *of Exercise 6 may be taken here.*)

(ii) Multiplication of an $m \times p$ matrix by a $p \times n$ matrix

To illustrate a further extension of matrix multiplication, consider the mappings $f:(x, y) \to (x', y')$ and $g:(x', y') \to (x'', y'')$ defined by

$$\left.\begin{aligned} x' &= px+qy \\ y' &= rx+sy \end{aligned}\right\} \quad \text{and} \quad \left.\begin{aligned} x'' &= ax'+by' \\ y'' &= cx'+dy' \end{aligned}\right\} \qquad \dots (1)$$

On substitution for x' and y',

$$\left.\begin{aligned} x'' &= a(px+qy)+b(rx+sy) \\ y'' &= c(px+qy)+d(rx+sy) \end{aligned}\right\}$$

$$\Rightarrow \quad \left.\begin{aligned} x'' &= (ap+br)x+(aq+bs)y \\ y'' &= (cp+dr)x+(cq+ds)y \end{aligned}\right\} \qquad \dots (2)$$

In matrix form, (2) can be written

$$\begin{pmatrix} x'' \\ y'' \end{pmatrix} = \begin{pmatrix} ap+br & aq+bs \\ cp+dr & cq+ds \end{pmatrix} \begin{pmatrix} x \\ y \end{pmatrix} \qquad \ldots (3)$$

Hence the linear mappings f followed by g map $(x, y) \rightarrow (x'', y'')$. Expressing each pair of equations in (1) in matrix form,

$$\begin{pmatrix} x' \\ y' \end{pmatrix} = \begin{pmatrix} p & q \\ r & s \end{pmatrix} \begin{pmatrix} x \\ y \end{pmatrix} \quad \text{and} \quad \begin{pmatrix} x'' \\ y'' \end{pmatrix} = \begin{pmatrix} a & b \\ c & d \end{pmatrix} \begin{pmatrix} x' \\ y' \end{pmatrix},$$

from which, on substitution for $\begin{pmatrix} x' \\ y' \end{pmatrix}$ in the second pair,

$$\begin{pmatrix} x'' \\ y'' \end{pmatrix} = \begin{pmatrix} a & b \\ c & d \end{pmatrix} \begin{pmatrix} p & q \\ r & s \end{pmatrix} \begin{pmatrix} x \\ y \end{pmatrix} \qquad \ldots (4)$$

(3) and (4) show that the result of the mapping f followed by the mapping g could have been written down from the following definition of multiplication of matrices:

$$\begin{pmatrix} a & b \\ c & d \end{pmatrix} \begin{pmatrix} p & q \\ r & s \end{pmatrix} = \begin{pmatrix} ap+br & aq+bs \\ cp+dr & cq+ds \end{pmatrix}$$

A little thought will show that the '*row into column*' rule for multiplication of matrices requires that the number of columns in the left-hand matrix is the same as the number of rows in the right-hand matrix. Hence it is only possible to multiply an $m \times p$ matrix by a $q \times n$ matrix if $q = p$, and the product matrix will be of order $m \times n$. The two matrices are then said to be *conformable for multiplication.*

Note. In checking whether or not a product exists, and also in working out the order of the product matrix, a comparison with matching dominoes may be helpful, as shown in Figure 3.

$$\begin{pmatrix} 3 & 2 & 1 \\ 5 & -3 & 7 \end{pmatrix} \begin{pmatrix} x \\ y \\ z \end{pmatrix} = \begin{pmatrix} 3x+2y+z \\ 5x-3y+7z \end{pmatrix}$$

2x3 Same 3x1

2x1

3

General definition of a matrix product. The product of an $m \times p$ matrix A and a $p \times n$ matrix B is the $m \times n$ matrix AB whose entry in the ith row and jth column is the sum of the products of corresponding entries in the ith row of A and the jth column of B.

Example 3. Given $P = \begin{pmatrix} 1 & 2 \\ 3 & 1 \end{pmatrix}$ and $Q = \begin{pmatrix} 4 & 5 \\ 2 & 0 \end{pmatrix}$, find PQ and QP.

$$PQ = \begin{pmatrix} 1 & 2 \\ 3 & 1 \end{pmatrix}\begin{pmatrix} 4 & 5 \\ 2 & 0 \end{pmatrix} = \begin{pmatrix} 4+4 & 5+0 \\ 12+2 & 15+0 \end{pmatrix} = \begin{pmatrix} 8 & 5 \\ 14 & 15 \end{pmatrix}$$

$$QP = \begin{pmatrix} 4 & 5 \\ 2 & 0 \end{pmatrix}\begin{pmatrix} 1 & 2 \\ 3 & 1 \end{pmatrix} = \begin{pmatrix} 4+15 & 8+5 \\ 2+0 & 4+0 \end{pmatrix} = \begin{pmatrix} 19 & 13 \\ 2 & 4 \end{pmatrix}$$

Notice that $PQ \neq QP$, so *multiplication of matrices is not commutative.*

To avoid ambiguity in the multiplication of matrices, PQ may be described as P *postmultiplied* by Q, or Q *premultiplied* by P.

Exercise 6

1 Find the following matrix products:

a $(1 \quad 2)\begin{pmatrix} 3 \\ 4 \end{pmatrix}$ ***b*** $(3 \quad 4)\begin{pmatrix} 2 \\ 5 \end{pmatrix}$ ***c*** $(5 \quad -2)\begin{pmatrix} 1 \\ -2 \end{pmatrix}$

d $(3 \quad 1 \quad 2)\begin{pmatrix} 2 \\ 1 \\ 3 \end{pmatrix}$ ***e*** $(2 \quad -3 \quad 4)\begin{pmatrix} 5 \\ 1 \\ 2 \end{pmatrix}$ ***f*** $(8 \quad -5 \quad -1)\begin{pmatrix} x \\ y \\ z \end{pmatrix}$

2 Find x in each of the following:

a $(x \quad 2)\begin{pmatrix} 1 \\ 3 \end{pmatrix} = (10)$ ***b*** $(3 \quad x)\begin{pmatrix} 5 \\ 2 \end{pmatrix} = (21)$

c $(x \quad 1)\begin{pmatrix} 4 \\ x \end{pmatrix} = (-30)$ ***d*** $(x \quad -3)\begin{pmatrix} x \\ -1 \end{pmatrix} = (12)$

3 Find each of the following products in its simplest form:

a $\begin{pmatrix} 3 & 1 \\ 1 & 2 \end{pmatrix}\begin{pmatrix} 2 \\ 3 \end{pmatrix}$ ***b*** $\begin{pmatrix} 2 & 1 \\ 0 & 1 \end{pmatrix}\begin{pmatrix} 3 \\ 4 \end{pmatrix}$ ***c*** $\begin{pmatrix} 2 & -1 \\ 1 & 2 \end{pmatrix}\begin{pmatrix} 2 \\ 3 \end{pmatrix}$

d $\begin{pmatrix} 1 & -1 \\ 1 & 0 \end{pmatrix}\begin{pmatrix} 5 \\ -2 \end{pmatrix}$ ***e*** $\begin{pmatrix} 2 & 3 \\ 1 & -2 \end{pmatrix}\begin{pmatrix} 1 \\ -3 \end{pmatrix}$ ***f*** $\begin{pmatrix} 0 & -1 \\ 1 & 0 \end{pmatrix}\begin{pmatrix} 6 \\ -4 \end{pmatrix}$

g $\begin{pmatrix} 3 & -2 \\ 4 & 5 \end{pmatrix}\begin{pmatrix} x \\ y \end{pmatrix}$ ***h*** $\begin{pmatrix} 1 & -3 \\ -2 & -1 \end{pmatrix}\begin{pmatrix} x \\ -2x \end{pmatrix}$ ***i*** $\begin{pmatrix} 4 & 2 \\ -1 & 1 \end{pmatrix}\begin{pmatrix} a \\ -2a \end{pmatrix}$

4 $A = (3 \quad 5)$, $B = \begin{pmatrix} 2 & -1 \\ -1 & 0 \end{pmatrix}$ and $C = \begin{pmatrix} 4 \\ -3 \end{pmatrix}$.

Which of the products AB, BA, BC, CB, AC, CA are possible? Simplify those products that exist.

5 Perform matrix multiplication, where possible, *first considering the order of the product.*

a $\begin{pmatrix}1 & -1\\2 & 3\end{pmatrix}\begin{pmatrix}5\\4\end{pmatrix}$ ***b*** $\begin{pmatrix}5\\4\end{pmatrix}\begin{pmatrix}1 & -1\\2 & 3\end{pmatrix}$ ***c*** $(4 \quad 3)\begin{pmatrix}5\\2\end{pmatrix}$

d $\begin{pmatrix}2 & 1 & 3\\3 & 0 & 1\\1 & 2 & 3\end{pmatrix}\begin{pmatrix}4\\0\\1\end{pmatrix}$ ***e*** $\begin{pmatrix}2 & 1\\3 & 0\\1 & 2\end{pmatrix}\begin{pmatrix}4\\0\\1\end{pmatrix}$ ***f*** $\begin{pmatrix}1 & -2 & 3\\-1 & 4 & 2\\3 & 1 & 0\end{pmatrix}\begin{pmatrix}2\\1\\3\end{pmatrix}$

g $\begin{pmatrix}1 & 2\\3 & 4\end{pmatrix}(2 \quad 5)$ ***h*** $(2 \quad 5)\begin{pmatrix}1 & 2\\3 & 4\end{pmatrix}$ ***i*** $(\cos\alpha \quad \sin\alpha)\begin{pmatrix}\cos\alpha & -\sin\alpha\\\sin\alpha & \cos\alpha\end{pmatrix}$

6 In each of the following, find a system of equations in x and y. Hence find x and y.

a $\begin{pmatrix}3 & 0\\0 & -2\end{pmatrix}\begin{pmatrix}x\\y\end{pmatrix}=\begin{pmatrix}12\\8\end{pmatrix}$ ***b*** $\begin{pmatrix}x & 0\\1 & y\end{pmatrix}\begin{pmatrix}2\\3\end{pmatrix}=\begin{pmatrix}6\\-1\end{pmatrix}$

c $\begin{pmatrix}3 & 1\\2 & -1\end{pmatrix}\begin{pmatrix}x\\y\end{pmatrix}=\begin{pmatrix}9\\1\end{pmatrix}$ ***d*** $\begin{pmatrix}2 & 1\\1 & 2\end{pmatrix}\begin{pmatrix}x\\y\end{pmatrix}=\begin{pmatrix}8\\1\end{pmatrix}$

e $\begin{pmatrix}2 & -1\\1 & -3\end{pmatrix}\begin{pmatrix}x\\y\end{pmatrix}=\begin{pmatrix}11\\-7\end{pmatrix}$ ***f*** $\begin{pmatrix}x & y\\y & x\end{pmatrix}\begin{pmatrix}3\\1\end{pmatrix}=\begin{pmatrix}5\\-1\end{pmatrix}$

7 Find the following products in their simplest form:

a $\begin{pmatrix}1 & 2\\4 & 1\end{pmatrix}\begin{pmatrix}2 & 1\\1 & 3\end{pmatrix}$ ***b*** $\begin{pmatrix}2 & 1\\1 & 5\end{pmatrix}\begin{pmatrix}1 & 2\\3 & 1\end{pmatrix}$

c $\begin{pmatrix}3 & 2\\1 & 4\end{pmatrix}\begin{pmatrix}1 & 4\\5 & 3\end{pmatrix}$ ***d*** $\begin{pmatrix}1 & 2\\2 & 4\end{pmatrix}\begin{pmatrix}3 & 2\\4 & 5\end{pmatrix}$

e $\begin{pmatrix}1 & 0\\-2 & 5\end{pmatrix}\begin{pmatrix}3 & -4 & 2\\0 & -2 & 1\end{pmatrix}$ ***f*** $\begin{pmatrix}0 & 1\\-1 & 1\end{pmatrix}\begin{pmatrix}4 & 5 & -3 & 1\\-3 & 2 & 1 & -2\end{pmatrix}$

8 Work out the following products:

a $\begin{pmatrix}1 & -2\\-3 & 6\end{pmatrix}\begin{pmatrix}1 & 0\\0 & 1\end{pmatrix}$ ***b*** $\begin{pmatrix}1 & 0\\0 & 1\end{pmatrix}\begin{pmatrix}1 & -2\\-3 & 6\end{pmatrix}$ ***c*** $\begin{pmatrix}a & b\\c & d\end{pmatrix}\begin{pmatrix}1 & 0\\0 & 1\end{pmatrix}$

d $\begin{pmatrix}1 & 0\\0 & 1\end{pmatrix}\begin{pmatrix}a & b\\c & d\end{pmatrix}$ ***e*** $\begin{pmatrix}1 & 0\\0 & 1\end{pmatrix}\begin{pmatrix}1 & 0\\0 & 1\end{pmatrix}$ ***f*** $\begin{pmatrix}1 & 0\\0 & 1\end{pmatrix}\begin{pmatrix}0 & 1\\1 & 0\end{pmatrix}$

The 2×2 matrix $\begin{pmatrix}1 & 0\\0 & 1\end{pmatrix}$ is called the *unit matrix* of order 2, and is denoted by I. It behaves like unity in the real number system. $(1.a = a.1 = a, a \in R)$.

If A is a 2×2 matrix, then $IA = AI = A$.

9 If $A = \begin{pmatrix} 1 & -2 \\ -3 & 6 \end{pmatrix}$ and $B = \begin{pmatrix} 2 & 2 \\ 1 & 1 \end{pmatrix}$, find AB and BA.

Note that $AB = \boldsymbol{O}$ does not necessarily mean that $A = \boldsymbol{O}$ or $B = \boldsymbol{O}$.

Powers of a square matrix A are defined as follows:

$$A^2 = A.A, \quad A^3 = A.A^2, \quad A^4 = A.A^3, \text{ and so on.}$$

10 Given that $A = \begin{pmatrix} 3 & -1 \\ 5 & 2 \end{pmatrix}$, find the matrices A^2 and A^3.

11 If $\begin{pmatrix} 3 & 1 \\ 2 & 1 \end{pmatrix}\begin{pmatrix} a & b \\ c & d \end{pmatrix} = \begin{pmatrix} 1 & 0 \\ 0 & 1 \end{pmatrix}$, find a, b, c, d.

12 If $\begin{pmatrix} p & q \\ r & s \end{pmatrix}\begin{pmatrix} 2 & 1 \\ 4 & 3 \end{pmatrix} = \begin{pmatrix} 1 & 0 \\ 0 & 1 \end{pmatrix}$, find p, q, r, s.

13 Figure 4 shows a network of roads connecting four towns A, B, C, D.

a Work out the total number of routes from A to C via B and via D.

b The tables summarize the number of routes from A to B and to D, and from B and D to C. Extract matrices from these, and interpret their product in terms of your answer to ***a***.

	B	D
A	2	3

	C
B	3
D	1

c If a new road is made from D to C, calculate by means of a matrix product the total number of routes from A to C now.

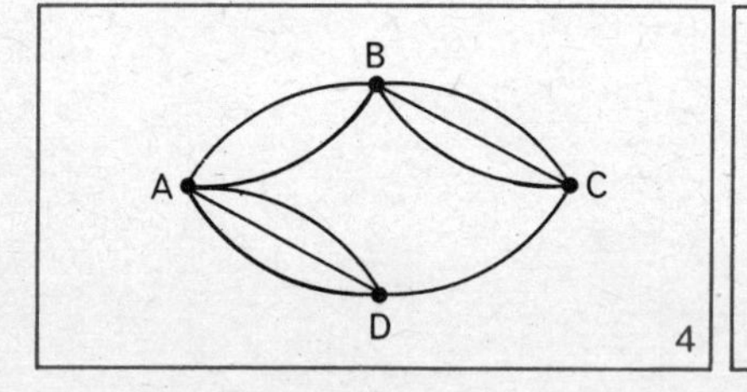

4

x_1 x_2 X

y_1 y_2 y_3 y_4 Y

z_1 z_2 z_3 Z

5

14 Figure 5 shows three countries X, Y and Z in which there are two, four and three airports. Travel agents supply the following tables of information about the number of routes connecting the airports.

	y_1	y_2	y_3	y_4
x_1	3	1	0	0
x_2	0	0	2	1

	z_1	z_2	z_3
y_1	1	1	0
y_2	1	1	0
y_3	0	1	2
y_4	0	0	1

a Copy Figure 5, and complete the network of routes between the airports.

b Calculate the number of routes from airports x_1 and x_2 in X to airports in Z: (*1*) from your diagram (*2*) using matrices.

Example. If $A = \begin{pmatrix} 1 & 2 \\ 3 & 4 \end{pmatrix}$, find p, q such that $A^2 = pA + qI$.

$$A^2 = pA + qI \quad \Leftrightarrow \quad \begin{pmatrix} 1 & 2 \\ 3 & 4 \end{pmatrix}\begin{pmatrix} 1 & 2 \\ 3 & 4 \end{pmatrix} = p\begin{pmatrix} 1 & 2 \\ 3 & 4 \end{pmatrix} + q\begin{pmatrix} 1 & 0 \\ 0 & 1 \end{pmatrix}$$

$$\Leftrightarrow \quad \begin{pmatrix} 7 & 10 \\ 15 & 22 \end{pmatrix} = \begin{pmatrix} p+q & 2p \\ 3p & 4p+q \end{pmatrix}$$

Hence $\left.\begin{aligned} p+q &= 7 \\ 2p &= 10 \\ 3p &= 15 \\ 4p+q &= 22 \end{aligned}\right\}$ from which $p = 5$, $q = 2$.

Exercise 6B

1 If $A = \begin{pmatrix} 2 & 1 \\ 1 & -1 \end{pmatrix}$, $B = \begin{pmatrix} 1 & -1 \\ 2 & 3 \end{pmatrix}$ and $C = \begin{pmatrix} 3 & 4 \\ -2 & 5 \end{pmatrix}$, find in simplest form:

a AB *b* BA *c* BC *d* CB *e* AC *f* CA

2 For the matrices A, B, C of question *1*, find:

a $(AB)C$ *b* $A(BC)$ *c* $(CB)A$ *d* $C(BA)$

What law appears to hold?

In general, for $n \times n$ matrices A, B and C, $(AB)C = A(BC)$.

3 For the matrices A, B, C of question *1*, find:

a $A(B+C)$ *b* $AB+AC$ *c* $(B+C)A$ *d* $BA+CA$

Which of these matrices are equal? What laws appear to hold?

In general, if matrices B and C can be *pre-multiplied* by A, then $A(B+C) = AB+AC$, and if B and C can be *post-multiplied* by A, then $(B+C)A = BA+CA$. These equalities are called the *left-hand and right-hand distributive laws* respectively.

4 Given that $P = \begin{pmatrix} 1 & 2 \\ 3 & 4 \\ 5 & 6 \end{pmatrix}$, $Q = \begin{pmatrix} 3 & 1 \\ 1 & 2 \\ 6 & 4 \end{pmatrix}$ and $R = \begin{pmatrix} 2 & 3 \\ 1 & 4 \end{pmatrix}$, find:

a PR *b* QR *c* $P+Q$

5 In question **4**, show that $PR+QR = (P+Q)R$.
Is it true that $R(P+Q) = RP+RQ$? Give a reason.

6 $A = \begin{pmatrix} 2 & 1 \\ -1 & 3 \end{pmatrix}$ and $B = \begin{pmatrix} 1 & 4 \\ 2 & -1 \end{pmatrix}$. Find:

a $A+B$ *b* $A-B$ *c* $(A+B)(A-B)$ *d* A^2 *e* B^2

Is it true that $(A+B)(A-B) = A^2 - B^2$?

7 $A = \begin{pmatrix} 1 & 2 \\ 3 & -1 \end{pmatrix}$ and $B = \begin{pmatrix} 2 & -3 \\ -2 & 1 \end{pmatrix}$. Find:

a $A+B$ *b* $(A+B)^2$ *c* A^2 *d* $2AB$ *e* B^2

Is it true that $(A+B)^2 = A^2+2AB+B^2$?

8 Assuming that A and B are matrices conformable for multiplication, use the distributive laws to show that in general

a $(A+B)^2 \neq A^2+2AB+B^2$ *b* $(A+B)(A-B) \neq A^2-B^2$

9 $A = \begin{pmatrix} k & 1 \\ 0 & k \end{pmatrix}$. Find A^2, A^3 and A^4 and hence deduce a formula for A^n, where n is a positive integer.

10 $A = \begin{pmatrix} 3 & -4 \\ 1 & -1 \end{pmatrix}$. Verify that $A^2-2A+I = \mathbf{O}$, where I is the unit matrix of order 2.

11 Show that the matrix $A = \begin{pmatrix} 2 & 3 \\ 3 & 2 \end{pmatrix}$ satisfies the equation

$$A^2-4A-5I = \mathbf{O}.$$

12 In the algebra of real numbers, $x^2-2xy-3y^2 = (x+y)(x-3y)$.
Given that $X = \begin{pmatrix} -1 & 2 \\ 0 & 3 \end{pmatrix}$ and $I = \begin{pmatrix} 1 & 0 \\ 0 & 1 \end{pmatrix}$, is it true that

$$X^2-2XI-3I^2 = (X+I)(X-3I)?$$

13 Find the two matrices of the form $X = \begin{pmatrix} x & 1 \\ 0 & y \end{pmatrix}$ such that $X^2 = I$.

14 If $A = \begin{pmatrix} 3 & -1 \\ 2 & -5 \end{pmatrix}$, find p, q such that $A^2 = pA + qI$.

15 Given $M = \begin{pmatrix} \cos\theta & \sin\theta \\ -\sin\theta & \cos\theta \end{pmatrix}$, calculate M^2 and M^3 when $\theta = 60°$.

16 $A = \begin{pmatrix} 1 & 1 \\ 3 & 2 \\ 1 & 2 \end{pmatrix}$ and $B = \begin{pmatrix} 2 & 1 & 3 \\ 3 & -1 & 2 \end{pmatrix}$. Find AB and BA.

17 $A = \begin{pmatrix} 1 & 0 & 2 \\ 1 & -1 & 1 \\ 2 & 1 & 3 \end{pmatrix}$ and $B = \begin{pmatrix} 2 & -1 & 3 \\ 0 & 1 & -1 \\ 1 & 1 & 1 \end{pmatrix}$. Find AB and BA.

7 *The inverse of a square matrix of order 2*

Let $A = \begin{pmatrix} a & b \\ c & d \end{pmatrix}$ and $I = \begin{pmatrix} 1 & 0 \\ 0 & 1 \end{pmatrix}$.

Pre-multiplying A by I, $IA = \begin{pmatrix} 1 & 0 \\ 0 & 1 \end{pmatrix}\begin{pmatrix} a & b \\ c & d \end{pmatrix} = \begin{pmatrix} a & b \\ c & d \end{pmatrix} = A$.

Post-multiplying A by I, $AI = \begin{pmatrix} a & b \\ c & d \end{pmatrix}\begin{pmatrix} 1 & 0 \\ 0 & 1 \end{pmatrix} = \begin{pmatrix} a & b \\ c & d \end{pmatrix} = A$.

Therefore $IA = AI = A$.

For this reason, the unit 2×2 matrix I is called the *identity matrix for multiplication* of 2×2 matrices. Note that A commutes with I, i.e. $IA = AI$.

Consider matrices $P = \begin{pmatrix} 3 & 2 \\ 7 & 5 \end{pmatrix}$ and $Q = \begin{pmatrix} 5 & -2 \\ -7 & 3 \end{pmatrix}$.

Pre-multiplying Q by P, $PQ = \begin{pmatrix} 3 & 2 \\ 7 & 5 \end{pmatrix}\begin{pmatrix} 5 & -2 \\ -7 & 3 \end{pmatrix} = \begin{pmatrix} 1 & 0 \\ 0 & 1 \end{pmatrix} = I$.

Post-multiplying Q by P, $QP = \begin{pmatrix} 5 & -2 \\ -7 & 3 \end{pmatrix}\begin{pmatrix} 3 & 2 \\ 7 & 5 \end{pmatrix} = \begin{pmatrix} 1 & 0 \\ 0 & 1 \end{pmatrix} = I$.

Therefore $PQ = QP = I$.

For this reason Q is called the *multiplicative inverse* of P and is denoted by P^{-1}. We can also say that P is the multiplicative inverse of Q and is therefore denoted by Q^{-1}.

It is customary to use the phrase *inverse of a matrix* to refer to its multiplicative inverse, since its additive inverse is usually called its negative.

In general, if A and B are square matrices of the same order such that $AB = BA = I$, then B is an inverse of A and A is an inverse of B. It can be shown that if these inverses exist, then they are unique, so we can talk about *the* inverse of A or *the* inverse of B.

Example. If $A = \begin{pmatrix} 5 & -2 \\ 3 & -1 \end{pmatrix}$ and $B = \begin{pmatrix} -1 & 2 \\ -3 & 5 \end{pmatrix}$, show that A and B are inverses of each other.

We have to show that $AB = I = BA$.

$$AB = \begin{pmatrix} 5 & -2 \\ 3 & -1 \end{pmatrix}\begin{pmatrix} -1 & 2 \\ -3 & 5 \end{pmatrix} = \begin{pmatrix} 1 & 0 \\ 0 & 1 \end{pmatrix} = I$$

$$BA = \begin{pmatrix} -1 & 2 \\ -3 & 5 \end{pmatrix}\begin{pmatrix} 5 & -2 \\ 3 & -1 \end{pmatrix} = \begin{pmatrix} 1 & 0 \\ 0 & 1 \end{pmatrix} = I.$$

Since $AB = I = BA$, A and B are inverses of each other.

Exercise 7

In questions ***1*** to ***6***, show that each matrix is the inverse of the other.

1 $\begin{pmatrix} 3 & 2 \\ 1 & 1 \end{pmatrix}$ and $\begin{pmatrix} 1 & -2 \\ -1 & 3 \end{pmatrix}$

2 $\begin{pmatrix} 1 & 1 \\ 1 & 2 \end{pmatrix}$ and $\begin{pmatrix} 2 & -1 \\ -1 & 1 \end{pmatrix}$

3 $\begin{pmatrix} 3 & 4 \\ 5 & 7 \end{pmatrix}$ and $\begin{pmatrix} 7 & -4 \\ -5 & 3 \end{pmatrix}$

4 $\begin{pmatrix} 5 & -7 \\ -2 & 3 \end{pmatrix}$ and $\begin{pmatrix} 3 & 7 \\ 2 & 5 \end{pmatrix}$

5 $\begin{pmatrix} 3 & -2 \\ -4 & 3 \end{pmatrix}$ and $\begin{pmatrix} 3 & 2 \\ 4 & 3 \end{pmatrix}$

6 $\begin{pmatrix} 2 & 5 \\ 3 & 8 \end{pmatrix}$ and $\begin{pmatrix} 8 & -5 \\ -3 & 2 \end{pmatrix}$

Study the patterns in the entries of the pairs of matrices in questions ***1*** to ***6***. Use this pattern to *write down* the inverse of each of the matrices in questions 7 to ***15***. Check by multiplication.

7 $\begin{pmatrix} 2 & 1 \\ 1 & 1 \end{pmatrix}$

8 $\begin{pmatrix} 2 & 3 \\ 3 & 5 \end{pmatrix}$

9 $\begin{pmatrix} 4 & 3 \\ 9 & 7 \end{pmatrix}$

10 $\begin{pmatrix} 2 & -3 \\ -1 & 2 \end{pmatrix}$ *11* $\begin{pmatrix} 9 & -5 \\ -7 & 4 \end{pmatrix}$ *12* $\begin{pmatrix} 5 & -7 \\ 3 & -4 \end{pmatrix}$

13 $\begin{pmatrix} 1 & -1 \\ 0 & 1 \end{pmatrix}$ *14* $\begin{pmatrix} \cos\theta & \sin\theta \\ -\sin\theta & \cos\theta \end{pmatrix}$ *15* $\begin{pmatrix} 11 & 12 \\ 10 & 11 \end{pmatrix}$

16a $I = \begin{pmatrix} 1 & 0 \\ 0 & 1 \end{pmatrix}$. Show that I is its own inverse, i.e. $I \,.\, I = I$.

b Noting that $A = \begin{pmatrix} 2 & 0 \\ 0 & 2 \end{pmatrix} = 2\begin{pmatrix} 1 & 0 \\ 0 & 1 \end{pmatrix}$, write down A^{-1}, the inverse of A.

c Write down the inverse of : (*1*) $\begin{pmatrix} 3 & 0 \\ 0 & 3 \end{pmatrix}$ (2) $\begin{pmatrix} a & 0 \\ 0 & a \end{pmatrix}$, $a \neq 0$.

17 $M = \begin{pmatrix} 3 & 5 \\ 1 & 2 \end{pmatrix}$. Find, as in question 7, M^{-1}.

Investigate whether or not the squares of M and M^{-1} are also inverses of one another.

18 $A = \begin{pmatrix} 4 & 2 \\ 1 & 1 \end{pmatrix}$ and $B = \begin{pmatrix} \frac{1}{2} & -1 \\ -\frac{1}{2} & 2 \end{pmatrix}$. Show that $AB = I = BA$, and so $B = A^{-1}$.

---

8 *More about inverses of square matrices of order 2*

From your answers to questions 7 to *15* of Exercise 7, did you notice that:

(i) *the difference of the 'cross-products' of the entries was always* 1?

For example, from $\begin{pmatrix} 2 & 3 \\ 3 & 5 \end{pmatrix}$, $(2 \times 5) - (3 \times 3) = 10 - 9 = 1$.

(Note the order—the main diagonal product first.)

(ii) *the inverse matrix could be found by interchanging the entries in the main diagonal, and changing the signs of the entries in the other diagonal?*

Does every 2×2 matrix have an inverse?

To answer this question, consider the 2×2 matrix $A = \begin{pmatrix} a & b \\ c & d \end{pmatrix}$.

Pre-multiplying A by $\begin{pmatrix} d & -b \\ -c & a \end{pmatrix}$,

$$\begin{pmatrix} d & -b \\ -c & a \end{pmatrix}\begin{pmatrix} a & b \\ c & d \end{pmatrix} = \begin{pmatrix} ad-bc & 0 \\ 0 & ad-bc \end{pmatrix} = (ad-bc)\begin{pmatrix} 1 & 0 \\ 0 & 1 \end{pmatrix}.$$

Hence $$\left[\frac{1}{ad-bc}\begin{pmatrix} d & -b \\ -c & a \end{pmatrix}\right]\begin{pmatrix} a & b \\ c & d \end{pmatrix} = \begin{pmatrix} 1 & 0 \\ 0 & 1 \end{pmatrix}.$$

Similarly, post-multiplying A by $\frac{1}{ad-bc}\begin{pmatrix} d & -b \\ -c & a \end{pmatrix}$ we obtain

$$\begin{pmatrix} a & b \\ c & d \end{pmatrix}\left[\frac{1}{ad-bc}\begin{pmatrix} d & -b \\ -c & a \end{pmatrix}\right] = \begin{pmatrix} 1 & 0 \\ 0 & 1 \end{pmatrix}.$$

It follows that if $ad-bc \neq 0$, the matrix $A = \begin{pmatrix} a & b \\ c & d \end{pmatrix}$ has inverse

$$A^{-1} = \frac{1}{ad-bc}\begin{pmatrix} d & -b \\ -c & a \end{pmatrix}.$$

$ad-bc$ is called the *determinant* of the matrix A, and is written det A. If det $A = 0$, A does not have an inverse, and is called a *singular matrix*. If det $A \neq 0$, then A is said to be non-singular.

Example. Given $P = \begin{pmatrix} 4 & 3 \\ -1 & -2 \end{pmatrix}$, find, if possible, P^{-1}.

$\det P = 4(-2)-(-1)3 = -8+3 = -5 \neq 0$ so P^{-1} exists.

$$P^{-1} = \frac{1}{\det P}\begin{pmatrix} -2 & -3 \\ 1 & 4 \end{pmatrix} = \frac{1}{-5}\begin{pmatrix} -2 & -3 \\ 1 & 4 \end{pmatrix} = \begin{pmatrix} \frac{2}{5} & \frac{3}{5} \\ -\frac{1}{5} & -\frac{4}{5} \end{pmatrix}$$

Exercise 8

State whether each of the matrices in question ***1*** to ***15*** has an inverse. If the inverse exists, find it.

1 $A = \begin{pmatrix} 4 & 3 \\ 2 & 2 \end{pmatrix}$ **2** $A = \begin{pmatrix} 2 & 1 \\ 4 & 3 \end{pmatrix}$ **3** $A = \begin{pmatrix} 3 & 2 \\ 1 & 2 \end{pmatrix}$

4 $A = \begin{pmatrix} 4 & 2 \\ 10 & 5 \end{pmatrix}$ **5** $A = \begin{pmatrix} 7 & 4 \\ 16 & 9 \end{pmatrix}$ **6** $A = \begin{pmatrix} 2 & 3 \\ 1 & 4 \end{pmatrix}$

7 $P = \begin{pmatrix} 2 & 1 \\ 4 & 2 \end{pmatrix}$ **8** $Q = \begin{pmatrix} 5 & 7 \\ 6 & 9 \end{pmatrix}$ **9** $B = \begin{pmatrix} 1 & 1 \\ 1 & 0 \end{pmatrix}$

10 $A = \begin{pmatrix} 2 & -3 \\ 1 & 5 \end{pmatrix}$ **11** $X = \begin{pmatrix} 1 & 6 \\ 3 & 4 \end{pmatrix}$ **12** $Y = \begin{pmatrix} 2 & 3 \\ -1 & 2 \end{pmatrix}$

13 $C = \begin{pmatrix} 3 & -1 \\ -1 & 2 \end{pmatrix}$ **14** $M = \begin{pmatrix} -2 & 4 \\ 1 & -1 \end{pmatrix}$ **15** $A = \begin{pmatrix} -1 & 2 \\ 0 & 3 \end{pmatrix}$

16 Given the matrices $A = \begin{pmatrix} 2 & 3 \\ 0 & 1 \end{pmatrix}$ and $B = \begin{pmatrix} 2 & 5 \\ 1 & 3 \end{pmatrix}$, calculate:

a AB *b* BA *c* A^{-1} *d* B^{-1}

e $(AB)^{-1}$ *f* $A^{-1}B^{-1}$ *g* $B^{-1}A^{-1}$ *h* $(BA)^{-1}$

List any equalities between pairs of these matrices.

17 $A = \begin{pmatrix} 2 & 1 \\ 4 & 3 \end{pmatrix}$ and $B = \begin{pmatrix} 1 & -1 \\ -2 & 3 \end{pmatrix}$. Verify that $(AB)^{-1} = B^{-1}A^{-1}$.

18 Assuming that the matrix $\begin{pmatrix} a & b \\ c & d \end{pmatrix}$ is the inverse of the matrix $\begin{pmatrix} 1 & 2 \\ 3 & 4 \end{pmatrix}$, $\begin{pmatrix} a & b \\ c & d \end{pmatrix}\begin{pmatrix} 1 & 2 \\ 3 & 4 \end{pmatrix} = \begin{pmatrix} 1 & 0 \\ 0 & 1 \end{pmatrix}$. Use this equation to find the entries a, b, c, d. Check your result by multiplication.

19 Using the method of question **18**, find the inverse of each of the following matrices:

a $\begin{pmatrix} 4 & 2 \\ 1 & 1 \end{pmatrix}$ *b* $\begin{pmatrix} 1 & 4 \\ 0 & 2 \end{pmatrix}$ *c* $\begin{pmatrix} 3 & 1 \\ 4 & 2 \end{pmatrix}$ *d* $\begin{pmatrix} 3 & -1 \\ -2 & 2 \end{pmatrix}$

20 Solve the matrix equation $\begin{pmatrix} 3 & 1 \\ 3 & 2 \end{pmatrix}X = \begin{pmatrix} 0 & 7 \\ 9 & 2 \end{pmatrix}$ for 2×2 matrix X by pre-multiplying each side by the inverse of matrix $\begin{pmatrix} 3 & 1 \\ 3 & 2 \end{pmatrix}$. Note that $IX = X$, where I is the identity matrix of order 2.

9 *Using matrices to solve systems of linear equations*

Consider the following system of equations in which x and y are variables on the set of real numbers.

$$\left.\begin{aligned} 3x+\ \ y &= 9 \\ 3x+2y &= 12 \end{aligned}\right\} \qquad \ldots (1)$$

Since $\begin{pmatrix} 3x+\ \ y \\ 3x+2y \end{pmatrix} = \begin{pmatrix} 3 & 1 \\ 3 & 2 \end{pmatrix}\begin{pmatrix} x \\ y \end{pmatrix}$, system (1) may be written as a single matrix equation:

$$\begin{pmatrix} 3 & 1 \\ 3 & 2 \end{pmatrix}\begin{pmatrix} x \\ y \end{pmatrix} = \begin{pmatrix} 9 \\ 12 \end{pmatrix} \qquad \ldots (2)$$

If we can find an equation equivalent to (2) of the form $\begin{pmatrix} x \\ y \end{pmatrix} = \begin{pmatrix} a \\ b \end{pmatrix}$, the solution of the system can be written down at once. To do this, we make use of the fact that the product of a matrix and its inverse is the identity matrix I, and proceed as follows.

For matrix $\begin{pmatrix} 3 & 1 \\ 3 & 2 \end{pmatrix}$, det $= (3\times 2)-(1\times 3) = 3$,

so $\begin{pmatrix} 3 & 1 \\ 3 & 2 \end{pmatrix}^{-1} = \frac{1}{3}\begin{pmatrix} 2 & -1 \\ -3 & 3 \end{pmatrix}$.

Pre-multiplying both sides of (2) by the inverse $\frac{1}{3}\begin{pmatrix} 2 & -1 \\ -3 & 3 \end{pmatrix}$,

$$\frac{1}{3}\begin{pmatrix} 2 & -1 \\ -3 & 3 \end{pmatrix}\begin{pmatrix} 3 & 1 \\ 3 & 2 \end{pmatrix}\begin{pmatrix} x \\ y \end{pmatrix} = \frac{1}{3}\begin{pmatrix} 2 & -1 \\ -3 & 3 \end{pmatrix}\begin{pmatrix} 9 \\ 12 \end{pmatrix}$$

$$\Leftrightarrow \quad \begin{pmatrix} 1 & 0 \\ 0 & 1 \end{pmatrix}\begin{pmatrix} x \\ y \end{pmatrix} = \frac{1}{3}\begin{pmatrix} 6 \\ 9 \end{pmatrix}$$

$$\Leftrightarrow \quad \begin{pmatrix} x \\ y \end{pmatrix} = \begin{pmatrix} 2 \\ 3 \end{pmatrix}$$

Hence $x = 2$ and $y = 3$, which gives $\{(2, 3)\}$ as the solution set of the system.

Replacing x by 2 and y by 3 in (1) readily verifies that $\{(2, 3)\}$ is the solution set of the system. This check is always worth making.

Exercise 9

Find the solution sets of the systems of equations in questions ***1–12*** by matrix methods; the variables are on the set of real numbers.

1 $\begin{aligned} x-y &= 5 \\ x+y &= 11 \end{aligned}$ *2* $\begin{aligned} x-y &= 0 \\ x+y &= 8 \end{aligned}$ *3* $\begin{aligned} 3x+y &= 7 \\ 3x+2y &= 5 \end{aligned}$

4 $\begin{aligned} 3x-y &= 5 \\ 2x+y &= 15 \end{aligned}$ *5* $\begin{aligned} 2x+y &= 5 \\ 2x+3y &= -1 \end{aligned}$ *6* $\begin{aligned} 3x-4y &= 18 \\ 5x+y &= 7 \end{aligned}$

7 $\begin{aligned} 2x+y &= 12 \\ 3x-2y &= 25 \end{aligned}$ *8* $\begin{aligned} x+2y &= 11 \\ 2x-y &= 2 \end{aligned}$ *9* $\begin{aligned} 2x+3y &= 5 \\ 4x-5y &= 21 \end{aligned}$

10 $\begin{aligned} 5x-3y &= 9 \\ 7x-6y &= 9 \end{aligned}$ *11* $\begin{aligned} 10x+5y+3 &= 0 \\ 5x+10y+9 &= 0 \end{aligned}$ *12* $\begin{aligned} 2x-y &= 10 \\ 5y-6x &= 10 \end{aligned}$

13 Find the inverse of the matrix $\begin{pmatrix} 2 & -1 \\ 1 & 3 \end{pmatrix}$, and use it to solve the following systems:

a $\begin{aligned} 2x-y &= 0 \\ x+3y &= 7 \end{aligned}$ *b* $\begin{aligned} 2x-y &= 9 \\ x+3y &= 1 \end{aligned}$ *c* $\begin{aligned} 4x-2y-5 &= 0 \\ 2x+6y+1 &= 0 \end{aligned}$

14 Try to solve the following systems of equations by matrices. Explain, with the aid of a Cartesian diagram, why you failed.

a $\begin{aligned} x+y &= 4 \\ 3x+3y &= 12 \end{aligned}$ *b* $\begin{aligned} 2x-y &= 3 \\ 6x-3y &= 15 \end{aligned}$

Let A, B, X be square matrices of the same order such that $AX = B$. Then if the inverse A^{-1} of A exists, we can find X as follows:

$$AX = B \Leftrightarrow A^{-1}AX = A^{-1}B \Leftrightarrow IX = A^{-1}B \Leftrightarrow X = A^{-1}B.$$

Use the fact that if $AX = B$ and A^{-1} exists, then $X = A^{-1}B$ to solve the matrix equations in questions ***15–19***.

15 $\begin{pmatrix} 2 & 1 \\ 1 & 1 \end{pmatrix} X = \begin{pmatrix} 1 & 1 \\ 2 & 0 \end{pmatrix}$ *16* $\begin{pmatrix} 2 & -3 \\ -1 & 2 \end{pmatrix} X = \begin{pmatrix} 1 & 2 \\ 3 & 4 \end{pmatrix}$

17 $\begin{pmatrix} 2 & 1 \\ 4 & 3 \end{pmatrix} X = \begin{pmatrix} 4 & -3 \\ -2 & 1 \end{pmatrix}$ *18* $\begin{pmatrix} -3 & 2 \\ -1 & 5 \end{pmatrix} X = \begin{pmatrix} 2 & 5 \\ 3 & -1 \end{pmatrix}$

19 If $ps \neq qr$, find the 2×2 matrix X such that

$$\begin{pmatrix} p & q \\ r & s \end{pmatrix} X = \begin{pmatrix} q & p \\ s & r \end{pmatrix}.$$

20a Find the inverse of the matrix $\begin{pmatrix} \cos\theta & \sin\theta \\ -\sin\theta & \cos\theta \end{pmatrix}$.

b Given that $\begin{pmatrix} \cos\theta & \sin\theta \\ -\sin\theta & \cos\theta \end{pmatrix}\begin{pmatrix} x \\ y \end{pmatrix} = \begin{pmatrix} x' \\ y' \end{pmatrix}$, find equations giving x, y in terms of x', y', $\sin\theta$ and $\cos\theta$.

Summary

1 *A matrix is a rectangular array of numbers arranged in rows and columns, the array being enclosed in round (or square) brackets.* The numbers are called *entries* or *elements.*

2 The *order of a matrix* is given by the number of rows followed by the number of columns,

e.g. $\begin{pmatrix} 3 & 1 & 7 \\ 4 & 2 & 5 \end{pmatrix}$ *Order*: 2×3 $\begin{pmatrix} 3 & 7 \\ 9 & 4 \end{pmatrix}$ *Order*: 2×2, *or* a square matrix of order 2.

3 *Two matrices are equal* if and only if they are of the same order and their corresponding entries are equal.

4 *A zero matrix*, $\boldsymbol{O}$, is a matrix whose elements are all zero.

5 *A unit matrix*, I, is a square matrix whose elements in the main diagonal are unity and whose other elements are all zero,

e.g. $I = \begin{pmatrix} 1 & 0 \\ 0 & 1 \end{pmatrix}$ and $I = \begin{pmatrix} 1 & 0 & 0 \\ 0 & 1 & 0 \\ 0 & 0 & 1 \end{pmatrix}$

6 *Addition of matrices*
If A and B are two matrices of the same order, the sum of A and B, denoted by $A+B$, is the matrix obtained by adding each entry of A to the corresponding entry of B,

$$\underset{A}{\begin{pmatrix} a & b \\ c & d \end{pmatrix}} + \underset{B}{\begin{pmatrix} p & q \\ r & s \end{pmatrix}} = \underset{A+B}{\begin{pmatrix} a+p & b+q \\ c+r & d+s \end{pmatrix}}$$

7 *Subtraction of matrices*
If A and B are two matrices of the same order, to subtract B from A, the negative of B is added to A, i.e.

$$A - B = A + (-B).$$

The *negative* of B is the matrix whose entries are the negatives of the entries in B, so that $B + (-B) = \boldsymbol{O}$.

$$\text{e.g.}\quad \overset{A}{\begin{pmatrix} a & b \\ c & d \end{pmatrix}} - \overset{B}{\begin{pmatrix} p & q \\ r & s \end{pmatrix}} = \overset{A}{\begin{pmatrix} a & b \\ c & d \end{pmatrix}} \overset{+}{+} \overset{(-B)}{\begin{pmatrix} -p & -q \\ -r & -s \end{pmatrix}}$$

$$\overset{A-B}{= \begin{pmatrix} a-p & b-q \\ c-r & d-s \end{pmatrix}}$$

8 *Multiplication of matrices by real numbers* (*scalars*)

To multiply a matrix by a real number k, we multiply each entry by that number,

e.g. $k\begin{pmatrix} a & b \\ c & d \end{pmatrix} = \begin{pmatrix} ka & kb \\ kc & kd \end{pmatrix}$ This operation is *scalar multiplication*.

9 *Multiplication of two matrices*

a $\begin{pmatrix} a & b \\ c & d \end{pmatrix}\begin{pmatrix} x \\ y \end{pmatrix} = \begin{pmatrix} ax+by \\ cx+dy \end{pmatrix}$

b $\begin{pmatrix} a & b \\ c & d \end{pmatrix}\begin{pmatrix} p & q \\ r & s \end{pmatrix} = \begin{pmatrix} ap+br & aq+bs \\ cp+dr & cq+ds \end{pmatrix}$

Rule. Multiply '*row into column and add the products*'. Multiplication of matrices is not in general commutative; it is associative, and distributive with respect to matrix addition.

10 *Inverse of a* 2×2 *matrix*

The *inverse* of the matrix $A = \begin{pmatrix} a & b \\ c & d \end{pmatrix}$ is the matrix

$A^{-1} = \dfrac{1}{ad-bc}\begin{pmatrix} d & -b \\ -c & a \end{pmatrix}$, provided that $ad-bc \neq 0$.

$ad-bc$ is the *determinant* of matrix A. If $\det A = 0$, A has no inverse, and is said to be a *singular* matrix.

Property. $A^{-1}A = AA^{-1} = I$.

11 *Matrix equations*

If A, B and X are square matrices of the same order such that $AX = B$ and A has an inverse A^{-1}, then

$$AX = B \Leftrightarrow A^{-1}AX = A^{-1}B \Leftrightarrow IX = A^{-1}B \Leftrightarrow X = A^{-1}B.$$

Note to the Teacher on Chapter 4

A relation from a set A to a set B may be

(i) many-many, e.g. the relation expressed by 'x is the uncle of y'
(ii) one-many, e.g. the relation expressed by 'x is the father of y'
(iii) many-one, e.g. the relation expressed by 'x has as father y'
(iv) one-one, e.g. the relation expressed by 'x has as capital y'.

The first two are of limited interest at school level, except where the sets A and B are equal, as for instance in order relations on the set of real numbers. But (iii) and (iv), which are mappings or functions, are of great interest and importance. Chapters in Books 3 and 5 introduced the ideas informally and laid the groundwork for a good understanding of function, an understanding which has been deepened through the study of transformations in geometry.

In this chapter, the concept of function is studied in more detail, and the new topics of *composition of functions* and *inverse functions* are carefully discussed. As before, technical language has been kept to a minimum in the text. However, it may be helpful to the teacher to mention some of the more advanced descriptions of function.

(*a*) $f: A \to B$ is said to be a mapping from A to B if the set of images of the elements of A is a subset of B. If the set of images of the elements of A is denoted by $f(A)$, then $f(A) \subset B$.

(*b*) $f: A \to B$ is said to be a mapping from A *onto* B if $f(A) = B$. Such a function is called a *surjection*.

(*c*) Of particular interest are functions which are *one-one*, i.e. those in which no two elements of the domain have the same image. This means that if $a, b \in A$, then $f(a) = f(b) \Rightarrow a = b$. A one-one function is called an *injection*.

(*d*) A function f which is *onto* and *one-one*, i.e. both surjective and injective, is called a *bijection*.

These four cases are illustrated in the following figure.

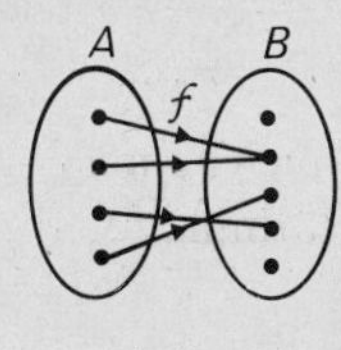

$f(A) \subset B$
'to'

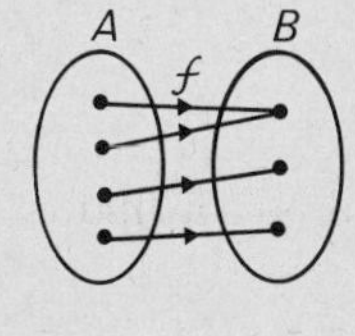

$f(A) = B$
'onto' (surjective)

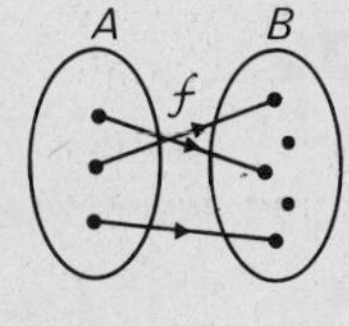

'one-one'
(injective)

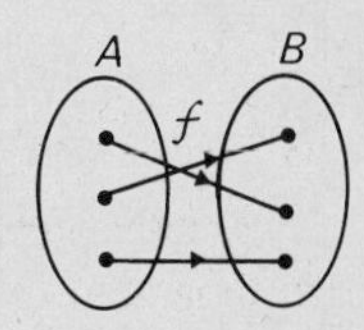

'one-one' and 'onto'
(bijective)

If $f:A \to B$ is a bijection, then there exists a function $g:B \to A$ such that $g \circ f = I_A$ and $f \circ g = I_B$, where I_A and I_B are the identity functions on the sets A and B respectively. Function g is said to be an *inverse* of the function f and is denoted by f^{-1}.

Section 1. In this Section the previous work on relations and mappings is reviewed, using a more precise approach. Exercise 1 is basic and should be taken carefully. Exercise 1B probes further and introduces the pupil to new functions which may be required at a later stage. Among these, there is the important *modulus function* and its graph.

Sections 2 and 3 are concerned with the composition of functions and the properties of composition. As suggested in the text, a simple way of introducing composition of functions is to consider the 'think of a number' type of problem. In the text example, we have the steps:

$$n \to 2n \to 2n+13$$

The sequence of steps can with advantage be illustrated by a flow chart.

The important facts to be thoroughly understood by the pupil are that:

(i) $g \circ f$ means 'Use f first and then follow it by g', or 'g after f'

(ii) $(g \circ f)(x)$ means $g(f(x))$.

In this connection, an arrow diagram or a flow chart is very helpful.

Although pupils are likely to accept without question the curious convention that '$g \circ f$' means 'f then g', it is worth noting the reason for this convention. It is simply that the custom of writing $f(a)$, rather than af, for the image of a under the mapping f, leads to writing $g(f(a))$, rather than afg, for the image of a under 'f then g'.

From Exercise 2, it is seen that composition of functions is not in general commutative. Other properties of composition such as associativity are investigated in Exercise 3. The associative property for composition of functions comes, roughly speaking, from the fact that both $(h \circ g) \circ f$ and $h \circ (g \circ f)$ mean 'f then g then h'. A formal proof of associativity is given in *Algebra and Number Systems*, Chapter 3 (Blackie/Chambers).

Sections 4 and 5 investigate the conditions under which a function has an inverse, and develop a method for finding a formula for the inverse function when it exists.

The work of Exercise 4 leads to the conclusions stated in colour at the end of the exercise. Pupils proceeding further with algebra will meet further sophistications, but the present content should give them a fair working knowledge of the subject, accompanied by a reasonable precision of language and ideas.

Section 5 suggests a method for finding a formula for an inverse function based on the fact that if f and f^{-1} are inverses, then

$$f^{-1}(y) = x \quad \Leftrightarrow \quad f(x) = y.$$

Figures 15 and 16 illustrate the ideas involved. Some manipulation is required in Exercise 5, but the basic principles involving functions and their inverses should not be lost sight of; questions **22** and **23**, in particular, should assist in this respect.

Functions, Composition of Functions, and Inverse Functions

1 Review of relations and functions

In Book 5 we looked at some relations and mappings from one set to another set. In the present chapter we develop these ideas further and study ways in which functions can be combined.

Example 1. Let $A = \{1, 4, 9\}$ and $B = \{1, 2, 3, 4\}$. Show in arrow diagrams the following relations from A to B:

(i) *is greater than* (ii) *is the square of* (iii) *is paired with* 2

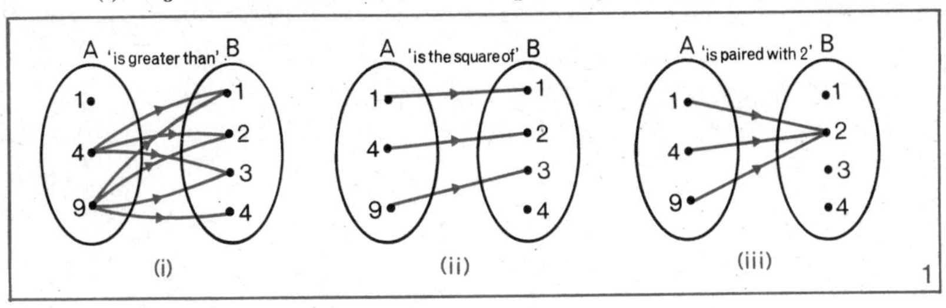

In (ii) the relation is a *mapping* since each element of A is related to exactly one element of B.

For the same reason the relation in (iii) is also a mapping.

Relations which are mappings are important in mathematics and are often called *functional relations*, or simply *functions*. A function is therefore another name for a mapping, but both terms are useful.

A *function f*, or a *mapping f*, from a set A to a set B is a relation in which each element of A is related to exactly one element of B. We write $f: A \rightarrow B$ ('f maps A to B').

If f maps an element $x \in A$ to an element $y \in B$, we say that 'y is the image of x under f', and denote this image by $f(x)$ (read 'f of x'); we also write $f: x \rightarrow f(x)$.

The set A is called the *domain* of the function and the set of images in B is called the *range* of the function (see Figure 2(i)).

In Worked Example 1(ii) above, the domain is $\{1, 4, 9\}$ and the range is $\{1, 2, 3\}$; in (iii) the domain is $\{1, 4, 9\}$ and the range is $\{2\}$, which is a subset of B.

In the special mapping illustrated in Figure 2(ii) in which the elements of the two sets are paired so that each element of A corresponds to one element of B, *and vice versa*, the sets A and B are said to be in *one-to-one correspondence*.

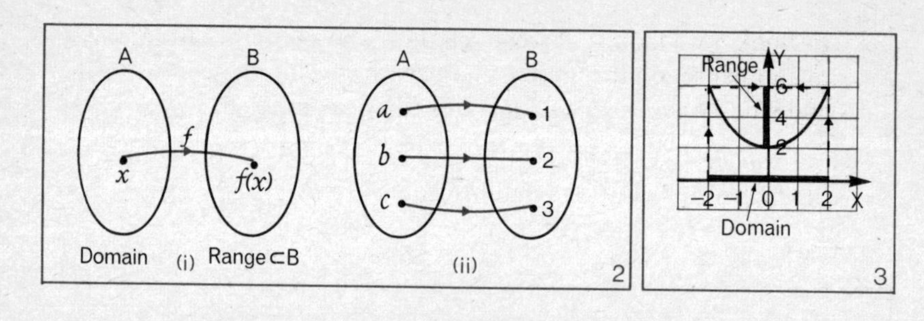

Example 2. $A = \{x: -2 \leqslant x \leqslant 2, x \in R\}$, and $f: A \rightarrow R$ is defined by $f(x) = x^2 + 2$. Find $f(-2)$, $f(0)$ and $f(2)$, and sketch the graph of f. State the range of f.

$$f(x) = x^2 + 2 \quad \Rightarrow \quad \begin{aligned} f(-2) &= (-2)^2 + 2 = 6 \\ f(0) &= 2 \\ f(2) &= 2^2 + 2 = 6 \end{aligned}$$

Hence the points $(-2, 6)$, $(0, 2)$ and $(2, 6) \in$ the graph of f.

The graph of f is shown in Figure 3.

The range of f is $\{y: 2 \leqslant y \leqslant 6, y \in R\}$.

Exercise 1

1 Relations from set $A = \{a, b, c\}$ to set $B = \{p, q, r\}$ are shown in Figure 4.

a Which of the relations are mappings?

b Which shows a one-to-one correspondence between the sets A and B?

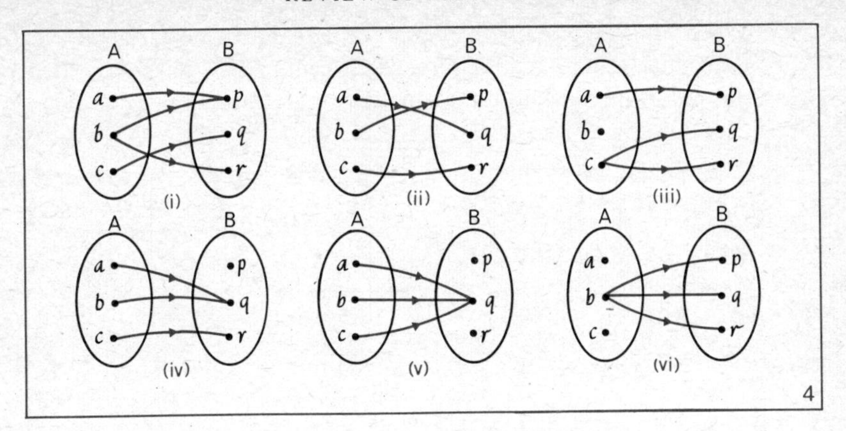

2 The function $f: R \to R$ is defined by $f(x) = x^2 + 1$.

a Calculate $f(2), f(-1), f(4)$ and $f(-3)$.
b Given $f(a) = 50$, find a.

3 The function $g: R \to R$ is defined by $g(x) = 2^x$.

a Find the values of g at 3, 2, 1, 0, -1, -2.
b What element of the domain has 64 as its image?

4 The function $h: S \to R$, where $S = \{0, 1, 2, 3, 4\}$, is given by $h(x) = x + 1$.

a Find the range of h, i.e. the image set.
b Illustrate h by a Cartesian graph.

5 The function $f: R \to R$ is defined by $f(x) = x^2$.

a Find the set of images of the elements 3, 2, 1, 0, -1, -2, -3 of the domain.
b Sketch the graph of $f (x \in R)$, and state the range of the function.

6 $A = \{x: -4 \leqslant x \leqslant 4, x \in R\}$, and $f: A \to R$ is defined by $f(x) = x^2 - 9$.

a Copy and complete this table of values of f:

x	-4	-3	-2	-1	0	1	2	3	4
$f(x)$									

b Sketch the graph of f, and state the range of the function.

7 Which of the graphs in Figure 5 could illustrate functions? (Remember that for a function each element of the domain is related to exactly one element of the range.)

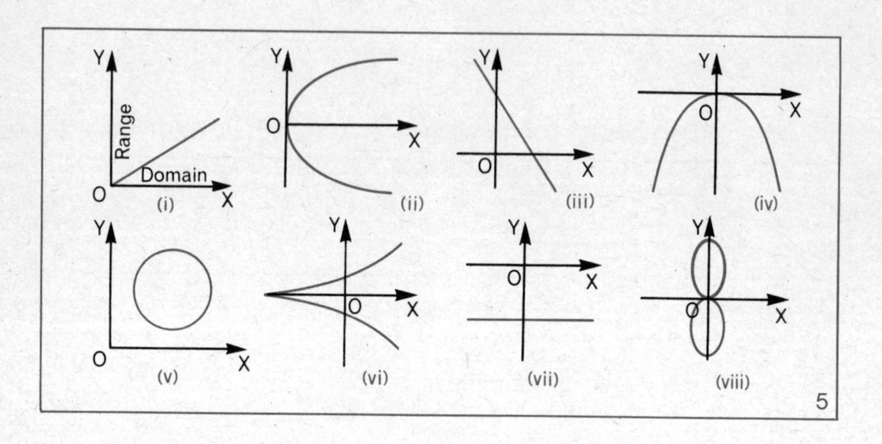

8 $A = \{0, 1, 2, \ldots, 10\}$ and $B = \{0, 1, 2, 3, 4\}$. A relation r from A to B is defined by the sentence 'x divided by 5 leaves remainder y'. ($x \in A$ and $y \in B$.)

a Show this relation by: (*1*) a set of ordered pairs $\{(0, 0), \ldots, (10, 0)\}$.
(*2*) a Cartesian graph.

b Is the relation r a function? Give a reason for your answer.

9 Under a mapping on the coordinate plane, the point P(x, y) is mapped to the point P$'(x', y')$, where $x' = x + 1$ and $y' = y - 3$. P$'$ is the image of P.

a Find the images of $(2, 4)$, $(-3, 3)$, $(-1, -3)$, $(3, -2)$ under the mapping.

b Give equations for the mapping which maps each image point P$'$ to P.

10 $A = \{x: -1 \leqslant x \leqslant 4, x \in R\}$, and $f: A \rightarrow R$ is defined by $f(x) = x^2 - 2x - 3$.

a Find the zeros of f, i.e. replacements of x such that $f(x) = 0$.

b Calculate $f(0), f(1), f(2), f(4)$.

c Hence sketch the graph of f, and state the range of f.

11 $A = \{x: -3 \leqslant x \leqslant 3, x \in R\}$, and $f: A \to R$ is defined by $f(x) = x(x+2)(x-2)$.

a Find the zeros of f.
b Calculate $f(1), f(-1), f(3), f(-3)$.
c Hence sketch the graph of f, and state the range of f.
d Solve the inequation $f(x) > 0$.

12 $A = \{x: -3 \leqslant x \leqslant 3, x \in R\}$, and $f: A \to R$ is defined by $f(x) = \sqrt{(9-x^2)}$.

a Calculate $f(0)$, $f(1)$, $f(2)$, $f(3)$, $f(-1)$, $f(-2)$, $f(-3)$, to two significant figures where necessary.
b Sketch the graph of f, and deduce the graph of $g: A \to R$, where $g(x) = -\sqrt{(9-x^2)}$.
c Complete: 'The union of the graph of f and the graph of g is ...'.

Exercise 1B (including some special functions)

(*Where the domain of a function is not stated, it is understood to be R.*)

1 The function $f: x \to c$, where c is a constant, is called a *constant function*. f assigns each real number to c, e.g. $f(3) = c, f(-1) = c$.

a Complete the following table for $f: x \to 4$ (i.e. $c = 4$).

x	-3	-2	-1	0	1	2	3
$f(x)$	4						

b Hence sketch the graph of f for domain R.

2 The function $I: A \to A$ defined by $I(x) = x$ is called the *identity function* on A. I assigns each member of the domain to itself, e.g. $3 \to 3$, $-1 \to -1$.

Complete a table similar to that in question ***1*** for the identity function on R, and hence sketch the graph of I.

3 $S = \{-2, -1, 0, 1, 2\}$. Show the following relations on S as

a a set of ordered pairs *b* a graph:

(1) *is greater than* (2) *is less than.* Compare the results.

These are called *inverse relations*. If (a, b) belongs to one relation, then (b, a) belongs to the inverse relation.

4 The relation $R_1 = \{(-1, -5), (0, -3), (1, -1), (2, 1), (3, 3), (4, 5)\}$.

a Write down the inverse relation R_2 as a set of ordered pairs.
b Show R_1 and R_2 on the same Cartesian diagram.
c Draw the axis of symmetry for the union of the graphs of R_1 and R_2, and state its equation.
d Is it true that R_1 and R_2 are functions?

The *modulus*, or *absolute value*, $|x|$, of a real number x is defined by: $|x| = x$ if $x \geqslant 0$, and $|x| = -x$ if $x < 0$.
For example, $|5| = 5$, $|-3| = 3$, $|0| = 0$.

5 The *modulus function* M defined by $M(x) = |x|$ assigns each real number to its absolute value. Complete this table, and hence draw the graph of M for $x \in R$.

x	-3	-2	-1	0	1	2	3
$\|x\|$	3						

6 Sketch the graphs of:

a $f: x \rightarrow -|x|$ *b* $g: x \rightarrow 2+|x|$ *c* $h: x \rightarrow x+|x|$

7 Sometimes a function is defined by different formulae over different intervals. Sketch the Cartesian graph of f defined by:

$$f(x) = \begin{cases} 2 \text{ when } x < 0 \\ x+2 \text{ when } x \geqslant 0 \end{cases}$$

8 Sketch the graph of the *step function* S defined by:

$$S(x) = \begin{cases} 0 \text{ when } 0 \leqslant x < 1 \\ 1 \text{ when } 1 \leqslant x < 2 \\ 2 \text{ when } 2 \leqslant x < 3 \end{cases}$$

9 The function $f: R \rightarrow R$ defined by a formula of the form $f(x) = ax+b$, where a and b are constant, is called the *linear function*.

a Show that $\dfrac{f(p)-f(q)}{p-q} = a$. Explain this result geometrically.

b Show also that $f(x+1)-f(x)$ is constant.

10 If a linear function $f: x \rightarrow ax+b$ is such that $f(1) = -1$ and $f(3) = 5$, find a and b. Hence find x for which $f(x) = 0$.

11 The function $q: R \to R$ defined by a formula of the form $q(x) = ax^2 + bx + c$, where a, b and c are constants, is called the *quadratic function.*

Show that if $f(x) = q(x+1) - q(x)$, then $f(x)$ defines a linear function f.

12 $f: R \to R$ is a given function, and a function f' is defined by $f'(x) = \lim_{h \to 0} \dfrac{f(x+h) - f(x)}{h}$. f' is called the *derived function.* Find the values of $f'(x)$, given:

a $f(x) = x$ ***b*** $f(x) = x^2$ ***c*** $f(x) = x^3$.

---

2 *Composition of functions*

In Book 2, Algebra, we met problems of the 'Think of a number' type. For example: 'n is a whole number. Double it, and add 13. The result is 23. Find n'. From the data,

$$2n + 13 = 23 \quad \Rightarrow \quad 2n = 10 \quad \Rightarrow \quad n = 5$$

This simple example illustrates an important kind of combination of two functions: (i) the '*multiply by* 2' function, and (ii) the '*add* 13' function.

The first function maps 5 to 2×5, i.e. 10, and the second function maps 10 to $10 + 13$, i.e. 23, as illustrated by the flow chart in Figure 6.

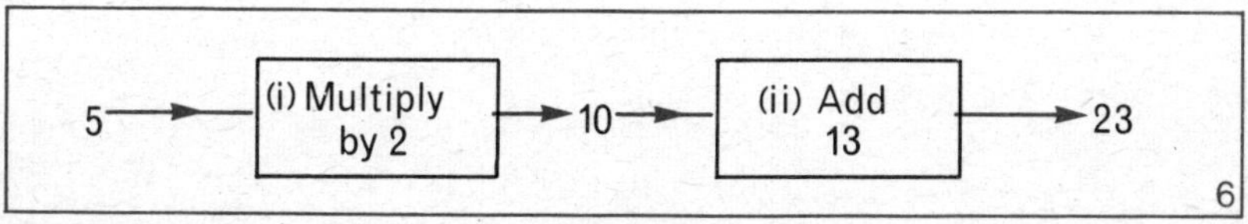

6

Combining the functions in this way suggests that there may be a third function which maps 5 directly to 23. Denote the 'multiply by 2' function by f and the 'add 13' function by g, so that

$$f(x) = 2x \quad \text{and} \quad g(x) = x + 13, \quad \text{for} \quad x \in R.$$

Using f first and then g.

$$f(5) = 2 \times 5 = 10 \quad \text{and} \quad g(10) = 10 + 13 = 23.$$

For $x \in R, f(x) = 2x$ and $g(2x) = 2x + 13$.

Now let a function $f: R \to R$ be defined by $h(x) = 2x + 13$.

Then $h(5) = (2 \times 5) + 13 = 23$, which checks, and so function h maps 5 directly to 23.

h is called the *composite* of f and g, and is denoted by $g \circ f$ (read 'g circle f'). Notice that h is defined by using f first and then g, as in the composition of transformations in geometry.

Formula for a composite function

Functions $f: A \to B$ and $g: B \to C$ are shown in Figure 7. Using f first and then g, f maps x to y, and g maps y to z.

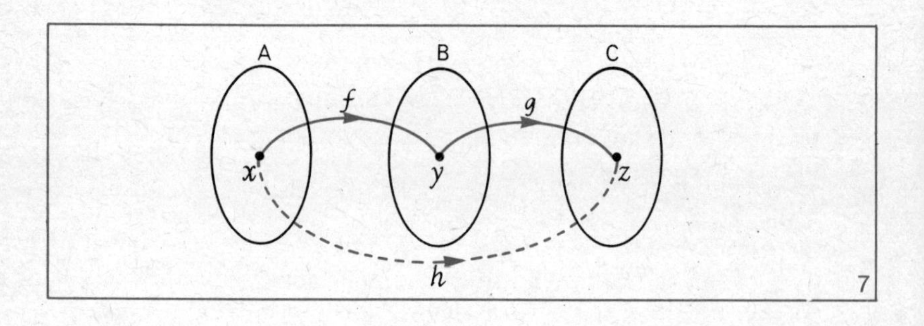

Since $y = f(x)$ and $z = g(y) = g(f(x))$, then the function $h: A \to C$ defined by the formula

$$h(x) = g(f(x)), \quad \text{(read '}g\text{ of }f\text{ of }x\text{')}$$

is the *composite* of f and g, and is denoted by

$$h = g \circ f.$$

Hence, $h(x) = (g \circ f)(x) = g(f(x))$, for all $x \in A$.

Note. $g \circ f$ is sometimes referred to as a '*function of a function*'.

Example. Functions $f: R \to R$ and $g: R \to R$ are defined by the formulae $f(x) = x^2$ and $g(x) = x - 1$.

Find (i) $(g \circ f)(3)$ (ii) $(f \circ g)(3)$ (iii) formulae for the functions $g \circ f$ and $f \circ g$.

(i) $(g \circ f)(3) = g(f(3)) = g(9) = 9-1 = 8$
(ii) $(f \circ g)(3) = f(g(3)) = f(2) = 2^2 = 4$
(iii) $g \circ f$ is the function from R to R defined by

$$(g \circ f)(x) = g(f(x)) = g(x^2) = x^2-1.$$

$f \circ g$ is the function from R to R defined by

$$(f \circ g)(x) = f(g(x)) = f(x-1) = (x-1)^2.$$

Note. In the above example the functions $g \circ f$ and $f \circ g$ are not equal, and so composition of functions is not commutative.

Exercise 2

1 Functions f: $A \to B$ and g: $B \to C$ are defined by Figure 8.

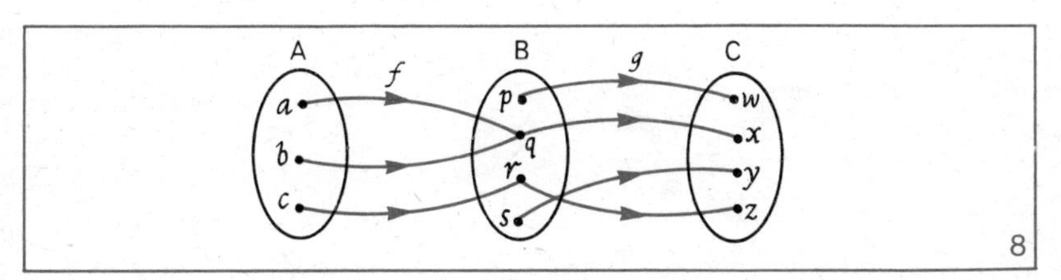

a Copy and complete the following:

$$\begin{array}{l} a \xrightarrow{f} q \xrightarrow{g} x \\ b \longrightarrow \ldots \longrightarrow \ldots \\ c \longrightarrow \ldots \longrightarrow \ldots \end{array}$$

b Hence give $(g \circ f)(a)$, $(g \circ f)(b)$ and $(g \circ f)(c)$.

c State the domain and range of the function $g \circ f$.

2 f and g are functions on the set of integers defined by the formulae $f(x) = x+1$ and $g(x) = 2x-1$.

a Copy and complete this table:

$$\begin{array}{l} x \xrightarrow{f} f(x) \xrightarrow{g} g(f(x)) \\ 2 \\ 1 \\ 0 \\ -1 \end{array}$$

b Show that $(g \circ f)(x) = 2x+1$, and use this formula for $g \circ f$ to check the entries in the third column. Illustrate the mappings in an arrow diagram.

3 Functions $f: R \to R$ and $g: R \to R$ are defined by $f(x) = 2x+1$ and $g(x) = 3x$.

a Knowing that $(g \circ f)(x) = g(f(x))$, calculate $(g \circ f)(3)$ and $(g \circ f)(-2)$ as in the worked example.
b Find $g(f(x))$ and use it to verify mentally the above values of $g \circ f$.

4 For the functions in question **3**, knowing that $(f \circ g)(x) = f(g(x))$, calculate $(f \circ g)(3)$, $(f \circ g)(-2)$ and $(f \circ g)(x)$.

5 The mappings $f: R \to R$ and $g: R \to R$ are defined by $f(x) = x+2$ and $g(x) = x^2$. Find:

a $(g \circ f)(1)$, $(g \circ f)(-3)$ and $(g \circ f)(x)$
b $(f \circ g)(1)$, $(f \circ g)(-3)$ and $(f \circ g)(x)$.

6 Repeat question **5** for the mappings $f: R \to R$ and $g: R \to R$ defined by $f(x) = x+1$ and $g(x) = x^3$.

7 For the functions $g: R \to R$ and $h: R \to R$ defined by $g(x) = 2x$ and $h(x) = x^2+4$, find, in simplest form:

a $(h \circ g)(x)$ *b* $(g \circ h)(x)$ *c* $(g \circ g)(x)$ *d* $(h \circ h)(x)$.

8 All of the following are mappings from R to R. Find a formula for $g \circ f$ in each case.

a $f(x) = x+1, g(x) = x^2$
b $f(x) = x+3, g(x) = 2x+1$
c $f(x) = x-1, g(x) = x^2+x+1$
d $f(x) = x^2, g(x) = 2x^2+1$
e $f(x) = 2x-1, g(x) = x^2-1$
f $f(x) = x^2+1, g(x) = \dfrac{1}{x^2+1}$
g $f(x) = -3x, g(x) = x^3-2x$
h $f(x) = x^2, g(x) = \sin x^\circ$.

9 For the functions in question **8**, find formulae for $f \circ g$.

10 Mappings $g: R \to R$ and $h: R \to R$ are defined by $g(x) = 1-3x$ and $h(x) = x^2-1$.

a Find the number $(h \circ g)(3)$. *b* If $(h \circ g)(x) = 3$, find x.

11 T_1 is the translation defined by $\begin{pmatrix} x \\ y \end{pmatrix} \to \begin{pmatrix} x+2 \\ y+3 \end{pmatrix}$, and T_2 is the translation defined by $\begin{pmatrix} x \\ y \end{pmatrix} \to \begin{pmatrix} x-1 \\ y+1 \end{pmatrix}$.

A(2, 4), B(4, 2) and C(2, 1) are the vertices of triangle ABC. Find the coordinates of the images of A, B and C under $T_2 \circ T_1$. Is $T_2 \circ T_1 = T_1 \circ T_2$?

12 Functions $g: R \to R$ and $h: R \to R$ are defined by $g(x) = 2x+1$ and $h(x) = x^2+2$.

a Find a formula defining $h \circ g$, giving your answer in the form px^2+qx+r.
b State the range of: *(1)* g and *(2)* h.
c What are the domain and range of $h \circ g$?
d What elements of the domain of $h \circ g$ have image 27?

---

3 *Some properties of composition of functions*

In earlier books we have studied the effects of each of the binary operations addition and multiplication on the various number systems. We inquired whether the sets were *closed* under the operation, whether the operation was *associative*, whether there was an *identity* element, whether each element had an *inverse* with respect to the operation and whether the operation was *commutative*. Since composition of functions is a binary operation, it is of interest to ask some of the same questions about this operation.

Exercise 3

1 Let f be a mapping from a set A to a set B, g be a mapping from the set B to a set C, and h be a mapping from the set C to a set D, as shown in Figure 9. Then $g \circ f$ is a mapping from A to C and $h \circ g$ is a mapping from B to D.

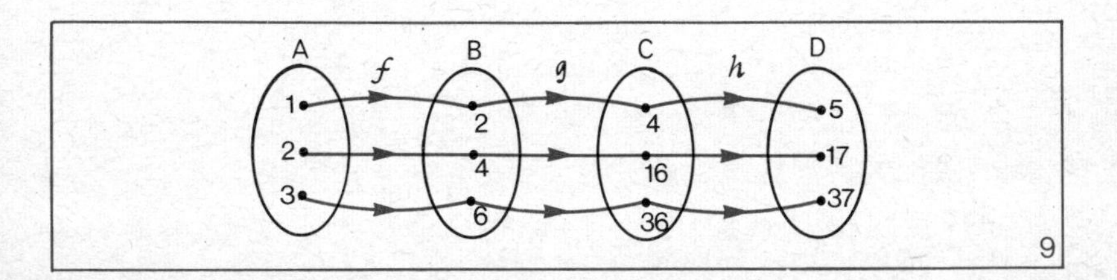

9

a Notice from the diagram that:

(1) $(g \circ f)(1) = 4$ (2) $h(4) = 5$ (3) $(h \circ (g \circ f))(1) = 5$
(4) $f(1) = 2$ (5) $(h \circ g)(2) = 5$ (6) $((h \circ g) \circ f)(1) = 5$

It follows that $(h \circ (g \circ f))(1) = ((h \circ g) \circ f)(1)$.

b Repeat ***a*** for the element 2 of set A.

c Repeat ***a*** for the element 3 of set A.

2 Let u be a mapping from a set A to a set B, v be a mapping from the set B to a set C, w be a mapping from the set C to a set D such that $u: x \to x^2$, $v: x \to x+5$ and $w: x \to 2x$. If $A = \{-2, 0, 1, 3\}$ show the mappings in a diagram similar to that in question ***1***.

a Find the values of $v \circ u$, and hence find the values of $w \circ (v \circ u)$.

b Find the values of u, and hence obtain the values of $(w \circ v) \circ u$.

c Hence show that $w \circ (v \circ u) = (w \circ v) \circ u$.

3 Let S be the set of all points in the coordinate plane. Let $f: S \to S$ be such that $f: (x, y) \to (x+1, y)$ and $g: S \to S$ be such that $g: (x, y) \to (y, x)$.

a Find the image of origin O(0, 0) under $g \circ f$ and under $f \circ g$.

b Find the image of the point (4, 3) under $g \circ f$ and under $f \circ g$.

c Find the image of the point (a, b) under $g \circ f$ and under $f \circ g$.

4 Let S be the set of all points in the coordinate plane. Let X, Y, Z be mappings: $S \to S$ such that $X: (x, y) \to (x, -y)$, $Y: (x, y) \to (-x, y)$ and $Z: (x, y) \to (-y, x)$. Find the image of the point (a, b) under the mappings:

a $Y \circ Z$ ***b*** $X \circ (Y \circ Z)$ ***c*** $X \circ Y$ ***d*** $(X \circ Y) \circ Z$

*Note. As questions **1**, **2**, **4** suggest, 'composition of functions' is in fact associative. Thus the brackets are not needed; in question **1** for instance, the function mapping A to D could be written $h \circ g \circ f$.*

5 $f: R \to R$, $g: R \to R$ and $h: R \to R$ are functions defined by $f(x) = 2x$, $g(x) = x+1$ and $h(x) = x^2$.

a Noting that $(h \circ g \circ f)(x) = (h \circ g)f(x) = (h \circ g)(2x) = h(2x+1) = (2x+1)^2$, show that $(f \circ g \circ h)(x) = 2(x^2+1)$.

b Calculate $(h \circ g \circ f)(2)$ and $(f \circ g \circ h)(2)$.

6 $f: R \to R$, $g: R \to R$ and $h: R \to R$ are functions defined by $f(x) = x-2$, $g(x) = x^3$ and $h(x) = 4x$.

a Show that $(h \circ g \circ f)(x) = 4(x-2)^3$, and $(f \circ g \circ h)(x) = 64x^3 - 2$.

b Calculate $(h \circ g \circ f)(1)$ and $(f \circ g \circ h)(1)$.

7 *a* The identity function I on R is defined by $I(x) = x$. Functions $u: R \rightarrow R$ and $v: R \rightarrow R$ are defined by $u(x) = 2x-3$ and $v(x) = x^2 - x + 5$.

Find $I \circ u, \quad u \circ I, \quad I \circ v, \quad v \circ I$.

b If f is any function from R to R, show that

$$I \circ f = f \circ I = f.$$

Note. This example shows that I has the identity property for composition of functions from R to R.

8 P and Q are transformations of the coordinate plane defined by $P: \begin{pmatrix} x \\ y \end{pmatrix} \rightarrow \begin{pmatrix} 2x+1 \\ y+3 \end{pmatrix}$ and $Q: \begin{pmatrix} x \\ y \end{pmatrix} \rightarrow \begin{pmatrix} 3-2x \\ 2y+3 \end{pmatrix}$. Complete:

a $Q \circ P: \begin{pmatrix} x \\ y \end{pmatrix} \rightarrow \begin{pmatrix} \quad \end{pmatrix}$ *b* $P \circ Q: \begin{pmatrix} x \\ y \end{pmatrix} \rightarrow \begin{pmatrix} \quad \end{pmatrix}$.

9 Linear functions $f: R \rightarrow R$ and $g: R \rightarrow R$ are defined by $f(x) = ax+b$ and $g(x) = cx+d$. Show that the composite of two linear functions, in either order, is a linear function.

10 Linear function $f: R \rightarrow R$ is defined by $f(x) = ax+b$ and a quadratic function $g: R \rightarrow R$ is defined by $g(x) = px^2+qx+r$. Show that the composite of a linear and quadratic function, in either order, is a quadratic function.

4 *Inverse functions*

Of the questions raised in Section 3, we now consider that involving inverses.

Let f be a function which maps a set A to a set B. Then each element $a \in A$ has an image $f(a) = a'$, say, in B. Can we find a function g which maps B to A so that $g(a') = a$? (See Figure 10.)

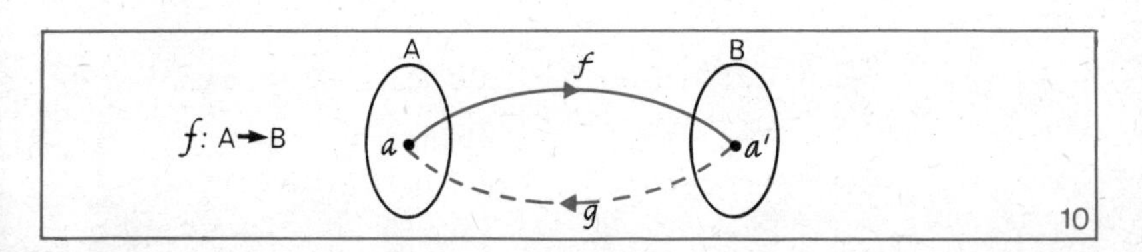

If the function g exists, then f and g are called *inverse functions*; g is the *inverse* of f. We can also say that f is the inverse of g.

Do all functions have inverses? We investigate this in Exercise 4.

Exercise 4

1 $f: A \to B$ is defined by the arrow diagram in Figure 11. Show in an arrow diagram the inverse $g: B \to A$.

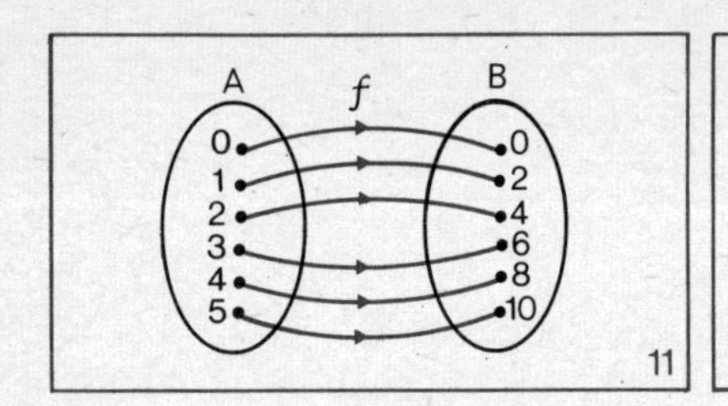

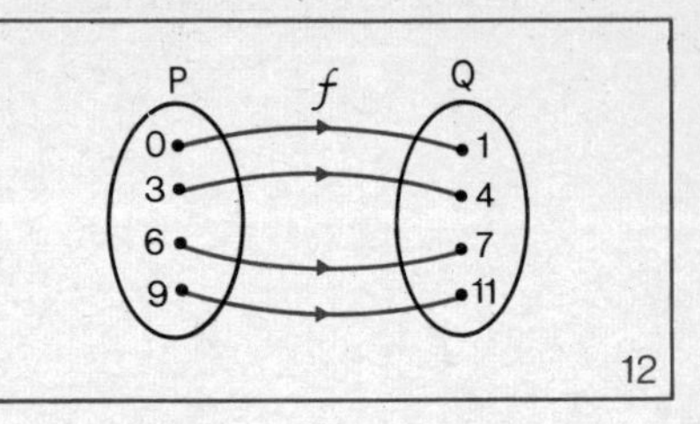

2 $f: P \to Q$ is defined in Figure 12. Show in an arrow diagram the inverse $g: Q \to P$.

3 $A = \{1, 2, 3, 4\}$ and $B = \{4, 5, 6, 7\}$. $f: A \to B$ maps x in A to $x+3$ in B.

a Show f in an arrow diagram.
b Show g, the inverse of f, in another arrow diagram, and complete the statement: g maps y in B to . . . in A.

4 $A = \{-2, -1, 0, 1, 2\}$ and $B = \{0, 1, 4\}$. $f: A \to B$ maps x in A to x^2 in B.

a Show f in an arrow diagram.
b Explain why an inverse function from B to A is not defined.

5 $A = \{-2, -1, 0, 1, 2\}$ and B is the set of cubes of the members of A. $f: A \to B$ maps x in A to x^3 in B.

a Show f in an arrow diagram.
b Has f an inverse function $g: B \to A$? If so, what is $g(y)$ for y in B?

6 In Figure 13, $f: A \to B$ and the inverse function $g: B \to A$.

a Find $(g \circ f)(a)$, $(g \circ f)(b)$, $(g \circ f)(c)$.
b Find $(f \circ g)(p)$, $(f \circ g)(q)$, $(f \circ g)(r)$.
c Complete: $g \circ f$ is the identity function on set ... and $f \circ g$ is the identity function on set ....

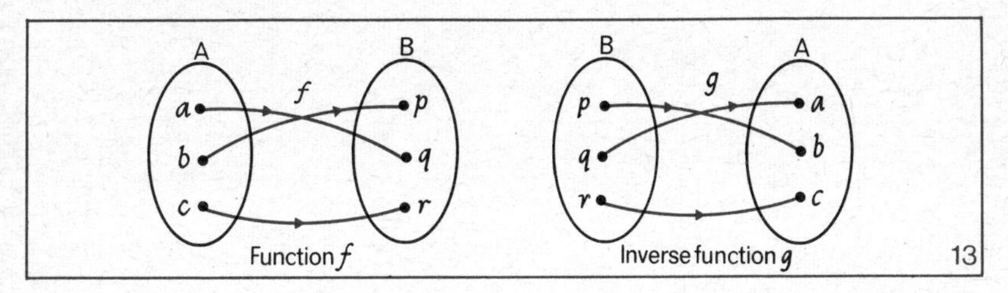

7 Figure 14 shows mappings (i) $f: A \to B$ (ii) $g: C \to D$ (iii) $h: E \to F$. Study the diagrams carefully and decide in which case reversing the arrows gives a mapping from the second set to the first set.

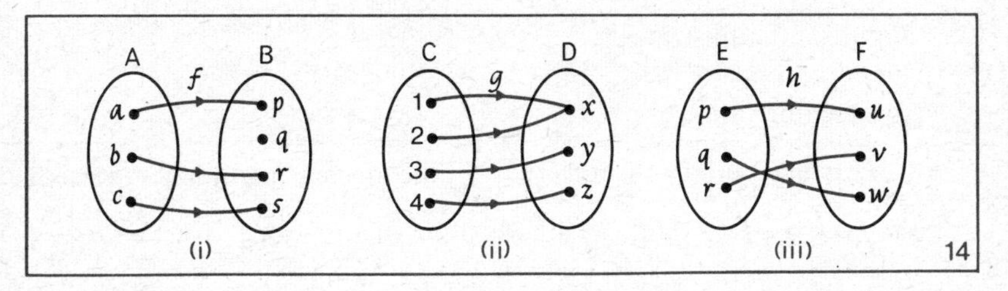

Exercise 4 shows that some functions do not have inverse functions.

A function $f: A \to B$ has an *inverse* function $g: B \to A$ if each member of B is the image of a unique member of A, i.e. if the sets A and B are in one-to-one correspondence.

When g exists, it is denoted by f^{-1} (read as 'f inverse'); the range of f is the domain of f^{-1}, and the domain of f is the range of f^{-1}

5 *Finding a formula for an inverse function*

If f and f^{-1} are inverse functions, then $f(x) = y \Leftrightarrow f^{-1}(y) = x$, as illustrated in Figure 15.

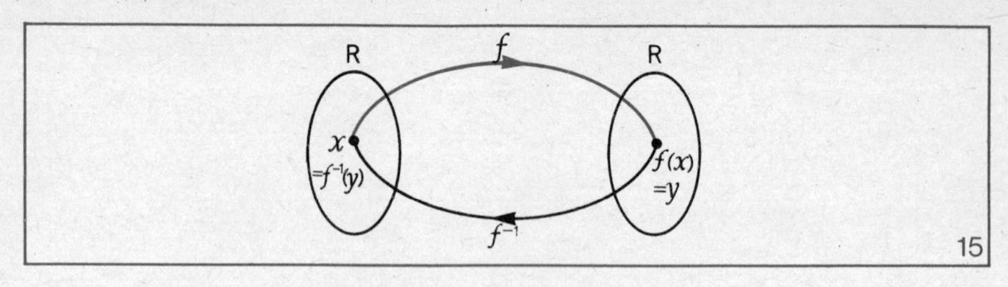

Let $f: R \rightarrow R$ be defined by $f(x) = 2x-3$, and let y be the image of x under f.

$$
\begin{aligned}
& f(x) = y \\
\Leftrightarrow \quad & 2x-3 = y \\
\Leftrightarrow \quad & 2x = y+3 \\
\Leftrightarrow \quad & x = \tfrac{1}{2}(y+3) \\
\Leftrightarrow \quad & f^{-1}(y) = \tfrac{1}{2}(y+3)
\end{aligned}
$$

i.e. the inverse function is $f^{-1}(x) = \frac{1}{2}(x+3)$.

Figure 16 illustrates these composite and inverse functions.

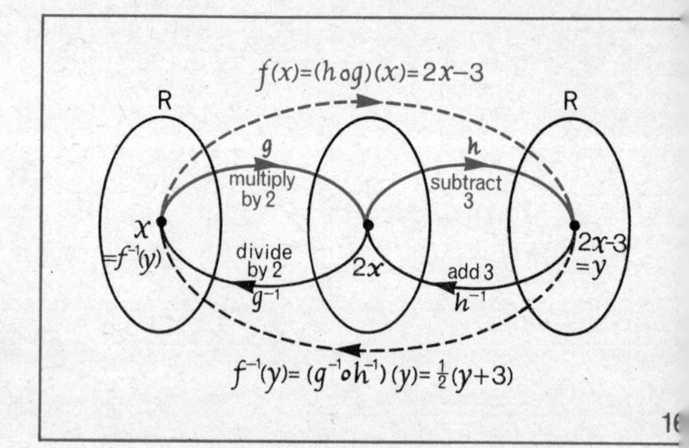

Example. Find a formula for f^{-1}, the inverse function of f defined by $f(x) = \dfrac{2}{3-4x}$. State a suitable domain and range for f.

Let y be the image of x under f.

$$
\begin{aligned}
& f(x) = y \\
\Leftrightarrow \quad & \frac{2}{3-4x} = y \\
\Leftrightarrow \quad & 2 = 3y-4xy \\
\Leftrightarrow \quad & 4xy = 3y-2 \\
\Leftrightarrow \quad & x = \frac{3y-2}{4y} \\
\Leftrightarrow \quad & f^{-1}(y) = \frac{3y-2}{4y}
\end{aligned}
$$

Hence the required formula is $f^{-1}(x) = \dfrac{3x-2}{4x}$.

Domain is $A = \{x : x \neq \frac{3}{4}, x \in R\}$, range is $B = \{x : x \neq 0, x \in R\}$.

Exercise 5

In questions ***1–12***, $f: R \to R$. In each case find a formula for the inverse function f^{-1}.

1	$f(x) = 2x+5$	2	$f(x) = 3x-1$	3	$f(x) = 1-3x$
4	$f(x) = 1-x$	5	$f(x) = x-\frac{1}{2}$	6	$f(x) = 1-\frac{1}{2}x$
7	$f(x) = \frac{1}{2}x+3$	8	$f(x) = \frac{1}{2}(x+9)$	9	$f(x) = \frac{1}{3}(2x-5)$
10	$f(x) = \frac{1}{2}(1-3x)$	*11*	$f(x) = \frac{2}{3}(5-x)$	*12*	$f(x) = x^3-4$

13 $A = \{x : x > 0, x \in R\}$ and f, g, h are functions defined on A by $f(x) = x+1$, $g(x) = 2x$, $h(x) = x^2$.

a Find formulae for the inverse functions f^{-1}, g^{-1}, h^{-1}.
b Evaluate: (*1*) $f^{-1}(5)$ (2) $g^{-1}(8)$ (3) $h^{-1}(4)$.

14 Functions $f: R \to R$ and $g: R \to R$ are defined by $f(x) = 2x$ and $g(x) = x+2$.

a Write down formulae for the inverse functions f^{-1} and g^{-1}.
b Find formulae for $g \circ f$, $f^{-1} \circ g^{-1}$ and $(g \circ f)^{-1}$.
c State the connection between $(g \circ f)^{-1}$ and $f^{-1} \circ g^{-1}$.

15 Functions $f: R \to R$ and $g: R \to R$ are defined by $f(x) = x^3$ and $g(x) = 3x-4$.

Find: ***a*** $f^{-1}(8)$ ***b*** $g^{-1}(8)$ **c** $(f^{-1} \circ g^{-1})(8)$ ***d*** $(g^{-1} \circ f^{-1})(8)$

In each of the following questions find a formula for f^{-1}, the inverse of f. State a suitable domain for f in each case.

16	$f(x) = \sqrt{(3-x)}$	*17*	$f(x) = \sqrt{x}-3$	*18*	$f(x) = 2\sqrt{x}+5$
19	$f(x) = \dfrac{1}{x+1}$	*20*	$f(x) = \dfrac{2}{3-x}$	*21*	$f(x) = \dfrac{x}{x-4}$

22a Sketch the graph of the function $f: R \to R$ defined by $f(x) = x^2$.
b Give a reason why no inverse exists.
c Define a suitably restricted domain for f which will permit the existence of an inverse function f^{-1}.
d Give a formula for f^{-1}, and sketch its graph.

23 Answer the following for each graph in Figure 17.

a Could the graph be the graph of a function?
b If so, does the function have an inverse?
c If the answer to ***b*** is 'no', can this be changed to 'yes' by restricting the domain of the function?
d If the answer to ***a*** is 'no', can this be changed to 'yes'?
e State in words, using the language of coordinate geometry, a condition which ensures that a graph is that of a function with an inverse.

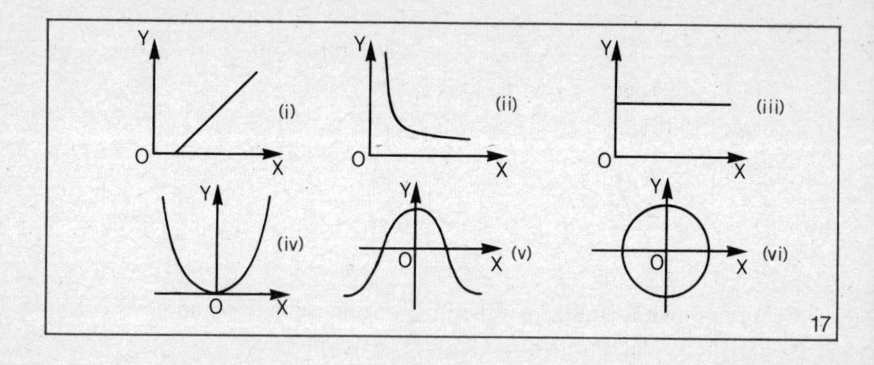

Summary

1 *A function, or mapping,* from a set A to a set B is a special kind of relation in which each element of A is related to exactly one element of B. (See Figure 1.)

2 *A function f may be described by*:

(i) a set of ordered pairs, e.g. $\{(2, 4), (3, 6)\}$

(ii) a formula, e.g. $f(x) = 2x$, and the domain and range of the function

(iii) an arrow diagram, or a Cartesian graph.

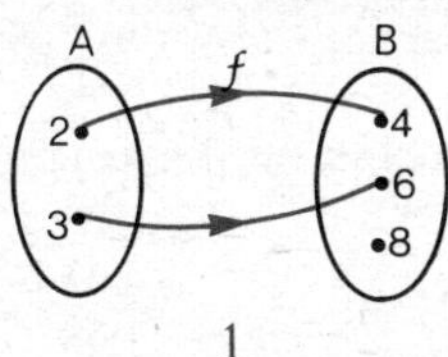

1

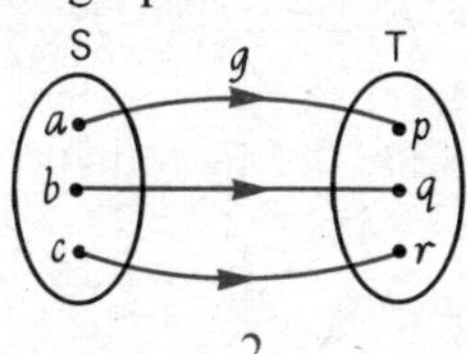

2

$f: A \to B; f(x) = 2x$, or $x \to 2x$. *Domain* $\{2, 3\}$, *range* $\{4, 6\}$.

S and T are in one-to-one correspondence.

3 *Composition of functions.* If f and g are functions such that $f: A \to B$ and $g: B \to C$, then the composite function $g \circ f: A \to C$ is defined by the formula $(g \circ f)(x) = g(f(x))$, $x \in A$. (See Figure 3.)

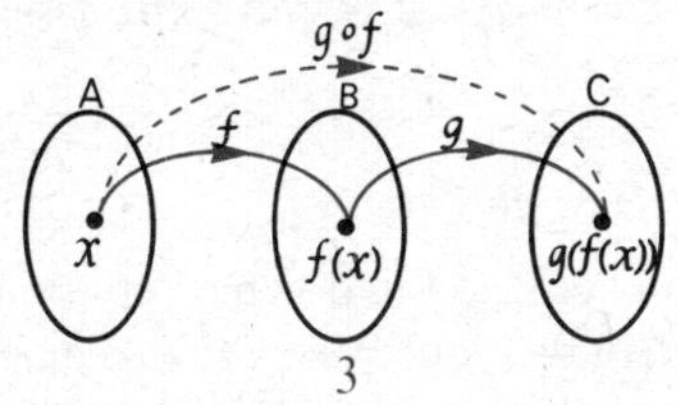

3

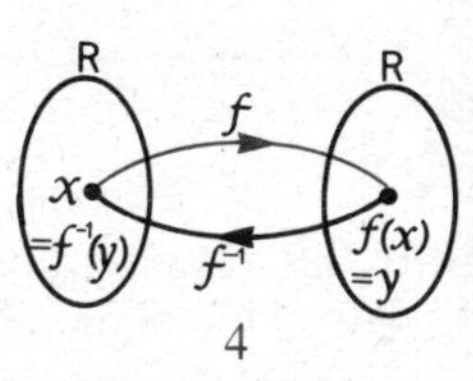

4

4 *Composition of functions* is not commutative, i.e. $g \circ f \neq f \circ g$ in general, but is associative, i.e. $[h \circ (g \circ f)](x) = [(h \circ g) \circ f](x) = (h \circ g \circ f)(x)$.

5 *Inverse functions.* A function $f: A \to B$ has an inverse function $g: B \to A$ if each member of B is the image of a unique member of A, i.e. if the sets A and B are in one-to-one correspondence.

When g exists, it is denoted by f^{-1}. (Figure 4.)

$$f^{-1}(y) = x \quad \Leftrightarrow \quad f(x) = y.$$

Topics to explore

(i) The Fibonacci sequence

Consider the ratios of successive terms of the Fibonacci sequence 1, 1, 2, 3, 5, 8, ... Let $R_1 = \frac{1}{1}, R_2 = \frac{2}{1}, R_3 = \frac{3}{2}$, etc.

a Plot $R_1, R_2, \ldots, R_6$ on graph paper. Show R_1, R_3, R_5 in red and the rest in blue.

b Given $R_7 = 1{\cdot}6154$, $R_8 = 1{\cdot}6190$, $R_9 = 1{\cdot}6176$, $R_{10} = 1{\cdot}6182$, $R_{11} = 1{\cdot}6180$, $R_{12} = 1{\cdot}6181$ plot $R_7, R_8, \ldots, R_{12}$ using 2 cm to 0·01 for R_n.

c On the diagram for ***b*** draw a line parallel to the n-axis at a distance $\frac{1+\sqrt{5}}{2}$ from it.

If you have drawn your diagram carefully it should *suggest* that as n gets larger and larger R_n has limit $\frac{1+\sqrt{5}}{2}$. This is in fact true.

The limit, or rather its reciprocal, is the ratio of the Golden Section and appears in many contexts. Verify that R_6 gives the ratio (length/breadth) for a sheet of foolscap paper, approximately. Find out what you can about the Golden Section in books about mathematics, art, biology, architecture.

(ii) Hire purchase

In this form of transaction you agree to put down a deposit and then pay off the remainder of the debt by regular instalments over a given period. The vendor or finance company naturally charges you interest on the debt outstanding.

Let us suppose you are paying £10 per month over 6 months, the rate of interest is 1% per month, and the original debt after paying the deposit is £x.

Then at the end of 1 month the debt has grown by 1%, i.e. by a factor of 1·01, and you have reduced it by £10, so that the remaining debt in £ is $(1{\cdot}01)x - 10$.

In the same way, after two months the debt in £ is

$$(1{\cdot}01)[(1{\cdot}01)x - 10] - 10 = (1{\cdot}01)^2x - 10(1{\cdot}01 + 1)$$

and after three months

$$(1{\cdot}01)[(1{\cdot}01)^2x-10(1{\cdot}01+1)]-10 = (1{\cdot}01)^3x-10[(1{\cdot}01)^2+(1{\cdot}01)+1]$$

The pattern is now clear. After six months the debt in £ is

$$(1{\cdot}01)^6x-10[(1{\cdot}01)^5+(1{\cdot}01)^4+\ldots+(1{\cdot}01)+1]$$

Since this is now zero,

$$(1{\cdot}01)^6x = 10[(1{\cdot}01)^5+(1{\cdot}01)^4+\ldots+(1{\cdot}01)+1]$$

To find x we need to sum the series on the right-hand side.

a Use long multiplication, or a machine, or Pascal's triangle, to calculate $(1{\cdot}01)^n$ for $n = 1, 2, \ldots, 6$.
b Use the formula for the sum of a geometric series to find x.
c How much more do you pay in this case than you would have paid in cash? Check by comparing with simple interest at 12% on £30 for 6 months.
d Compare with Hire Purchase advertisements in the newspapers.

(iii) The invention of logarithms

Napier, the inventor of logarithms, made use of a striking one-to-one correspondence between geometric and arithmetic sequences. Let us consider a simpler case than Napier's:

1,	2,	4,	8,	16,	32,	64,	128,	256
NIL,	I,	II,	III,	IV,	V,	VI,	VII,	VIII

a Check that the *product* of two arabic numbers always corresponds to the *sum* of the corresponding roman numbers.
b Explain in your own words how this enables a multiplication problem to be *transformed* into an addition problem.
c Choose a different geometric sequence, and use it to replace 1, 2, 4, ..., etc., above. Does the property still hold?
d Choose from Chapter 2 a geometric sequence with positive terms and form an arithmetic sequence from it by taking the logarithm of each term from your three-figure tables.
e How could you draw a graph of a geometric sequence in such a way that the points lay on a straight line? Notice that this is often done in serious newspapers for graphs of population growth, inflation, etc.

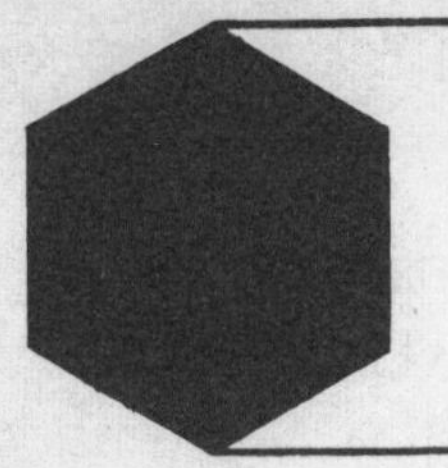

Revision Exercises

Revision Exercise on Chapter 1

Systems of Equations

Revision Exercise 1

(Assume that the variables are on R.)

1 By finding their solution sets, show that the following systems of equations are equivalent:

$$\left.\begin{aligned} 3x-2y &= 17 \\ 2x-5y &= 4 \end{aligned}\right\} \quad \text{and} \quad \left.\begin{aligned} 4x-5y &= 18 \\ 3x+2y &= 25 \end{aligned}\right\}$$

2 AB is the line $x+y+1=0$ and AC is the line $2x-7y+11=0$. Calculate the coordinates of the point A.

3 In the formula $Q = a+bt$, $Q = 4$ when $t = 2$ and $Q = -8$ when $t = 6$. Find a and b, and hence calculate Q when $t = 3$.

4 The sides of a triangle are formed by the lines $x-y+3=0$, $x+y-7=0$ and $x-2y+5=0$. Find the coordinates of the three vertices of the triangle.

Solve the systems of equations in questions **5** to ***10***.

5 $\begin{aligned} x+y+z &= 5 \\ 3x-y+2z &= 11 \\ 5x-3y+z &= 13 \end{aligned}$ *6* $\begin{aligned} x+y &= -1 \\ y+z &= 1 \\ z+x &= 8 \end{aligned}$ *7* $\begin{aligned} 3x+2y+5z &= 1 \\ -x+y+z &= 3 \\ 4x+3y+3z &= 2 \end{aligned}$

8 $\begin{aligned} 2x-y+3z &= 20 \\ 5x+2y+z &= 4 \\ 7x-3y-2z &= 6 \end{aligned}$ *9* $\begin{aligned} 2x+3y+5z &= 13 \\ 6x-5y+4z &= -3 \\ 4x-8y+15z &= -16 \end{aligned}$ *10* $\begin{aligned} 3x-4y &= -1 \\ 5y+4z &= 12 \\ 3z+2x &= 4 \end{aligned}$

11 The circle $x^2+y^2+2gx+2fy+c=0$ passes through the points $(0, -1)$, $(2, 3)$ and $(1, 6)$. Find g, f and c.

12 Find a, b and c such that the expression x^3+ax^2+bx+c is equal to zero when x is replaced by 1, 3 or -2.

13 The parabola $y=ax^2+bx+c$ passes through the points $(1, -2)$, $(-2, 7)$ and $(3, 12)$. Find a, b and c.

14 Find the solution set of each of the following systems of equations and illustrate by sketch graphs:

a $y=x^2-4$, $y=x+2$ **b** $x^2+y^2=25$, $x+y+1=0$ **c** $y=1-x^2$, $y=1-x$

Solve the following systems of equations:

15 $x^2-2xy+3=0$, $2x+y-4=0$ *16* $x^2+3y^2=21$, $x-3y=9$ *17* $2x^2-3xy-2y^2=0$, $x+2y=4$

18 $2x^2-3y^2=5$, $2x-3y=1$ *19* $x^2-2y^2+1=0$, $2x+3y=1$ *20* $2x+3y+1=0$, $x^2-xy-y^2+2x=1$

21
$$x+y+z=0$$
$$x-y+z=4$$
$$xyz=48$$

22
$$x+y+2z+6u=3$$
$$x+2y+3z+7u=6$$
$$x-y+z+u=3$$
$$x+3y-z+2u=0$$

23 The circle $x^2+y^2+2gx+2fy+c=0$ passes through the points $(0,0)$, $(4,0)$ and $(0,-3)$. Show that the circle has equation $x^2+y^2-4x+3y=0$.

Find the coordinates of the points in which the line $x+y=1$ cuts the circle.

24a Find the solution set of the system: $y=x^2$, $y=x+2$, and illustrate the solution in a sketch graph.

b Shade that part of the graph which shows the solution set of the system of inequations: $y \geqslant x^2$, $y \leqslant x+2$.

Revision Exercise on Chapter 2

Sequences and Series

Revision Exercise 2

1 Find the first three terms of the sequences given by the formulae:

a $u_n=4n-1$ **b** $u_n=4(n-1)$ **c** $u_n=n^2-1$ **d** $u_n=(n-1)^2$

2 In the sequences given by question ***1a*** and ***c***, which terms are 143, and in the sequences given by ***b*** and ***d*** which terms are 64?

3 Find a simple formula for the nth terms of the sequences starting:

a 5, 10, 15, 20, ... ***b*** 100, 10, 1, 0·1, ... ***c*** $-1, 2, -3, 4, \ldots$

4 Find the first four terms of the sequences with nth terms:

a $u_n = 2n-1$ ***b*** $u_n = 2(n-1)(n-2)(n-3)-\frac{1}{2}(2-4n)$.
What do you notice?

5 For the arithmetic sequence 1, 5, 9, 13, ..., find:

a the tenth term ***b*** which term is 101.

6 ***a*** Find the sum of 20 terms of the arithmetic series $18+15+12+9+\ldots$
b How many terms must be taken so that their sum is 60? Explain.

7 Prove that the sum of the first n odd whole numbers is n^2. Find the sum of the first 100 odd whole numbers.

8 Show that there are two geometric sequences with first term $\frac{1}{2}$ and seventh term 32.

9 The first term of a geometric series is 3, and the common ratio is 2.

a Write down the nth term, and find the sum of n terms.
b How many terms must be taken to give a sum of 381?

10 The third term of a geometric sequence is 32, and the sixth term is 2048. Find the common ratio and the first term.

11 Show that the nth term of the geometric sequence in question ***10*** can be expressed as 2^{2n-1}, and that the *product* of the first n terms is 2^{n^2}.

12 A line is divided into six parts whose lengths form a geometric sequence. If the shortest length is 3 cm and the longest 96 cm, find the length of the whole line.

13 The first term of a geometric series is 28, and its sum to infinity is 16. Find the second and third terms.

14 The second term of a geometric series is -9, and its sum to infinity is 12. Find a quadratic equation in r, the common ratio, and explain why one root of this equation must be rejected.

15 For each of the following series, state whether it could be an arithmetic series or a geometric series. In the case of an arithmetic series find the sum of 101 terms; in the case of a suitable geometric series find the sum to infinity.

a $2, 4, 6, 8, \ldots$ *b* $2, -4, 6, -8, \ldots$ *c* $2, 4, 8, 16, \ldots$
d $2, -4, 8, -16, \ldots$ *e* $2, -1, \frac{1}{2}, -\frac{1}{4}, \ldots$

16 The first two terms of a series are $1+\sqrt{2}$ and $1+\frac{1}{\sqrt{2}}$.

a If the series is arithmetic, show that the common difference is $-\frac{1}{2}\sqrt{2}$, and that the sum of 10 terms is $\frac{5}{2}(4-5\sqrt{2})$.
b If the series is geometric, show that it has a sum to infinity, and that this sum is $4+3\sqrt{2}$.

17a £100 is invested at 10% per annum compound interest.
(*1*) Find the total amount after 1, 2 and 3 years.
(*2*) State the growth factor, and calculate the amount after 6 years.
b Machinery depreciates each year by $12\frac{1}{2}\%$ of its value at the beginning of the year. Show that after six years its value is less than half the original value.

18 Find the smallest value of n for which the sum of n terms and the sum to infinity of the series $1+\frac{1}{2}+\frac{1}{4}+\ldots$ differ by less than 0·01.

Revision Exercise on Chapter 3

Matrices — 2

Revision Exercise 3

1 $A = \begin{pmatrix} 1 & 4 & 5 \\ 3 & 2 & -3 \end{pmatrix}$ and $B = \begin{pmatrix} -2 & 3 & -4 \\ 0 & 6 & -5 \end{pmatrix}$.

a Find a 2×3 matrix C such that $A+B = C$.

b Find a matrix D such that $D+C = \boldsymbol{O}$.

2 For the matrix $A = \begin{pmatrix} a_{11} & a_{12} & a_{13} & a_{14} \\ a_{21} & a_{22} & a_{23} & a_{24} \end{pmatrix}$, $a_{ij} = 2i+j-1$. Find the numerical values of the entries of A.

3 The element in the ith row and jth column of the 3×3 matrix C is c_{ij}.

a Write down the matrix C using the notation of question **2**.
b If $c_{ij}=i+2j$, write down matrix C with numerical entries.

4 $P=\begin{pmatrix}2 & -1\\ 1 & 0\end{pmatrix}$, $Q=\begin{pmatrix}3 & -2\\ -1 & 4\end{pmatrix}$ and $R=\begin{pmatrix}-3 & 3\\ 2 & 1\end{pmatrix}$. Find in simplest form:

a $P+Q+R$ *b* $P-Q-R$ *c* $3P-2Q$ *d* $2P-Q+3R$

5 Given that $2\begin{pmatrix}a & 3\\ -1 & b\end{pmatrix}-3\begin{pmatrix}1 & -c\\ d & -4\end{pmatrix}=\begin{pmatrix}0 & 0\\ 0 & 0\end{pmatrix}$, find a, b, c and d.

6 $A=\begin{pmatrix}5 & 3\\ -2 & 1\end{pmatrix}$ and $B=\begin{pmatrix}4 & 1\\ 3 & -7\end{pmatrix}$. Find:

a $A+B$ *b* $(A+B)'$ *c* A' *d* B' *e* $A'+B'$

Which two matrices are equal?

7 $A=(2\quad 3)$, $B=\begin{pmatrix}1 & 4\\ -1 & 0\end{pmatrix}$ and $C=\begin{pmatrix}5\\ 2\end{pmatrix}$.

Which of the products AB, BA, BC, CB, AC, CA exist? Simplify those products that do exist.

8 $M=\begin{pmatrix}\frac{1}{\sqrt{2}} & \frac{1}{\sqrt{2}}\\ -\frac{1}{\sqrt{2}} & \frac{1}{\sqrt{2}}\end{pmatrix}$ and $N=\begin{pmatrix}\frac{1}{2}\sqrt{3} & -\frac{1}{2}\\ \frac{1}{2} & \frac{1}{2}\sqrt{3}\end{pmatrix}$. Find M^2, N^2 and N^3, where $N^3=N.N^2$.

9 $P=\begin{pmatrix}1 & 2\\ -1 & 0\end{pmatrix}$ and $Q=\begin{pmatrix}-1 & 2\\ 3 & -2\end{pmatrix}$. Which of the following are true?

a $P+Q=Q+P$ *b* $PQ=QP$ *c* $(P+Q)(P-Q)=P^2-Q^2$

10 Repeat question **9** for the matrices $P=\begin{pmatrix}2 & 1\\ -1 & 2\end{pmatrix}$ and $Q=\begin{pmatrix}3 & 2\\ -2 & 3\end{pmatrix}$.

11 M is the matrix $\begin{pmatrix}\frac{1}{2} & \frac{1}{2}\sqrt{3}\\ -\frac{1}{2}\sqrt{3} & \frac{1}{2}\end{pmatrix}$.

a Write down the inverse M^{-1} of the matrix M.
b Hence, or otherwise, solve the system of equations

$$x'=\tfrac{1}{2}x+\tfrac{1}{2}\sqrt{3}y$$
$$y'=-\tfrac{1}{2}\sqrt{3}x+\tfrac{1}{2}y$$

for x and y in terms of x' and y'.

12 Given that $\begin{pmatrix} p & q \\ r & s \end{pmatrix}\begin{pmatrix} 4 & -6 \\ -1 & 2 \end{pmatrix} = \begin{pmatrix} 1 & 0 \\ 0 & 1 \end{pmatrix}$, find p, q, r and s.

13 $S = \begin{pmatrix} a & 1 \\ -2 & b \end{pmatrix}$, $T = \begin{pmatrix} 2 & 1 \\ 0 & -1 \end{pmatrix}$ and $ST = \begin{pmatrix} 2 & 0 \\ -4 & 2 \end{pmatrix}$.

a Find a and b. ***b*** Using these values of a and b, find the matrix TS and the matrix R such that $RST = TS$.

14 $A = \begin{pmatrix} -1 & 2 \\ -3 & 4 \end{pmatrix}$, $B = \begin{pmatrix} x \\ y \end{pmatrix}$ and $C = \begin{pmatrix} -2 \\ 1 \end{pmatrix}$. If $AB = C$, find x and y.

15 Find $V\begin{pmatrix} 3 \\ 4 \end{pmatrix}$, where V is the inverse of the matrix $\begin{pmatrix} 2 & 1 \\ 3 & 1 \end{pmatrix}$.

Hence solve the system of equations $2x+y = 3$, $3x+y = 4$, $x, y \in R$.

16 The matrix $M_\alpha = \begin{pmatrix} \cos\alpha & -\sin\alpha \\ \sin\alpha & \cos\alpha \end{pmatrix}$ gives the mapping $P(x, y) \rightarrow P_1(x_1; y_1)$ under a rotation α about origin O.

a Write down similar matrices which map:

(1) $P_1(x_1, y_1)$ to $P_2(x_2, y_2)$ under a rotation β about O (M_β),

(2) $P(x, y)$ to $P_2(x_2, y_2)$ under a rotation $(\alpha+\beta)$ about O $(M_{\alpha+\beta})$.

b Explain why $M_\beta M_\alpha \begin{pmatrix} x \\ y \end{pmatrix} = M_{\alpha+\beta}\begin{pmatrix} x \\ y \end{pmatrix}$.

c Use the matrix equation in ***b*** to derive formulae for $\cos(\alpha+\beta)$ and $\sin(\alpha+\beta)$.

17 Copy and complete the tables below for the number of routes from A to B and from B to C.

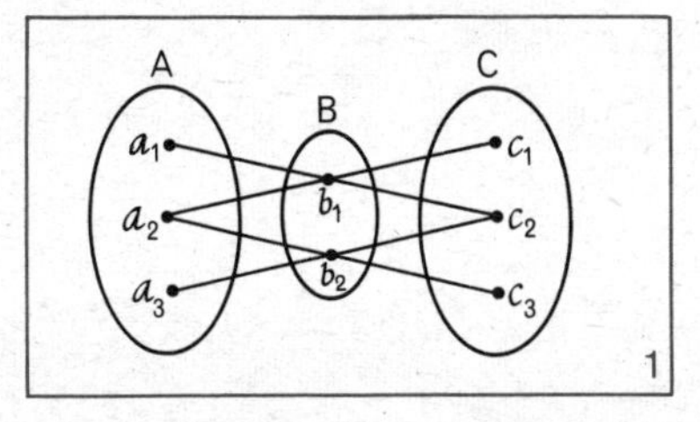

	b_1	b_2
a_1		
a_2		
a_3		

	c_1	c_2	c_3
b_1			
b_2			

Express the entries in the tables in matrix form, and hence calculate the total number of routes from A to C.

18a Under the mapping whose matrix is $\begin{pmatrix} 3 & 2 \\ 1 & 4 \end{pmatrix}$, find the coordinates of the image of the point $(-1, 3)$.

b If the image of the point (x, y) under this mapping is the point $(6, -8)$, find x and y.

19a $A = \begin{pmatrix} 2 & 1 \\ 4 & 3 \end{pmatrix}$ and $B = \begin{pmatrix} 2 & 3 \\ 1 & 4 \end{pmatrix}$. Find A^{-1} and B^{-1}.

b Solve each of the following equations for the 2×2 matrix X, making use of your results in ***a***:

(*1*) $\begin{pmatrix} 2 & 1 \\ 4 & 3 \end{pmatrix} X = \begin{pmatrix} 2 & 5 \\ 8 & 9 \end{pmatrix}$ (2) $\begin{pmatrix} 2 & 3 \\ 1 & 4 \end{pmatrix} X = \begin{pmatrix} -4 & 6 \\ -7 & 8 \end{pmatrix}$

20a If $A = \begin{pmatrix} 5 & -3 \\ 4 & -2 \end{pmatrix}$ find any real numbers c for which $A - cI$ is singular.

b Repeat ***a*** when $A = \begin{pmatrix} 1 & -2 \\ 3 & 4 \end{pmatrix}$.

21 $P = \begin{pmatrix} a & b \\ c & d \end{pmatrix}$. Write down the inverse matrix P^{-1}. Show that the squares of P and P^{-1} are also inverses of one another.

Revision Exercise on Chapter 4

Functions, Composition of Functions, and Inverse Functions

Revision Exercise 4

1 *a* Which of the diagrams in Figure 2 define functions from set A to set B? Give the range of each function.

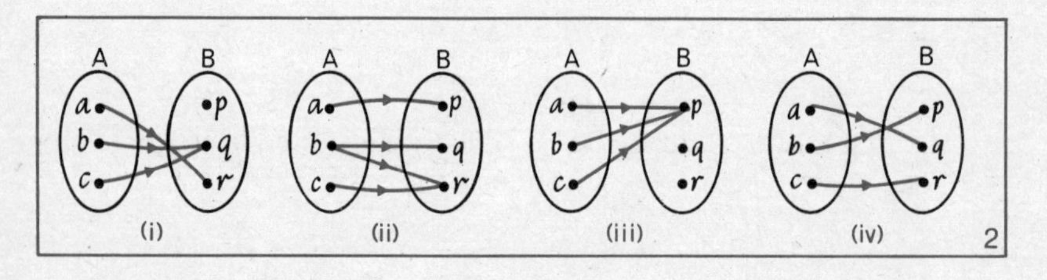

b Which function puts A and B in one-to-one correspondence?

2 Let $A = \{a, b\}$ and $B = \{1, 2\}$. Use arrow diagrams to show all the possible mappings from A to B.

3 $A = \{-4 \leqslant x \leqslant 6\}$ and B is the set of non-negative real numbers. Function $f\colon A \to B$ is defined by $f(x) = 24 + 2x - x^2$.

a Find, if possible, $f(5), f(-5), f(0), f(1)$.
b Find a for which $f(a) = 21$.

4 A function $g\colon R \to R$, defined by $g(x) = ax + b$, is such that $g(-3) = 6$ and $g(5) = -6$. Find the constants a and b, and hence calculate $g(3)$.

5 *a* State the largest subset of the set of real numbers which can be a domain for a function f defined by $f(x) = \dfrac{3}{x^2 - 4}$.

b Find the values of f at $x = -1, 5, \frac{1}{2}$.
c What elements of the domain have image $\frac{1}{4}$?

6 $h\colon x \to 1 - 2x$ and $k\colon x \to x^2$ are functions from R to R.

a Find formulae for $h \circ k$ and $k \circ h$.
b Find x for which $(h \circ k)(x) = -7$.
c What elements in the domain of $k \circ h$ have image 4?

7 $A = \{1, 2, 3, 4\}$, and functions $f\colon A \to A$ and $g\colon A \to A$ are defined by: $f(1) = 3,\ f(2) = 1,\ f(3) = 4,\ f(4) = 2$
$g(1) = 4,\ g(2) = 2,\ g(3) = 1,\ g(4) = 3$.
Define the composite functions $g \circ f$ and $f \circ g$ as sets of ordered pairs.

8 The functions $f\colon R \to R$ and $g\colon R \to R$ are defined by $f(x) = x^2 + 2$ and $g(x) = 2x - 1$.

a Find the numbers $(g \circ f)(-3)$ and $(f \circ g)(-3)$.
b Obtain formulae for $g \circ f$ and $f \circ g$ and hence find the number $(f \circ g)(a) - (g \circ f)(a)$, giving your answer in factorised form.

9 Functions $f\colon R \to R$ and $g\colon R \to R$ are defined by $f(x) = 3x - 2$ and $g(x) = x^3$. Find formulae for the composite functions:

a $g \circ f$ *b* $f \circ g$ *c* $f \circ f$ *d* $g \circ g$

10 $A = \{x : 0 \leqslant x \leqslant 2\pi, x \in R\}$ and function $f\colon A \to R$ is defined by $f(x) = 1 + 2\sin(\frac{1}{3}\pi + x)$. Calculate:

a $f(\frac{2}{3}\pi)$ *b* x for which $f(x) = 0$.

11 $f: R \to R, g: R \to R$ and $h: R \to R$ are functions defined by $f(x) = 2x$, $g(x) = x^2$ and $h(x) = 3 - x$.

a Calculate $(h \circ g \circ f)(2)$ and $(h \circ g \circ f)(\frac{1}{2})$.
b Find a formula for $h \circ g \circ f$.
c Find $(f \circ g \circ h)(5)$ and $(f \circ g \circ h)(x)$.

12 $f: R \to R$, $g: R \to R$ and $h: R \to R$ are functions defined by $f(x) = 2 - x$, $g(x) = x^2 + 1$ and $h(x) = 3x$.

a Calculate the numbers $(h \circ g \circ f)(3)$ and $(f \circ g \circ h)(3)$.
b Find, in simplest form, $(h \circ g \circ f)(x)$ and $(f \circ g \circ h)(x)$; use ***a*** to check your results.

13 In each of the following find a formula for f^{-1}, the inverse of f. State a suitable domain and range for f in each case.

a $f(x) = 2 - 5x$ *b* $f(x) = \frac{1}{2}(x - 1)$ *c* $f(x) = \frac{1}{4}(1 - x)$

d $f(x) = \dfrac{1}{2x}$ *e* $f(x) = \dfrac{1}{4 - x}$ *f* $f(x) = 1 + \dfrac{1}{x^3}$

14 $f: A \to A$ and $g: A \to A$, where $A = \{x: x > 0, x \in R\}$, are functions defined by $f(x) = x + 2$ and $g(x) = \dfrac{2}{x}$. Find

a $f^{-1}(x)$ *b* $g^{-1}(x)$ *c* $(g \circ f)(2)$ *d* $(f \circ g)(1)$

e $(f^{-1} \circ g^{-1})(2)$ *f* $(g^{-1} \circ f^{-1})(5)$

15 f and g are mappings from R to R defined by $f(x) = 2x$ and $g(x)$ is such that $f(g(x)) = -x$. Find $g(x)$.

Geometry

Note to the Teacher on Chapter 1

There is little in *Sections* 1, 2 and 3 that is new to pupils, except the arrangement and development of known facts in a more formal way. Teachers should emphasise that the presentation of proofs as set equalities really includes a proof of the converse theorem as well. For example, $L = \{A\} \cup \{P: m = m_{AP}\}$ states that all points on the line L are included and also that no other points are included.

In forming the equation of a line the most useful form is $y-b = m(x-a)$. Able pupils may write down in one step the corresponding form $y-b = \dfrac{y_1-b}{x_1-a}(x-a)$ for the equation of the line determined by the points (a, b) and (x_1, y_1). The forms $y = mx+c$ and $ax+by = c$ are usually the most useful forms for obtaining information about lines by algebraic methods.

In Exercises 2, 4 and 6, where the emphasis is on practice in using the various forms of equation of a straight line, the answers are given in the most easily obtained simplified form, i.e. $y = mx+c$, or $py = qx+r$ to remove fractions. However, in Exercise 7 the method suggested is to solve the system of equations $\left.\begin{array}{r} ax+by = c \\ px+qy = r \end{array}\right\}$, so that the equations must first be arranged in this form.

Teachers may wish to point out that the intersection of two straight lines is only the simplest case of a general problem in coordinate geometry. If we can solve the appropriate system of equations, we can find the intersections of lines and of lines and curves.

For the condition $m_1 m_2 = -1$, it should be noted that only directions are important and hence consideration of any two lines can be reduced to consideration of two lines through the origin.

In Exercise 8 the questions have been grouped under four headings in order to simplify selection and to provide some suitable revision and extension of earlier work.

1 The Gradient and Equations of a Straight Line

In the geometry you have studied so far nearly all the figures have consisted of straight lines or circles. Parabolas, ellipses and more complicated curves are met from time to time in two dimensions, and there are many interesting surfaces in three dimensions, such as those on the sphere, cylinder, cone and paraboloid. But straight lines occur everywhere. For this reason it is necessary to study them in detail, and to investigate their gradients and equations.

1 *The gradient of a straight line*

In Book 5 (page 101) we defined the gradient of the line AB shown in Figure 1 as:

$$\frac{y\text{-component of AB}}{x\text{-component of AB}} = \frac{\text{CB}}{\text{AC}}$$

Denoting the gradient of AB by m_{AB},

we have $m_{\text{AB}} = \dfrac{\text{CB}}{\text{AC}}$

$$= \frac{y_2 - y_1}{x_2 - x_1}, \quad x_2 \neq x_1.$$

$m_{\text{AB}} = \dfrac{y_2 - y_1}{x_2 - x_1}$ is called the *gradient formula* for AB.

Note. Assuming that OX is perpendicular to OY, and that AB makes an angle $\theta°$ with OX, then $\angle$ CAB $= \theta°$ (corresponding angles)

and $\tan \theta° = \dfrac{\text{CB}}{\text{AC}} = m_{\text{AB}}$.

Thus (i) the gradient of a line is the tangent of the angle it makes with OX

(ii) the gradient of a line parallel to the y-axis is undefined
(iii) parallel lines have the same gradient.

Exercise 1

1 Use the gradient formula to find the gradients of the lines joining:

a (3, 4), (6, 7) *b* (6, 7), (3, 4) *c* (1, 1), (5, 5)

d (0, 2), (2, 6) *e* (5, 3), (6, 7) *f* (0, 0), (4, 3)

g (3, 0), (0, 4) *h* (1, 4), (4, 1) *i* (3, 6), (0, 0)

2 In Figure 2, find the gradient of each line, where possible. Which of the lines have:

a positive *b* negative *c* zero
d undefined gradients?

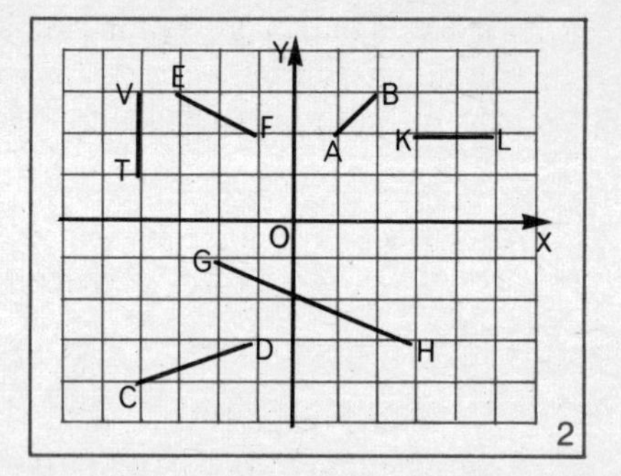

3 Plot the following pairs of points on squared paper, and find the gradient of the line through each pair, where possible:

a (4, 1), (5, 3) *b* (5, 5), (10, 0) *c* (8, −3), (6, 1)

d (−5, 3), (−6, 0) *e* (8, −8), (−8, 8) *f* (0, −2), (−4, −6)

g (10, 10), (−2, −2) *h* (5, 3), (5, −4) *i* (−5, −3), (−10, −3)

4 Look back over question **3**, and state how the *slope of a line* is related to the *sign of its gradient.*

5 K is the point (0, 5), L is (2, −1), M is (−7, −3). Without drawing a diagram, use gradients to find whether the following line segments slope up or down from left to right: KL, LM, KM, OM.
Verify by means of a diagram.

6 Show that the opposite sides of the quadrilateral with vertices A(−2, −4), B(5, −1), C(6, 4), D(−1, 1) are parallel.

7 P is the point (1, 1), Q is (−1, 0), R is (−2, 3), and PQRS is a parallelogram. What are the gradients of RS and PS?

8 A is the point (0, 1), B(8, 8), C(5, 3), D(3, 6). Use gradients to show that ACBD is a parallelogram.

9 Use gradients to show that the points P(−6, 1), Q(2, −7), R(−2, −3) lie in the same straight line.

10 Prove that only one of the points C(−2, −1) and D(−1, −2) lies on the straight line joining A(−6, −4) and B(0, $\frac{1}{2}$).

11 The tangent of the angle which a straight line makes with the x-axis is −2. If the straight line passes through the point (2, 3) write down the coordinates of other three points on the line.

12 Find the angles which the lines joining the following pairs of points make with OX:

a (0, 0), (1, 1) *b* (−2, −2), (4, 4) *c* (0, 0), (1, $\sqrt{3}$)

d (0, 0), ($\sqrt{3}$, 1) *e* (3, 4), (5, 6) *f* (3, 6), (5, 4)

2 *The equation of the line through the point C (0, c), with gradient m; y=mx+c*

We saw in Book 3 (page 129) and again in Book 5 (page 103), that the coordinates of every point on the straight line through the point (0, c), with gradient m, satisfy an equation of the form $y = mx + c$. We now give a general proof.

The line L cuts the y-axis at C(0, c) and has gradient m.

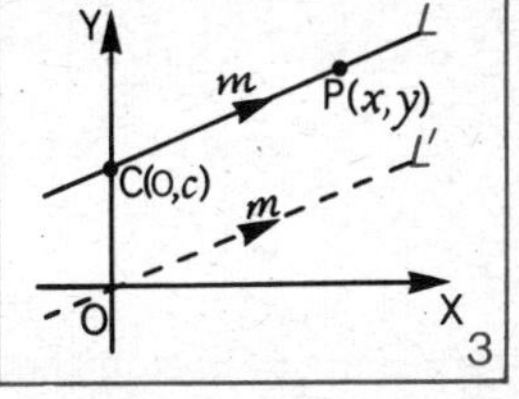

P(x, y) is any point on L distinct from C.

$$L = \{C\} \cup \{P: m_{CP} = m\}$$

$$= \{C(0, c)\} \cup \{P(x, y): \frac{y-c}{x} = m, x \neq 0\}$$

$= \{(x, y): y - c = mx\}$, since C(0, c) is the member of this set given by $x = 0$.

$$= \{(x, y): y = mx + c\}$$

The equation of the line is $y = mx + c$.

Note

(i) The use of equality of sets in this proof has allowed us to combine a 'theorem and converse' situation in one proof. Thus:

If the point (a, b) belongs to the set L, then $(a, b) \in \{(x, y): y = mx + c\}$. Conversely, if $(a, b) \in \{(x, y): y = mx + c\}$, then the point (a, b) belongs to the set L.

In other words if the point (a, b) lies on the line L then $b = ma + c$. Conversely, if $b = ma + c$, then the point (a, b) lies on the line L.

This explanation can be summed up very neatly, thus:

$$\text{The point } (a, b) \text{ lies on the line } L \quad \Leftrightarrow \quad b = ma + c.$$

(ii) The form $y = mx + c$ is useful when finding the gradient of a line. For example, $2x + 3y = 12 \quad \Leftrightarrow \quad y = -\frac{2}{3}x + 4$.

So the gradient of the line with this equation is $-\frac{2}{3}$, and the line cuts the y-axis at $(0, 4)$.

(iii) For the parallel line through the origin, L' in Figure 3, $c = 0$; so the equation of the line through O, with gradient m, is $y = mx$.

Exercise 2

1 Write down the equations of the lines through O with gradients:

a 6 *b* $\frac{1}{3}$ *c* $-\frac{3}{4}$ *d* 0

2 Find the equations of the lines through O and the points:

a $(1, 1)$ *b* $(2, 4)$ *c* $(-5, 3)$ *d* $(1, -\frac{1}{2})$

3 Which of the following are true and which are false?

a $(-1, 1)$ lies on the line with equation $y = x$.
b $(-1, -1)$ lies on the line with equation $y = x$.
c $(6, 4)$ lies on the line with equation $2y = 3x$.
d $(\frac{1}{3}, \frac{1}{4})$ lies on the line with equation $4y = 3x$.
e $(-5, 10)$ lies on the line through the origin and the point $(\frac{1}{2}, -1)$.

4 Write down the equations of the lines with gradient 2 which pass through the following points. Sketch the lines on one diagram and label them.

a $(0, 4)$ *b* $(0, 0)$ *c* $(0, 1)$ *d* $(0, -4)$

5 Write down the equations of the lines with gradient -1 which pass through the points given in question ***4***. Sketch and label the lines.

6 Your sketch in question ***4*** shows four members of the set $\{y = 2x+c, \text{ where } c \in R\}$.
Write down the corresponding set for question **5**.

7 Sketch, on separate diagrams, several members of the sets of lines:

a $\{y = x+c, c \in R\}$ *b* $\{y = mx+4, m \in R\}$

8 Give the gradients of the following lines and the coordinates of the points where they cut the y-axis:

a $y = 2x-3$ *b* $y = \frac{1}{2}x+1$ *c* $2y = 4x+5$

d $3y = 12x-4$ *e* $2y = x+6$ *f* $x+y = 4$

g $2x+3y-9 = 0$ *h* $x+y+1 = 0$ *i* $3y+9 = 0$

9 Sketch each of the lines in question ***8***, either by using the gradient and point, or by joining the points of intersection of the line with the two axes (where this applies).

10 Give the equation of the image of the line through the origin with gradient 10 under a translation: ***a*** 6 units parallel to the y-axis *b* -2 units parallel to the y-axis.

11 What translation would map the line with equation $y = \frac{1}{2}x-5$ onto the parallel line through the origin?

12 Which of the following are true and which are false?

a The point $(\frac{1}{2}, -\frac{1}{3})$ lies on the line with equation $6x+3y = 5$.
b The point $(4, -15)$ does not lie on the line with equation $y = 3x-27$.
c A translation of 1 unit parallel to the y-axis maps the line with equation $y = -x-1$ onto a parallel line through the origin.

---

3 The linear equation $Ax+By+C=0$

We can show that $Ax+By+C = 0$, where A and B are not both zero, is the equation of a straight line.

(i) Suppose that $A \neq 0, B = 0$.

Then we have $\{(x, y): Ax + C = 0\} = \left\{(x, y): x = -\frac{C}{A}\right\}$.

This defines a line parallel to the y-axis.

(ii) Suppose that $A = 0, B \neq 0$.

Then we have $\{(x, y): By + C = 0\} = \left\{(x, y): y = -\frac{C}{B}\right\}$.

This defines a line parallel to the x-axis.

(iii) Suppose that $A \neq 0, B \neq 0$.

Then we have $\{(x, y): Ax + By + C = 0\}$

$$= \left\{(x, y): y = \left(\frac{-A}{B}\right)x + \left(\frac{-C}{B}\right)\right\}$$

$$= \{(x, y): y = mx + c\}, \text{ where } m = \frac{-A}{B} \text{ and } c = \frac{-C}{B}.$$

This defines the line through the point $\left(0, \frac{-C}{B}\right)$ *with gradient* $\frac{-A}{B}$.

So $Ax + By + C = 0$ is the equation of a straight line, and for this reason is called a 'linear' equation: the equation is of the first degree in x and y.

Note. Both theorem and converse are involved here. Thus:

(i) a straight line has an equation of the form $Ax + By + C = 0$, where A and B are not both zero;

(ii) an equation of the form $Ax + By + C = 0$, where A and B are not both zero, is the equation of a straight line.

Exercise 3

1 Rearrange the following equations of straight lines in the form $Ax + By + C = 0$ where A, B, C are integers:

a $y = \frac{2}{3}x - 1$ *b* $x + \frac{1}{3}y = 2$ *c* $y = -\frac{3}{4}x$

2 Which of the following are equations of straight lines?

a $x + y = 0$ *b* $x^2 + y^2 = 1$ *c* $\frac{1}{2}x = \frac{1}{3}y$

d $5x = 3y$ *e* $x + y^2 + 1 = 0$ *f* $\frac{x}{y} + \frac{y}{x} = 0$

g $5 = 2x$ *h* $\frac{1}{2} = \frac{2}{3}x - \frac{1}{4}y$ *i* $xy = 1$

3 Find in the form $Ax+By+C = 0$ the equations of the lines:

a through (0, 4), with gradient 5
b through the points (0, 8) and $(-4, -2)$.

4 Give the equations of the lines through the point (2, 3), which are parallel to: ***a*** the x-axis ***b*** the y-axis.

5 A(1, 1) and C(5, -2) are opposite vertices of a rectangle ABCD whose sides are parallel to the axis. Write down the equations of the four sides of the rectangle.

6 If $(a, 2)$ lies on the line with equation $3x-2y+1 = 0$, find a.

7 If $(-5, b)$ is on the line with equation $2x-4y = 3$, find b.

8 If (p, q) lies on the line with equation $3x+4y+5 = 0$, write down a relation between p and q.

---

4 *The equation of the line through the point* (a, b)*, with gradient* m*;* $y-b=m(x-a)$

The line L passes through the point A(a, b) and has gradient m.

P(x, y) is any point on L distinct from A.

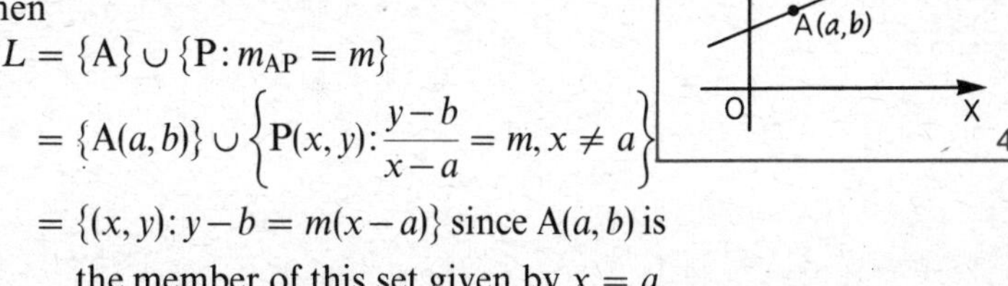

Then

$$L = \{A\} \cup \{P: m_{AP} = m\}$$

$$= \{A(a, b)\} \cup \left\{P(x, y): \frac{y-b}{x-a} = m, x \neq a\right\}$$

$$= \{(x, y): y-b = m(x-a)\}$$ since A(a, b) is the member of this set given by $x = a$.

So the line L through the point (a, b) with gradient m has equation $y-b = m(x-a)$.

Example. Find the equation of the line through the point $(-3, 1)$ which is parallel to the line with equation $2x+4y-1 = 0$.

$2x+4y-1 = 0 \Leftrightarrow y = -\frac{1}{2}x+\frac{1}{4}$, so the required gradient is $-\frac{1}{2}$.

Using '$y-b = m(x-a)$'
the required equation is $y-1 = -\frac{1}{2}(x-(-3))$

$$\Leftrightarrow \quad 2y-2 = -1(x+3)$$
$$\Leftrightarrow \quad 2y-2 = -x-3$$
$$\Leftrightarrow \quad 2y = -x-1$$
$$\text{or } x+2y = -1$$

Exercise 4

(Express the equations in your answers in the form $y = mx+c$, or equivalent.)

1 Find the equations of the lines through the following points, with given gradients:

a $(2, 3), 4$ *b* $(1, -1), 2$ *c* $(0, 4), -2$

d $(4, 1), -2$ *e* $(-3, -3), 1$ *f* $(-4, 2), \frac{1}{2}$

2 Find the equations of the lines through the following points, parallel to the given lines:

a $(1, 2), y = 6x-4$ *b* $(-5, 2), x+y = 5$

c $(1, 0), x-y = 0$ *d* $(-3, 2), 2x+3y+4 = 0$

3 A is the point (4, 2), B is (3, −1), C is (0, 3). Find the equations of the lines through:

a A, parallel to BC *b* B, parallel to AC *c* C, parallel to AB.

4 Find the equations of the lines through the following pairs of points:

a $(3, 4), (1, 2)$ *b* $(0, 1), (-5, -2)$

c $(-1, -3), (4, 0)$ *d* $(0, 0), (2, -3)$

e $(1, -4), (3, -6)$ *f* $(-5, -2), (-8, -1)$

5 P is the point (4, 0), Q is (0, −3), R is (−5, −1). Given that PQRS is a parallelogram, obtain the equations of RS and PS.

6 Choose a point on the line with equation $y = -\frac{1}{2}x$. Find the equations of the lines through the point, and

a parallel to the y-axis *b* parallel to the x-axis

c with gradient -2 *d* with gradient $\frac{1}{2}$.

7 Find the equations of the medians (lines from vertices to midpoints of opposite sides) of $\triangle$ABC, in which A is the point (6, 8), B is (−4, 0) and C is (2, −2).

8 EFGH is a square. E is (3, 0), F(7, 3), G(4, 7). Find the coordinates of H, and the equations of the sides and diagonals of the square.

9 The points A(6, 8), B(0, 6), C(−2, −2) and D(10, 2) are vertices of a quadrilateral.

a Find the coordinates of L, M, P and Q, the midpoints of AB, BC, CD and DA respectively.

b Find the equations of LM, AC and QP. What can you state about their gradients?

10 An isosceles triangle PQR has PQ = PR. The base QR lies on the x-axis, P lies on the y-axis and $2x-3y+9=0$ is the equation of PQ. Find the equation of PR.

---

5 *Perpendicular lines*

Exercise 5

1 a Plot the points A(1, 2), B(3, 2), C(4, 5) and their images A′, B′, C′ under an anticlockwise rotation of 90° of the points of the plane about O. State the coordinates of A′, B′, C′.

b If OA and OA′ have gradient m_1 and m_2 respectively, verify that $m_1 m_2 = -1$.

c Find the corresponding results for the pairs of lines OB, OB′ and OC, OC′.

2 *a* Write down the coordinates of the images D′, E′, F′ of the points D(−3, 1), E(−2, −1), F(5, −2) under the rotation given in ***1a***.

b If OD, OD′ have gradients m_1, m_2 respectively, verify that $m_1 m_2 = -1$.

c Find the corresponding results for the pairs of lines OE, OE′ and OF, OF′.

Note. Let OP, OQ be lines of gradient m_1, m_2 respectively. We have shown in the numerical cases considered in questions *1* and *2* that

if OQ *is perpendicular to* OP *then* $m_1 m_2 = -1$.

We now give general proofs for the above result and its converse.

(i) Under a rotation of $+90°$ about O,

$$P(a, b) \to Q(-b, a).$$

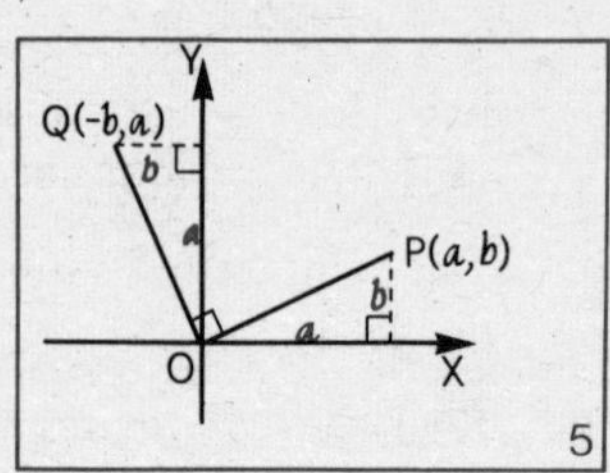

Let the gradient of $OP = m_1$,
and the gradient of $OQ = m_2$.

Then $m_1 = \dfrac{b}{a}$, and $m_2 = \dfrac{a}{-b}$.

$$m_1 m_2 = \frac{b}{a} \times \frac{a}{-b} = -1, a \neq 0, b \neq 0.$$

If OQ is perpendicular to OP, then $m_1 m_2 = -1$.

(ii) *Proof of converse.* Let OR, a line of gradient m_3, be perpendicular to OP in the plane of OP and OQ.

Since OR is perpendicular to OP,
then $m_1 m_3 = -1$.
But $m_1 m_2 = -1$.
Hence $m_2 = m_3$.

The 'two' lines OQ, OR have the same gradient and must coincide. Thus OQ is perpendicular to OP.

If $m_1 m_2 = -1$, then OQ is perpendicular to OP.

Note. (i) If the two lines under consideration do not pass through the origin, take the two lines through the origin which are parallel to the given lines.

(ii) The case of perpendicular lines which are parallel to the axes is not included in the above proof.

Example. Find the equation of the line through the point $(-5, 1)$ which is perpendicular to the line with equation $2x+4y+3=0$.

$$2x+4y+3=0 \quad \Leftrightarrow \quad y = -\tfrac{1}{2}x-\tfrac{3}{4}$$

So the gradient of the given line is $-\frac{1}{2}$, and the gradient of the perpendicular line is 2.

Using '$y-b = m(x-a)$'
the required equation is

$$\begin{aligned} & y-1 = 2(x+5) \\ \Leftrightarrow\ & y-1 = 2x+10 \\ \Leftrightarrow\ & y = 2x+11 \end{aligned}$$

Exercise 6

1 Which of the following pairs of numbers could be gradients of perpendicular lines?

a $4, \frac{1}{4}$ *b* $3, -\frac{1}{3}$ *c* $\frac{2}{5}, \frac{5}{2}$ *d* $\frac{2}{3}, -\frac{2}{3}$ *e* $-\frac{3}{4}, \frac{4}{3}$

2 Write down the gradients of the lines perpendicular to the lines with gradients:

a $\frac{1}{3}$ *b* -4 *c* 6 *d* $-\frac{5}{2}$ *e* 1

3 Write down the equations of the lines through the origin perpendicular to the lines whose gradients are given in question **2**.

4 Find the equations of the lines perpendicular to the line with equation $y = 3x$, which pass through the points:

a $(2, 1)$ *b* $(3, -1)$ *c* $(-4, -4)$

5 Show that one line in each of the following pairs is perpendicular to the other:

a $y = 4x+3, y = -\frac{1}{4}x-1$ *b* $2x+y = 3, x-2y = 4$

6 Obtain the equations of the lines through $(4, 1)$ perpendicular to the lines with equations: *a* $y = 5$ *b* $x = 6$.

7 Find the equations of the lines through the origin perpendicular to the lines with equations:

a $2x-4y+1 = 0$ *b* $5x-3y+2 = 0$ *c* $6x+5y-10 = 0$

8 Find the equations of the lines through the origin perpendicular to the lines with equations:

a $12x-4y+1 = 0$ *b* $5x-8y+2 = 0$ *c* $6x+2y+11 = 0$

9 Find the equations of the lines perpendicular to the line with equation $2x+4y+3 = 0$, and passing through the points:

a $(1, 1)$ *b* $(-5, -5)$ *c* $(0, 3)$

10 Find the equation of the line through $(2, -1)$, perpendicular to the line with the equation $x = 3$.

11 A is the point $(-3, 6)$. OA is the diagonal of a rhombus. Find the gradients of both diagonals of the rhombus.

12 The line joining P(-1, -1) and Q(3, 1) is one side of a rectangle. Find the equations of the other two sides through P and Q. Find also the equation of the fourth side, given that it passes through (1, 5).

13 A is (1, 2), B(3, -4), C(-1, -1), D(1, b). Find b, given that AB is perpendicular to CD.

14 Find the equations of the altitudes of the triangle with vertices A(4, 0), B(0, 4) and C(-2, -2).

15 A is the point (5, 5), B is (-8, 8), C(-2, -2), D(1, -1). Show by means of gradients that AC is perpendicular to BD.

16 Show that P(0, -2), Q(4, 2), R(2, 4), S(-2, 0) are vertices of a rectangle.

17 K is (1, 2), L(6, 0), M(3, 7). Prove that triangle KLM is right-angled: ***a*** by means of gradients ***b*** by using the distance formula $d^2 = (x_2 - x_1)^2 + (y_2 - y_1)^2$, and the converse of Pythagoras' theorem. What further information about the triangle does ***b*** give?

18 Prove that the lines with equations $ax + by + c = 0$ and $bx - ay + d = 0$ are perpendicular to each other.

---

6 *The intersection of two straight lines*

In Algebra, Chapter 1 we solved systems of linear equations; we can now make further use of these methods.

Example 1. Find the coordinates of the point of intersection of the lines with equations $2x + y = 5$ and $3x - 4y = 13$.

$$\begin{array}{rcl} 2x+y = 5 & \times 4 & \Leftrightarrow & 8x+4y = 20 \\ 3x-4y = 13 & \times 1 & \Leftrightarrow & 3x-4y = 13 \\ & & \textit{Add} & 11x \quad = 33 \\ & & \Leftrightarrow & x \quad = 3 \end{array}$$

Substituting $x = 3$ in the first equation, $\quad 6 + y = 5$

$$\Leftrightarrow \quad y = -1$$

The point of intersection is (3, -1).
(Alternatively, we could have started by *substituting* $y = 5 - 2x$ in the second equation.)

Example 2. $\triangle ABC$ has vertices $A(\frac{1}{2}, 4)$, $B(-6, -1)$, $C(4, -3)$. The median AM and the altitude BP intersect at K. Find the coordinates of K.

(i) M is the point

$$\left(\frac{-6+4}{2}, \frac{-1-3}{2}\right)$$

i.e. $(-1, -2)$

$$m_{AM} = \frac{4-(-2)}{\frac{1}{2}-(-1)} = \frac{6}{1\frac{1}{2}} = 4.$$

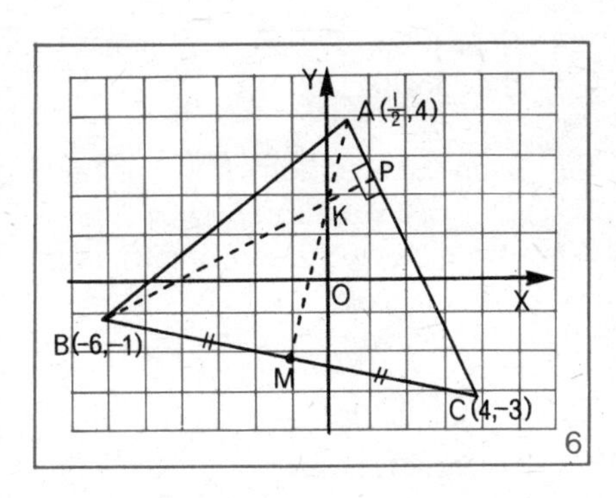

The equation of AM is

$$y-4 = 4(x-\tfrac{1}{2})$$
$$\Leftrightarrow \quad y-4 = 4x-2$$
$$\Leftrightarrow \quad 4x-y = -2 \qquad \text{.................... (1)}$$

(ii) $$m_{AC} = \frac{4+3}{\frac{1}{2}-4} = \frac{7}{-3\frac{1}{2}} = -2, \text{ so } m_{BP} = \tfrac{1}{2}$$

The equation of BP is $y+1 = \frac{1}{2}(x+6)$

$$\Leftrightarrow \quad 2y+2 = x+6$$
$$\Leftrightarrow \quad x-2y = -4 \qquad \text{.................... (2)}$$

$$\begin{array}{llcl} 4x-y=-2 & \times 2 & \Leftrightarrow & 8x-2y=-4 \\ x-2y=-4 & \times 1 & \Leftrightarrow & x-2y=-4 \\ & \textit{Subtract} & & 7x \quad\; = 0 \\ & & \Leftrightarrow & x \quad\;\; = 0 \end{array}$$

Substituting $x = 0$ in equation (1), $y = 2$. K is $(0, 2)$.

(Note the form in which we put the equations (1) and (2) in order to solve the system.)

Exercise 7

1 Find the coordinates of the point of intersection of each of these pairs of lines:

a $x+y = 2$ and $x-y = -1$ *b* $2x+5y = 1$ and $x-3y = -5$

2 Find the coordinates of the vertices of the triangle whose sides have equations $x = 3$, $x - y = 2$ and $2x + y + 5 = 0$.

3 The line through A $(-1, 3)$ perpendicular to the line BC with equation $2x - y = 10$ meets BC at D. Find:

a the equation of the perpendicular line through A
b the coordinates of D.

4 *a* Find the equation of the line through P $(3, 2)$ which is parallel to the line with equation $2x + 3y = 6$.
b Write down the coordinates of the points where this line through P cuts the axes.

5 K is the point $(-2, -3)$, L $(-4, 1)$, M $(5, -2)$. Find:

a the equation of the line through M parallel to KL
b the equation of the line through K perpendicular to LM
c the coordinates of the point of intersection of these two lines.

6 Sketch the parallelogram with vertices A$(-4, -2)$, B$(4, 0)$, C$(6, 4)$, D$(-2, 2)$. Find:

a the equations of the diagonals AC and BD
b the coordinates of the point of intersection of the diagonals.

7 P is the point $(0, 4)$, Q$(-2, -6)$, R$(6, -2)$. Find:

a the equation of the perpendicular from P to QR
b the equation of the perpendicular bisector of PR
c the coordinates of the point of intersection of these two lines.

8 Find the coordinates of the vertices of the triangle whose sides have equations $x + 3y = 10$, $x - 4y = -4$ and $x - 2y = -4$.

9 Try to find the coordinates of the point of intersection of the lines with equations $2x - y + 5 = 0$ and $4x - 2y - 7 = 0$. Explain.

10 Sketch the square with vertices A$(-1, -1)$, B$(1, -5)$, C$(5, -3)$, D$(3, 1)$. The diagonal BD meets the line from A to the midpoint of BC at T. Find the coordinates of T.

7 *Miscellaneous applications*

The Exercises that follow make use of the methods of earlier sections of this chapter (summarised on page 109), and also the following results for two points $A(x_1, y_1)$ and $B(x_2, y_2)$:

(i) The *distance formula* $AB = \sqrt{(x_2-x_1)^2+(y_2-y_1)^2}$ (from Book 4, page 117).

(ii) The *midpoint* of AB is $[\frac{1}{2}(x_1+x_2), \frac{1}{2}(y_1+y_2)]$ (from Book 5, pages 124 and 125).

In most questions rough sketches will be helpful.

Exercise 8

(i) Equations and intersections of lines

1 Find the equations of the lines through A(1, 2) which are parallel and perpendicular to the line through the points B(4, 1) and C(1, 4).

2 Find the equations of the perpendicular bisectors of the lines joining the points: *a* D(8, 4) and E(2, 6) *b* F(−1, 3) and G(1, −3).

3 Find the equations of the sides of the triangle with vertices P(−2, 3), Q(2, 3), R(0, 8).

4 a Find the equation of the line through the origin perpendicular to the line with equation $4x+y=34$.

b Find the point of intersection of these two lines, and hence calculate the shortest distance from O to the given line.

5 a Find the equations of the lines through A(−1, 5) parallel and perpendicular to the line with equation $4x-3y=6$.

b Find the point of intersection of the given line and the line through A perpendicular to it, and hence calculate the length of this perpendicular from A.

(ii) Lines associated with triangles

6 The vertices of a triangle are A(−6, 1), B(4, 5) and C(8, −3).

a Find the coordinates of the point of intersection of the medians AM and BN of the triangle.

b Prove that the three medians are concurrent by showing that this point lies on the third median CP.

7 The vertices of a triangle are D(6, 6), E(−3, 0) and F(0, −3).

a Find the coordinates of the point of intersection of the altitudes of the triangle through D and E.
b Prove that the three altitudes are concurrent by showing that this point lies on the third altitude.

8 The vertices of a triangle are P(−3, 1), Q(5, 5) and R(6, −2).

a Find the point of intersection, S, of the perpendicular bisectors of PQ and PR.
b Prove that the perpendicular bisectors of the three sides of the triangle are concurrent by showing that S lies on the perpendicular bisector of QR.
c Use the distance formula to show that a circle can be drawn with centre S to pass through P, Q and R (the circumcircle of $\triangle$PQR).

(iii) Parallelograms

9 O is the origin, A is (4, 1), C is (2, 5), and OABC is a parallelogram. Find the coordinates of B in each of the following ways:

a by noting that $\overrightarrow{OA} = \overrightarrow{CB}$
b by finding the equations of AB and CB
c by using properties of the diagonals of a parallelogram.

10 Repeat question **9** for parallelogram EFGH, where E is (4, 0), F(−2, −2) and G(0, 2), to find H.

11 P is the point (−1, −1), Q(7, −3), R(9, 5), S(1, 7). Prove that:

a the opposite sides of PQRS are parallel; so what shape is PQRS?
b the angles of PQRS are right angles; so what shape is PQRS?
c the sides of PQRS are equal; so what shape must it be?

(iv) Transformations

12 Write down the gradients and equations of the images under reflection in the x-axis of the lines with equations: *a* $y = x$ *b* $y = 2x$ *c* $5x + 2y = 0$.

13 Find the equations of the images of the line $y = x$ under:

a a half turn about O *b* a half turn about (1, 1)
c a quarter turn about O *d* a quarter turn about (1, 1)
e a translation of 3 units in the direction: (*1*) OY (*2*) OX.

14a Find the images A_1 and B_1 of the points A(2, 3) and B(3, −1) under the dilatation [O, 3].
b Find the gradients and equations of AB and A_1B_1.
c Use the distance formula to show that $A_1B_1 = 3AB$.

15 Parallelogram KLMN has vertices K(−2, −1), L(3, 1), M(4, 3), N(−1, 1).

a Find the coordinates of the images L_1, M_1, N_1 of L, M, N under an anticlockwise quarter turn about K.
b Verify by using gradients that L_1M_1, M_1N_1 and N_1L_1 are perpendicular to LM, MN and NL respectively.

Exercise 8B

1 The vertices of △ABC are A(−4, 10), B(10, 3), C(0, −10). Find the point of intersection of the median BM and the altitude CP.

2 P is the point (1, 9), Q(−2, 3), R(4, 7). Show that the midpoint of PR lies on the line through the midpoint of QR which is parallel to PQ.

3 A line l has equation $x+3y = 13$, and a line m passing through A(3, 10) and perpendicular to l meets l at T. Find the coordinates of T.

Show that l, m and the line with equation $2x-y+2 = 0$ are concurrent.

Find the equations of the images of these three lines under an anticlockwise rotation of 90° about T.

4 A is the point (4, −2), and AO is produced its own length to B. Give the coordinates of B.

OC, equal in length to OA, is perpendicular to OA, and C lies in the third quadrant. Find the coordinates of C.

Find the coordinates of E, if AEBC is a square, and of F if OAFE is a square.

5 A parallelogram has sides parallel to the lines $y = \frac{1}{3}x$ and $y = -\frac{12}{5}x$,

and two of its opposite vertices are $(-5, 10)$ and $(15, 3)$. Find the equations of its sides, and the coordinates of its other two vertices.

6 ABCD is a square, with A the point $(2, 3)$ and $C(6, 5)$. Find the point of intersection of the diagonals and the coordinates of B and D.

7 a (*1*) Show in a diagram the set of four lines $\{mx+y=2: m=0, 1, 2, 3\}$.
(*2*) Show in a diagram the set of four lines $\{2x+y=c: c=0, 1, 2, 3\}$.

b Find the point of intersection of the lines $mx+y=2$ and $2x+y=c$. Discuss the significance of: (*1*) $c=2$ and (*2*) $m=2$, relating your answers to part ***a***.

8 a Show that the point $(2, 4)$ lies on the line with equation $y=2x$. Find the equation of the image l' of this line in the line with equation $y=x$.

b Under a translation $\begin{pmatrix} 4 \\ 3 \end{pmatrix}$ the image of the line with equation $3x+y=12$ is m'. Find the equation of m' and the point of intersection of l' and m'.

9 $A(-3, 2)$, $B(0, 3)$, $C(-1, 10)$ are three vertices of a parallelogram ABCD. Find the coordinates of D.

Under a half turn H_1 about D, the image of this parallelogram is A′B′C′D′, where $A \to A'$ and so on. Find the equations of the images of the sides of the parallelogram ABCD.

Under a half turn H_2 about a point Q, the image of A′B′C′D′ is such that $B' \to A$. Find the coordinates of Q.

Replace $H_2 \circ H_1$ by a single transformation and describe it fully.

10 Under a transformation K the point (x, y) maps onto the point $(x+2, 2y+1)$. Find the images O_1, A_1, B_1, C_1 of the points $O(0, 0)$, $A(1, 0)$, $B(1, 1)$, $C(0, 1)$.

Find the equations of the lines OB, AC, O_1B_1 and A_1C_1. Find the coordinates of the point of intersection of O_1B_1 and A_1C_1.

The line with equation $y=\frac{1}{2}x$ intersects AB in the point D. Find the coordinates of the image of D under the transformation K.

11 P is the point (a, b), Q is (c, d) and OP is perpendicular to OQ. Use Pythagoras' theorem in $\triangle OPQ$ to prove that $bd+ac=0$. *Deduce* that if the gradient of OP is m_1 and the gradient of OQ is m_2, then $m_1m_2=-1$.

Summary

1 *The gradient of the line joining the points* $A(x_1, y_1)$ *and* $B(x_2, y_2)$ *is*

$$m_{AB} = \frac{y_2 - y_1}{x_2 - x_1}, \quad \text{provided that } x_2 \neq x_1.$$

2 (i) A line sloping up from left to right has a *positive gradient.*
(ii) A line sloping down from left to right has a *negative gradient.*
(iii) A line parallel to the x-axis has *zero gradient.*
(iv) A line parallel to the y-axis has *no defined gradient.*

3 *The equation of the line through the point* $(0, c)$, *with gradient* m, *is*

$$y = mx + c$$

4 *The equation of the line through the origin, with gradient* m, *is*

$$y = mx$$

5 *The equation of the line through the point* (a, b), *with gradient* m, *is*

$$y - b = m(x - a)$$

6 *The general equation of a line is* $Ax + By + C = 0$ *a linear equation.*

7 m_1, m_2 are the *gradients of perpendicular lines* $\Leftrightarrow m_1 m_2 = -1$.

8 *Also, for two points* $A(x_1, y_1)$ *and* $B(x_2, y_2)$, *we have:*

(i) the distance formula, $AB = \sqrt{(x_2 - x_1)^2 + (y_2 - y_1)^2}$
(ii) the midpoint of AB is $[\frac{1}{2}(x_1 + x_2), \frac{1}{2}(y_1 + y_2)]$.

2 Vectors 2

1 *Revision*

The following reminders of the work on vectors in Book 5 will help you with the revision examples in Exercise 1.

1 *The set of all directed line segments with the same magnitude and direction is a geometrical vector.*

The directed line segments are representatives of the vector.

The vector may be written $\boldsymbol{u}$, or, in components, $\begin{pmatrix} a \\ b \end{pmatrix}$.

2 *Addition.* Vectors may be added by the triangle or parallelogram rules.

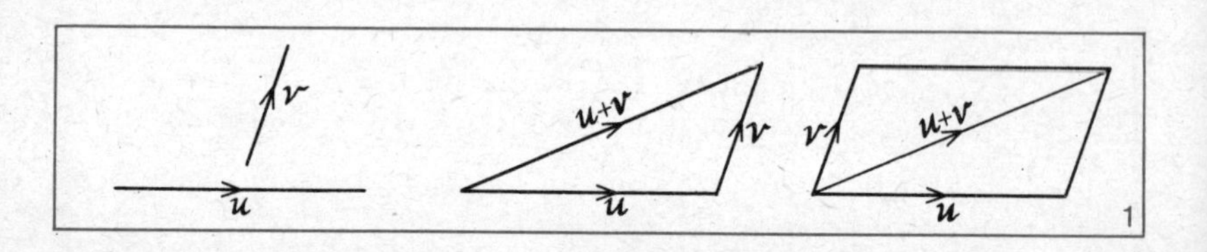

1

$$\boldsymbol{u}+\boldsymbol{v} = \boldsymbol{v}+\boldsymbol{u} \text{ (commutative law)}$$
$$(\boldsymbol{u}+\boldsymbol{v})+\boldsymbol{w} = \boldsymbol{u}+(\boldsymbol{v}+\boldsymbol{w}) \text{ (associative law)}$$
$$\boldsymbol{u}+\boldsymbol{0} = \boldsymbol{u} = \boldsymbol{0}+\boldsymbol{u} \text{ (identity element)}$$
$$\boldsymbol{u}+(-\boldsymbol{u}) = \boldsymbol{0} = (-\boldsymbol{u})+\boldsymbol{u} \text{ (additive inverse)}$$

All of the above can be expressed in component form.

3 *Subtraction.* Subtraction of one vector from another is defined by the rule

$$\boldsymbol{u}-\boldsymbol{v} = \boldsymbol{u}+(-\boldsymbol{v}); \begin{pmatrix} a \\ b \end{pmatrix} - \begin{pmatrix} c \\ d \end{pmatrix} = \begin{pmatrix} a \\ b \end{pmatrix} + \begin{pmatrix} -c \\ -d \end{pmatrix} = \begin{pmatrix} a-c \\ b-d \end{pmatrix}.$$

Note to the Teacher on Chapter 2

Chapter 2 takes up the study of geometrical vectors from the end of Book 5, Geometry, Chapter 2. The main objectives of the chapter are to emphasise the ease with which vector theorems, already familiar in two dimensions, can be extended to three dimensions, and to introduce the scalar product of two vectors. As in Vectors-1 in Book 5, vectors are presented in bold type. In written work, pupils will find it best to underline a letter by a wavy line to indicate a vector.

Section 1 is a revision section in two dimensions which leads on, in *Sections* 2 and 3, to the discussion of vectors in three dimensions, first in terms of directed line segments and then in components. Teachers may wish to indicate, particularly to good classes, that it is possible to have vectors with any number of components and that their rules of combination are similar to those with two or three components. This elementary course continues to deal with vectors which can be represented geometrically, as the geometrical medium helps pupils to make sense of the manipulations in which they are involved by actual drawing or by mental pictures.

Section 3 includes a warning about the distinction between co-ordinates and components as only when the position vector $\boldsymbol{a}$ is represented by $\overrightarrow{OA}$ are the *coordinates* of A the same numbers as the *components* of $\boldsymbol{a}$. If A is the point (x_1, y_1, z_1) and B is (x_2, y_2, z_2), then the components of the vector represented by $\overrightarrow{AB}$ must be found by subtraction and are $x_2 - x_1$, $y_2 - y_1$, and $z_2 - z_1$.

Section 4 deals with the magnitude of a vector and the distance formula as simple extensions of Pythagoras' theorem.

Position vectors will be found most helpful in geometrical demonstrations. For geometry, one fundamental result that must be got over to pupils is that $\boldsymbol{a} = \boldsymbol{b}$ implies that $\overrightarrow{OA} = \overrightarrow{OB}$ and hence A and B coincide. On page 127, teachers may choose one or both methods depending on their inclinations and the quality of the classes they are dealing with. The method beginning with $n\,\overrightarrow{AP} = m\,\overrightarrow{PB}$ is more sophisticated and illustrates the neatness of vector methods, while that beginning $\overrightarrow{OP} = \overrightarrow{OA} + \overrightarrow{AP}$ may be more convincing to less able pupils. The Section Formula in coordinates is closely linked with the Section Formula in terms of position vectors.

Careful note should be taken of the remarks on choice of origin for position vectors. Teachers will have to assess the advantages or disadvantages of displaying the origin in an actual diagram where O is not essential. Does it make the argument clearer? Does it complicate the figure too much? The natural end point for the mathematician is the omission of O from the actual drawing.

It will be obvious that the use of models will facilitate the introduction of number triples; even the use of a corner of the classroom as origin can be useful for illustration. At the same time it is perhaps better to postpone emphasis on rectangular axes until pupils are clear that a great many vector theorems, e.g. the Section Formula, can be referred to any basis, before the introduction of magnitude and angle. Teachers will see the advantage of using triples (x_1, y_1, z_1), etc. for results that are to be transferred to coordinate geometry.

Scalar product is the most important new topic. $\boldsymbol{a}.\boldsymbol{b}$ is defined, for non-zero vectors, as $|\boldsymbol{a}||\boldsymbol{b}|\cos\theta$, and its value $a_1b_1+a_2b_2+a_3b_3$ in terms of components is derived by means of the cosine rule. It should be noted that the first form is independent of any basis used for components. The component form is then used to demonstrate the commutative and distributive laws. In such a development it should be noted that it would then be illogical to use the distributive law to 'prove' the cosine rule or to 'prove' that

$$(a_1\boldsymbol{i}+a_2\boldsymbol{j}+a_3\boldsymbol{k}).(b_1\boldsymbol{i}+b_2\boldsymbol{j}+b_3\boldsymbol{k}) = a_1b_1+a_2b_2+a_3b_3.$$

Throughout discussion of the scalar product great stress should be put on correct notation. The dot should be used *only* to indicate scalar product and *not* in numerical products like ab or $|\boldsymbol{a}||\boldsymbol{b}|$.

4 *Multiplication by a number*. If k is positive, $k\boldsymbol{u}$ has the same direction as $\boldsymbol{u}$ and k times the magnitude of $\boldsymbol{u}$.

If k is negative, $k\boldsymbol{u}$ has the opposite direction to $\boldsymbol{u}$ and k times the magnitude of $\boldsymbol{u}$.

If $k = 0$, $0\boldsymbol{u} = \boldsymbol{0}$. $\quad k\begin{pmatrix} a \\ b \end{pmatrix} = \begin{pmatrix} ka \\ kb \end{pmatrix}$.

5 *Distributive laws*. $\quad k\boldsymbol{u} + k\boldsymbol{v} = k(\boldsymbol{u} + \boldsymbol{v})$.
$k\boldsymbol{u} + m\boldsymbol{u} = (k+m)\boldsymbol{u}$.

6 *Position vectors*. If O is the origin and P is the point (h, k), the vector represented by $\overrightarrow{OP}$ is the *position vector* of P. The position vector of P is denoted by $\boldsymbol{p}$, and $\boldsymbol{p} = \begin{pmatrix} h \\ k \end{pmatrix}$ in component, or number pair, form.

The coordinates of the point are the components of its position vector.

$\overrightarrow{PQ} = \overrightarrow{OQ} - \overrightarrow{OP}$, and represents $\boldsymbol{q} - \boldsymbol{p}$.

If M is the midpoint of PQ, $\boldsymbol{m} = \frac{1}{2}(\boldsymbol{p} + \boldsymbol{q})$.

If G is the centroid of $\triangle$ABC, $\boldsymbol{g} = \frac{1}{3}(\boldsymbol{a} + \boldsymbol{b} + \boldsymbol{c})$.

7 *Magnitude of a vector*. If $\overrightarrow{AB}$ represents the vector $\boldsymbol{u} = \begin{pmatrix} a \\ b \end{pmatrix}$, then the magnitude of $\boldsymbol{u} = |\boldsymbol{u}| = |\overrightarrow{AB}| = \sqrt{(a^2 + b^2)}$.

Exercise 1

1 State in component form the vectors represented by $\overrightarrow{AB}$, given:

a A(4, 4), B(8, 5) *b* A(5, 3), B(9, 2) *c* A(-5, -4), B(3, -4)

2 The position vectors of P, Q, R and S are $\begin{pmatrix} 1 \\ 2 \end{pmatrix}$, $\begin{pmatrix} 4 \\ 4 \end{pmatrix}$, $\begin{pmatrix} -2 \\ -5 \end{pmatrix}$ and $\begin{pmatrix} 4 \\ 0 \end{pmatrix}$. Find, in terms of components, the vectors represented by:

a $\overrightarrow{PQ}$ *b* $\overrightarrow{QR}$ *c* $\overrightarrow{RS}$ *d* $\overrightarrow{PS}$

3 If $\overrightarrow{CD}$ represents the vector $\begin{pmatrix} 4 \\ 2 \end{pmatrix}$, find the coordinates of D when C is the point: *a* (0, 0) *b* (2, 3) *c* (-5, 4) *d* (-1, -2)

4 Calculate $|\boldsymbol{u}|$ if: *a* $\boldsymbol{u} = \begin{pmatrix} 6 \\ 8 \end{pmatrix}$ *b* $\boldsymbol{u} = \begin{pmatrix} 0 \\ -1 \end{pmatrix}$ *c* $\boldsymbol{u} = \begin{pmatrix} -4 \\ -4 \end{pmatrix}$

5 Find p and q in each of the following:

a $\begin{pmatrix}p\\q\end{pmatrix}+\begin{pmatrix}3\\5\end{pmatrix}=\begin{pmatrix}9\\6\end{pmatrix}$ *b* $\begin{pmatrix}p\\-3\end{pmatrix}+\begin{pmatrix}4\\q\end{pmatrix}=\begin{pmatrix}10\\-1\end{pmatrix}$

6 Find the vector $\boldsymbol{v}$, given that:

a $2\boldsymbol{v}+\begin{pmatrix}7\\4\end{pmatrix}=\begin{pmatrix}3\\10\end{pmatrix}$ *b* $4\boldsymbol{v}+\begin{pmatrix}-3\\-4\end{pmatrix}=\boldsymbol{v}+\begin{pmatrix}12\\17\end{pmatrix}$

7 Solve the following vector equations for $\boldsymbol{x}$:

a $\frac{1}{2}(\boldsymbol{x}+\boldsymbol{a})=\frac{1}{3}(\boldsymbol{x}+\boldsymbol{b})$ *b* $\boldsymbol{x}+3(\boldsymbol{a}+\boldsymbol{b})=4(\boldsymbol{a}-\boldsymbol{b})-3\boldsymbol{x}$

8 Using Figure 2, fill in the missing directed line segments in the following statements.

a $\overrightarrow{AC}=\overrightarrow{AB}+\ldots$ *b* $\overrightarrow{AD}=\overrightarrow{AB}+\overrightarrow{BC}+\ldots$
c $\overrightarrow{AC}=\overrightarrow{AB}+\ldots+\overrightarrow{DC}$ *d* $\overrightarrow{EB}+\overrightarrow{BC}=\ldots+\overrightarrow{DC}$
e $\overrightarrow{BE}=\overrightarrow{BC}+\overrightarrow{CD}-\ldots$ *f* $\overrightarrow{AD}+(-\overrightarrow{DA})=\ldots$

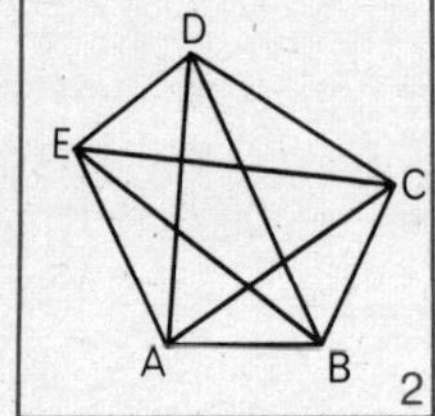

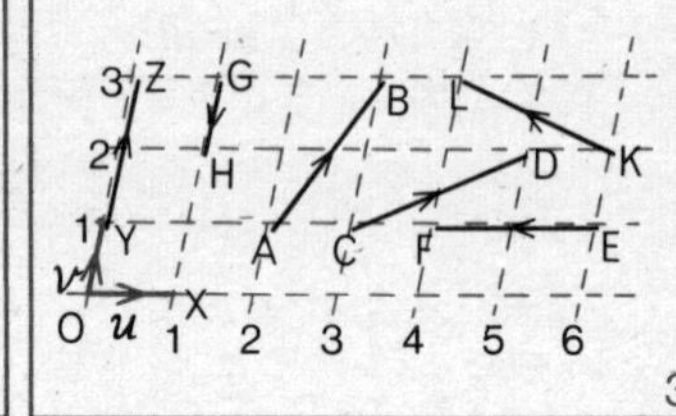

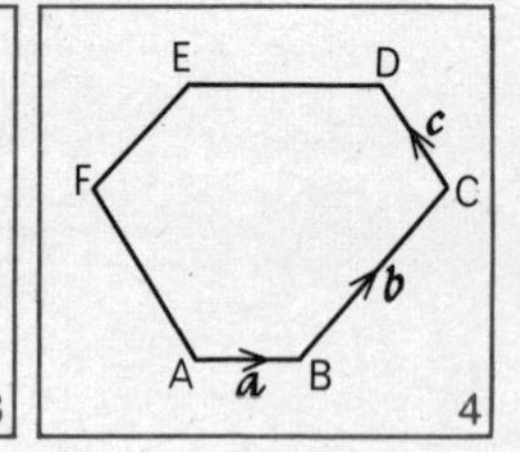

9 From Figure 3, express the vectors represented by $\overrightarrow{AB}$, $\overrightarrow{CD}$, $\overrightarrow{EF}$, $\overrightarrow{GH}$, $\overrightarrow{KL}$ and $\overrightarrow{YZ}$ in number pair form.

10 In Figure 3 $\overrightarrow{OX}$ represents $\boldsymbol{u}$ and $\overrightarrow{OY}$ represents $\boldsymbol{v}$. Express the vectors represented by $\overrightarrow{AB}$, $\overrightarrow{CD}$, $\overrightarrow{EF}$, $\overrightarrow{GH}$, $\overrightarrow{KL}$ and $\overrightarrow{YZ}$ in terms of $\boldsymbol{u}$ and $\boldsymbol{v}$.

Note. Check that in question ***10*** all your answers are of the form $l\boldsymbol{u}+m\boldsymbol{v}$. This illustrates a very important general property of vectors in a plane. Provided that $\boldsymbol{u}$ and $\boldsymbol{v}$ are not parallel, every vector in the plane can be expressed in the form $l\boldsymbol{u}+m\boldsymbol{v}$ where l and m are real numbers (including zero). It is also true that this form in $\boldsymbol{u}$ and $\boldsymbol{v}$ is unique for each vector, for, if a vector representative is given, there is only one parallelogram that can be drawn with the representative as diagonal and with sides parallel to $\boldsymbol{u}$ and $\boldsymbol{v}$.

Note. Vectors are called *coplanar* when they can all be represented in the same plane.

This is an extension of the meaning of coplanar in its usual sense. Three coplanar vectors, for example, might be represented by directed line segments in three different *parallel* planes.

11 In Figure 4, ABCDEF is a hexagon with three pairs of opposite sides parallel. If $\overrightarrow{AB}$, $\overrightarrow{BC}$ and $\overrightarrow{CD}$ represent $\boldsymbol{a}$, $\boldsymbol{b}$ and $\boldsymbol{c}$ respectively, state why it is possible for $\overrightarrow{DE}$, $\overrightarrow{EF}$ and $\overrightarrow{FA}$ to represent $k\boldsymbol{a}$, $l\boldsymbol{b}$ and $m\boldsymbol{c}$ where k, l and m are real numbers.

Express the vectors represented by $\overrightarrow{AC}$ and $\overrightarrow{DF}$ in terms of $\boldsymbol{a}$, $\boldsymbol{b}$, k and l. If AC is parallel to FD, find a relation between k and l.

12 With respect to the usual coordinate axes, P is the point $(-1, -1)$ Q is $(3, -1)$ and R is $(3, 2)$.

a Express the vectors represented by $\overrightarrow{PQ}$, $\overrightarrow{QR}$ and $\overrightarrow{RP}$ in number pair form.

b Use ***a*** to verify that $\overrightarrow{PQ} = \overrightarrow{RQ} - \overrightarrow{RP}$.

c Calculate the magnitudes of the vectors represented by $\overrightarrow{PQ}$, $\overrightarrow{QR}$ and $\overrightarrow{RP}$. What can you say about $\triangle PQR$?

13 With oblique axes and unequal scales as in Figure 3, $\overrightarrow{PQ}$, $\overrightarrow{QR}$ and $\overrightarrow{RS}$ represent $\begin{pmatrix}5\\4\end{pmatrix}$, $\begin{pmatrix}-8\\10\end{pmatrix}$ and $\begin{pmatrix}3\\-14\end{pmatrix}$ respectively.

a By considering $\overrightarrow{PQ} + \overrightarrow{QR} + \overrightarrow{RS}$, show that P and S coincide.

b The midpoint of QR is L. Express the vector represented by $\overrightarrow{PL}$ in number pair form.

c G is the point of trisection of PL nearer to L. Give the vector represented by $\overrightarrow{PG}$ in number pair form.

d M is the midpoint of RP, and H is the point of trisection of QM nearer to M. Give the vectors represented by $\overrightarrow{QM}$ and $\overrightarrow{QH}$ in number pair form.

---

2 *Vectors in three dimensions*

So far we have been considering relations between those vectors which can all be represented in one plane, i.e. vectors in two dimensions. Vectors, however, cannot always be represented by lines in the same plane and they can be used equally well in three dimensions. In this Section we shall consider the more general case of vectors in three dimensions, beginning with addition.

Definition of addition of two vectors

This is exactly the same as in two dimensions, since representatives of the two vectors may be chosen which have a point in common and so lie in the same plane.

Thus in Figure 5, $\overrightarrow{AB}$ and $\overrightarrow{BC}$ are representatives of the vectors $\boldsymbol{u}$ and $\boldsymbol{v}$, chosen to have the point B in common; then the sum of $\boldsymbol{u}$ and $\boldsymbol{v}$, denoted by $\boldsymbol{u}+\boldsymbol{v}$, is the vector represented by $\overrightarrow{AC}$.

In the two-dimensional case we verified that we had a *proper* definition, i.e. that it made no difference which pair of representatives we chose. We shall now see that this is true in three dimensions also.

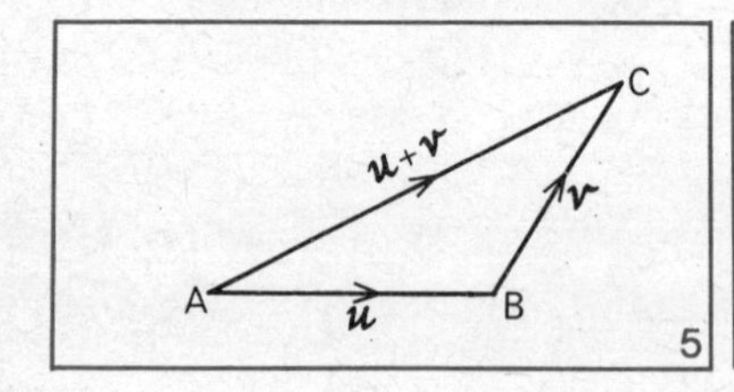

5

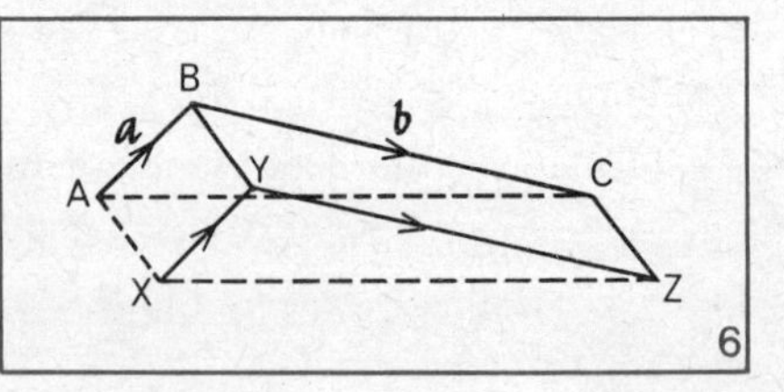

6

In Figure 6, $\overrightarrow{AB}$ and $\overrightarrow{XY}$ are representatives of vector $\boldsymbol{a}$, and $\overrightarrow{BC}$ and $\overrightarrow{YZ}$ are representatives of vector $\boldsymbol{b}$. We have to prove that $\overrightarrow{AC}$ and $\overrightarrow{XZ}$ are representatives of the same vector.

Since $\overrightarrow{AB}$ and $\overrightarrow{XY}$ are representatives of the same vector, ABYX is a parallelogram, and so AX is equal and parallel to BY.

Similarly BY is equal and parallel to CZ, so that $\overrightarrow{AX}$, $\overrightarrow{BY}$ and $\overrightarrow{CZ}$ are all representatives of the same vector.

Thus AXZC is a parallelogram.

Hence AC and XZ are equal, parallel and in the same direction; so $\overrightarrow{AC}$ and $\overrightarrow{XZ}$ represent the same vector.

Properties of addition

As in the case of two-dimensional work we shall be able to treat the symbols $\boldsymbol{a}$, $\boldsymbol{b}$, $\boldsymbol{c}$, etc., like the symbols of ordinary algebra as far as addition and subtraction are concerned, provided we are sure they follow the same rules. For this reason we now verify that vectors have the same properties as ordinary numbers under addition.

Associative Law

In Figure 7, let $\overrightarrow{PQ}$, $\overrightarrow{QR}$ and $\overrightarrow{RS}$ represent vectors $\boldsymbol{a}$, $\boldsymbol{b}$ and $\boldsymbol{c}$ respectively

$(\boldsymbol{a}+\boldsymbol{b})+\boldsymbol{c}$ is represented by
$(\overrightarrow{PQ}+\overrightarrow{QR})+\overrightarrow{RS}$
$=\overrightarrow{PR}+\overrightarrow{RS}$ by the addition of two vectors
$=\overrightarrow{PS}$ by the addition of two vectors.
In the same way $\boldsymbol{a}+(\boldsymbol{b}+\boldsymbol{c})$ is represented by
$\overrightarrow{PQ}+(\overrightarrow{QR}+\overrightarrow{RS})=\overrightarrow{PQ}+\overrightarrow{QS}=\overrightarrow{PS}$.

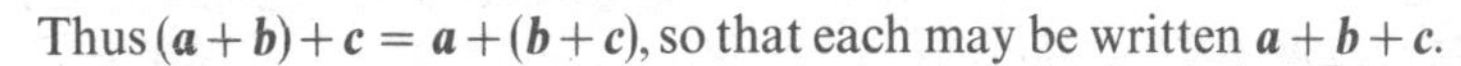

Thus $(\boldsymbol{a}+\boldsymbol{b})+\boldsymbol{c}=\boldsymbol{a}+(\boldsymbol{b}+\boldsymbol{c})$, so that each may be written $\boldsymbol{a}+\boldsymbol{b}+\boldsymbol{c}$.

Note. $\overrightarrow{PR}$ is shown as a broken line to emphasise the three-dimensional nature of the configuration being illustrated. (If the configuration is regarded as being two-dimensional, the proof still applies.) For the three-dimensional case, PQRS is a tetrahedron and the above additions have been carried out in four different planes.

Identity element

As in the two-dimensional case there is a zero vector $\boldsymbol{0}$ defined by

$$\boldsymbol{u}+\boldsymbol{0}=\boldsymbol{u}=\boldsymbol{0}+\boldsymbol{u}, \text{ for all } \boldsymbol{u}.$$

It is convenient to use $\boldsymbol{0}$ also as the zero for directed line segments.

Additive inverse (negative of a vector)

Consider the sum of the two directed line segments $\overrightarrow{AB}$ and $\overrightarrow{BA}$. Since $\overrightarrow{AB}+\overrightarrow{BA}=\overrightarrow{AA}$, which represents the zero vector $\boldsymbol{0}$, we say that each of the directed line segments is the *additive inverse* or *negative* of the other, and write $\overrightarrow{AB}=-\overrightarrow{BA}$ and $\overrightarrow{BA}=-\overrightarrow{AB}$. If $\overrightarrow{AB}$ represents $\boldsymbol{u}$ and $\overrightarrow{BA}$ represents $\boldsymbol{v}$, then $\boldsymbol{u}+\boldsymbol{v}=\boldsymbol{0}$; each of $\boldsymbol{u}$, $\boldsymbol{v}$ is the additive inverse of the other, and we write $\boldsymbol{u}=-\boldsymbol{v}$ and $\boldsymbol{v}=-\boldsymbol{u}$.

Subtraction of vectors is defined by $\boldsymbol{u}-\boldsymbol{v}=\boldsymbol{u}+(-\boldsymbol{v})$.

Commutative Law

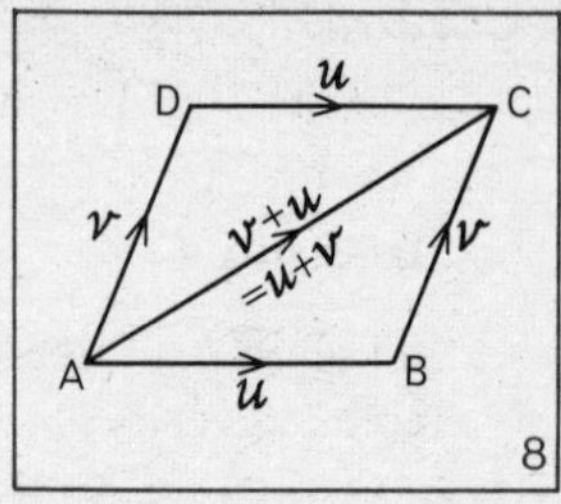

As with two-dimensional vectors it follows from a parallelogram in the plane of ABC (Figure 8) that

$$\boldsymbol{u}+\boldsymbol{v} = \boldsymbol{v}+\boldsymbol{u}.$$

Multiplication by a number

We have seen that two vectors can always be represented by directed line segments with a point in common, i.e. in the same plane. It follows that all the work in Book 5 about multiplication of a vector by a number holds good. Notice that this 'multiplication' is not at all the same as ordinary multiplication in which the two factors and the product belong to the same set R. Here we multiply a *vector* by a *number* to obtain a *vector*; compare multiplying a sum of money by a number to obtain a sum of money (see Reminders **4** and **5** on page 111).

Exercise 2

1 In Figure 9, ABCD is a tetrahedron. BL, CM and DN are concurrent at P.

- ***a*** Express $\overrightarrow{AP}$ as a sum of *two* directed line segments in three ways.
- ***b*** Express $\overrightarrow{AP}$ as a sum of *three* directed line segments named in the figure in three ways.
- ***c*** Try to express $\overrightarrow{AP}$ as a sum of four named line segments, and also as a sum of five.
- ***d*** Do you think that there is a limit to the number of possible segments in space that could be added together to give $\overrightarrow{AP}$?

2 In Figure 9, supply one missing directed line segment in each of the following statements.

a $\overrightarrow{AP}+\overrightarrow{PN}+\ldots = \overrightarrow{AB}$ ***b*** $\overrightarrow{MP}+\overrightarrow{PD} = \overrightarrow{MB}+\ldots+\overrightarrow{ND}$

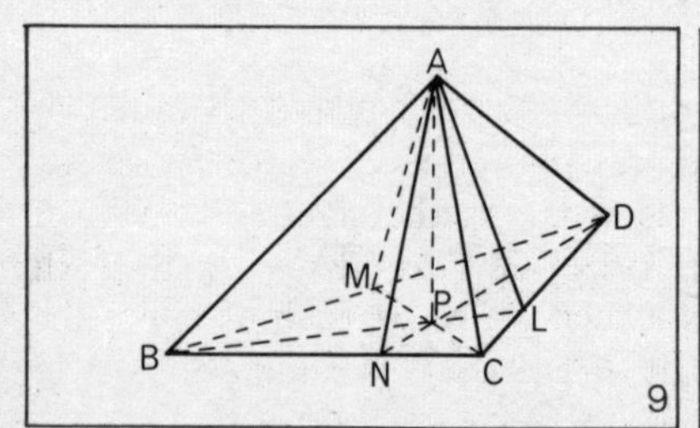

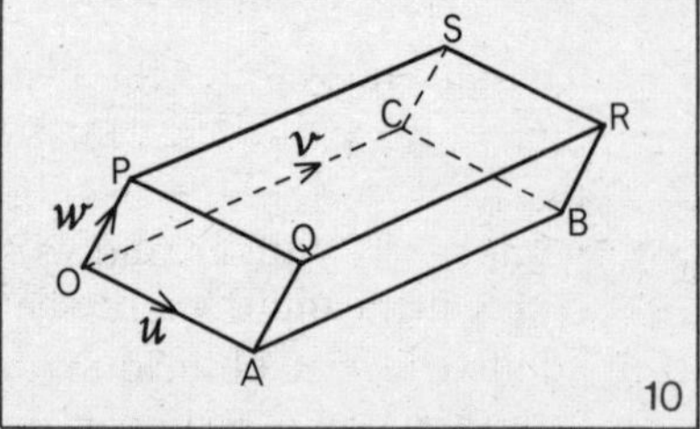

3 *a* Draw a straight line EF 2 cm long. Construct directed line segments equal to: (*1*) $4\overrightarrow{EF}$ (*2*) $-2\overrightarrow{EF}$ (*3*) $\frac{1}{2}\overrightarrow{EF}$.

b Where would you expect to find X in Figure 9 if:

(*1*) $\overrightarrow{AX} = 2\overrightarrow{AP}$ (*2*) $\overrightarrow{AX} = \frac{1}{3}\overrightarrow{AP}$ (*3*) $\overrightarrow{AX} = \overrightarrow{AP}$?

4 In Figure 9, express the following differences as single line segments.

a $\overrightarrow{AP}-\overrightarrow{AC}$ *b* $\overrightarrow{AC}-\overrightarrow{AP}$ *c* $\overrightarrow{NP}-\overrightarrow{ND}$

5 In the same figure express each of the following as a difference.

a $\overrightarrow{PL}$ *b* $\overrightarrow{PD}$ *c* $\overrightarrow{AN}$ *d* $\overrightarrow{AP}$ *e* $\overrightarrow{AB}$

6 Figure 10 shows a figure called a *parallelepiped*; it is a solid bounded by six faces, each of which is a parallelogram. The edges $\overrightarrow{OA}$, $\overrightarrow{OC}$ and $\overrightarrow{OP}$ represent the vectors $\boldsymbol{u}$, $\boldsymbol{v}$ and $\boldsymbol{w}$ respectively.

a Write down all the representatives in the diagram of $\boldsymbol{u}$, $\boldsymbol{v}$ and $\boldsymbol{w}$.

b Use the figure to show that the sum of $\boldsymbol{u}$, $\boldsymbol{v}$ and $\boldsymbol{w}$ is represented by $\overrightarrow{OR}$, no matter in which of the six possible orders the sum is taken.

7 Simplify:

a $2(\boldsymbol{p}+\boldsymbol{q})-3(2\boldsymbol{p}-\boldsymbol{q})+5(3\boldsymbol{q}-2\boldsymbol{r})$ *b* $3[2(\boldsymbol{p}-\boldsymbol{q})-3(\boldsymbol{p}+\boldsymbol{q})]$.

8 In $\triangle$ABC, $\overrightarrow{BC}$ represents $\boldsymbol{a}$, $\overrightarrow{CA}$ represents $\boldsymbol{b}$ and $\overrightarrow{AB}$ represents $\boldsymbol{c}$. L and M are the midpoints of BC and CA respectively.

a Show that $\overrightarrow{LM}$ represents $\frac{1}{2}(\boldsymbol{a}+\boldsymbol{b})$.

b Express the vector represented by $\overrightarrow{BA}$ in terms of $\boldsymbol{a}$ and $\boldsymbol{b}$ and hence state a relation between $\overrightarrow{LM}$ and $\overrightarrow{BA}$.

c State in words the relation between the lines LM and BA, beginning 'The line joining the midpoints of two sides of a triangle is...........' (This result is often useful and is usually known as the *midpoint theorem.*)

9 With the same diagram as in question **8**, let the line through M parallel to CB cut AB at the point N.

a Show that $\overrightarrow{MN}$ represents $\frac{1}{2}\boldsymbol{b}+k\boldsymbol{c}$ for some k.

b Show that $\overrightarrow{MN}$ represents $l\boldsymbol{a}$ for some l.

c Use $\boldsymbol{a}+\boldsymbol{b}+\boldsymbol{c}=\boldsymbol{0}$ to show that $(l+\frac{1}{2})\boldsymbol{b}+(l+k)\boldsymbol{c}=\boldsymbol{0}$, and hence that $k=\frac{1}{2}$ and $l=-\frac{1}{2}$. Interpret the result in words.

3 *Basis for vectors in space; vectors as number triples; unit vectors*

We saw in Section 1, page 112, that if $\overrightarrow{OX}$ represents vector $\boldsymbol{u}$ and $\overrightarrow{OY}$ represents vector $\boldsymbol{v}$, then every vector in the XOY plane can be expressed uniquely in the form $l\boldsymbol{u}+m\boldsymbol{v}$, where l and m are real numbers. In Figure 11, for example, $\overrightarrow{OB}$ represents $4\boldsymbol{u}+3\boldsymbol{v}$.

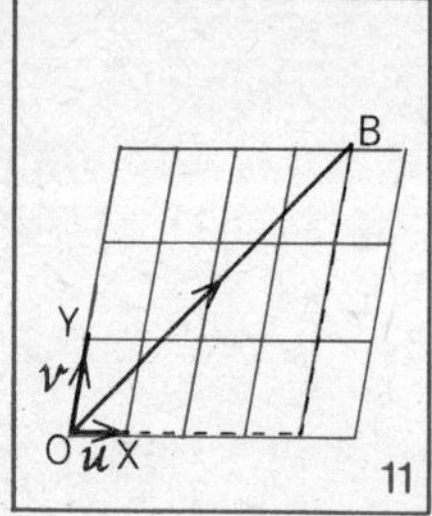

11

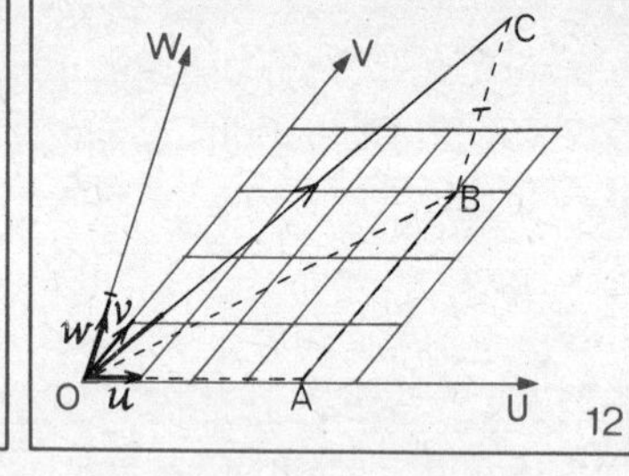

12

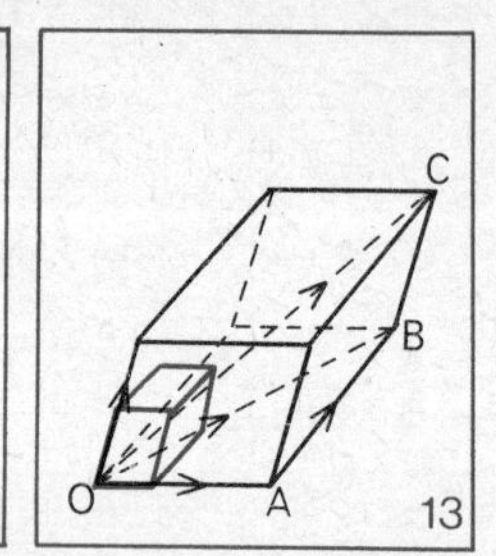

13

We now extend this idea to three dimensions. Figure 12 shows representatives of three vectors $\boldsymbol{u}$, $\boldsymbol{v}$ and $\boldsymbol{w}$ which are used as units for the three axes OU, OV and OW. (Think of OW in a different plane from UOV.)

In the figure, $\overrightarrow{OC} = \overrightarrow{OA}+\overrightarrow{AB}+\overrightarrow{BC}$.

Hence $\overrightarrow{OC}$ represents $4\boldsymbol{u}+3\boldsymbol{v}+2\boldsymbol{w}$.

In the same way any vector in space can be expressed as a linear combination of $\boldsymbol{u}$, $\boldsymbol{v}$ and $\boldsymbol{w}$. We say that $\boldsymbol{u}$, $\boldsymbol{v}$ and $\boldsymbol{w}$ form a *basis* for vectors in three dimensions.

Instead of the parallelograms with sides $\boldsymbol{u}$ and $\boldsymbol{v}$ in Figure 11, we have a framework of parallelepipeds with edges representing $\boldsymbol{u}$, $\boldsymbol{v}$ and $\boldsymbol{w}$, one of which is shown in colour in Figure 13.

Note. $\boldsymbol{u}$, $\boldsymbol{v}$ and $\boldsymbol{w}$ need not be of the same length or at right angles to one another.

In fact, every vector $\boldsymbol{r}$ in space can be expressed in terms of three non-coplanar base vectors $\boldsymbol{u}$, $\boldsymbol{v}$ and $\boldsymbol{w}$ in the form

$$\boldsymbol{r} = l\boldsymbol{u}+m\boldsymbol{v}+n\boldsymbol{w}.$$

l, m and n are called the *components* of the vector $\boldsymbol{r}$. We can express $\boldsymbol{r}$ as a *number triple* in the form of a *row vector* $(l \quad m \quad n)$ or a *column vector* $\begin{pmatrix} l \\ m \\ n \end{pmatrix}$.

This method of defining a vector is unique since there is only one parallelepiped with a space diagonal representing $\boldsymbol{r}$ and its edges parallel to $\boldsymbol{u}$, $\boldsymbol{v}$ and $\boldsymbol{w}$. (Compare the note on page 112 about two-dimensional vectors.)

(i) Whether the axes are rectangular or not, if O is the origin and A is the point (x_1, y_1, z_1), the vector $\boldsymbol{a}$ represented by $\overrightarrow{OA}$ may be written $\begin{pmatrix} x_1 \\ y_1 \\ z_1 \end{pmatrix}$ (see Figure 14(i)).

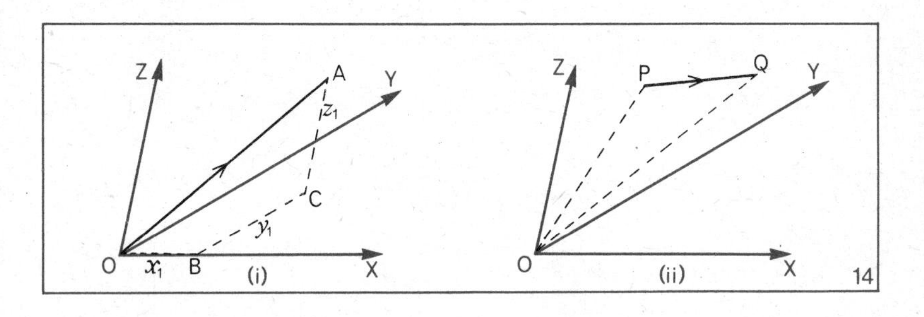

14

(ii) Care must be taken not to confuse *coordinates* with *components*.

In Figure 14(ii), let $\overrightarrow{PQ}$ represent a vector $\boldsymbol{a}$ where P is (x_1, y_1, z_1) and Q is (x_2, y_2, z_2).

Since $\overrightarrow{PQ} = \overrightarrow{PO} + \overrightarrow{OQ}$

$= \overrightarrow{OQ} - \overrightarrow{OP}$,

$$\boldsymbol{a} = \begin{pmatrix} x_2 \\ y_2 \\ z_2 \end{pmatrix} - \begin{pmatrix} x_1 \\ y_1 \\ z_1 \end{pmatrix} = \begin{pmatrix} x_2 - x_1 \\ y_2 - y_1 \\ z_2 - z_1 \end{pmatrix}$$

Hence $x_2 - x_1$, $y_2 - y_1$, $z_2 - z_1$ are the *components* of $\boldsymbol{a}$.

If $\boldsymbol{u}$, $\boldsymbol{v}$ and $\boldsymbol{w}$ are used as a basis, then

$$\boldsymbol{a} = (x_2 - x_1)\boldsymbol{u} + (y_2 - y_1)\boldsymbol{v} + (z_2 - z_1)\boldsymbol{w}.$$

Example. A is the point (0, 1, 2), B is (1, 3, −1) and C is (3, 7, −7). Show that A, B and C are collinear, and find the value of AB : BC.

Let $\boldsymbol{a} = \begin{pmatrix} 0 \\ 1 \\ 2 \end{pmatrix}$, $\boldsymbol{b} = \begin{pmatrix} 1 \\ 3 \\ -1 \end{pmatrix}$ and $\boldsymbol{c} = \begin{pmatrix} 3 \\ 7 \\ -7 \end{pmatrix}$.

Then $\overrightarrow{AB}$ represents $\boldsymbol{b} - \boldsymbol{a} = \begin{pmatrix} 1 \\ 3 \\ -1 \end{pmatrix} - \begin{pmatrix} 0 \\ 1 \\ 2 \end{pmatrix} = \begin{pmatrix} 1 \\ 2 \\ -3 \end{pmatrix}$,

and $\overrightarrow{BC}$ represents $\boldsymbol{c} - \boldsymbol{b} = \begin{pmatrix} 3 \\ 7 \\ -7 \end{pmatrix} - \begin{pmatrix} 1 \\ 3 \\ -1 \end{pmatrix} = \begin{pmatrix} 2 \\ 4 \\ -6 \end{pmatrix} = 2\begin{pmatrix} 1 \\ 2 \\ -3 \end{pmatrix}$.

Hence $\overrightarrow{BC} = 2\overrightarrow{AB}$. It follows that A, B and C are collinear, and that BC = 2AB, or AB : BC = 1 : 2.

Exercise 3

1 $\boldsymbol{s} = \begin{pmatrix} 3 \\ 4 \\ 2 \end{pmatrix}$, $\boldsymbol{t} = \begin{pmatrix} 1 \\ -1 \\ 3 \end{pmatrix}$ and $\boldsymbol{u} = \begin{pmatrix} -2 \\ 3 \\ -2 \end{pmatrix}$.

a Express as column vectors: (*1*) $\boldsymbol{s}+\boldsymbol{t}$ (*2*) $\boldsymbol{t}+\boldsymbol{u}$.
b Verify that $(\boldsymbol{s}+\boldsymbol{t})+\boldsymbol{u} = \boldsymbol{s}+(\boldsymbol{t}+\boldsymbol{u})$.

2 $\boldsymbol{p} = \begin{pmatrix} -2 \\ 4 \\ 0 \end{pmatrix}$. Express as column vectors: *a* $3\boldsymbol{p}$ *b* $\frac{1}{2}\boldsymbol{p}$ *c* $-\boldsymbol{p}$.

3 $\boldsymbol{a} = \begin{pmatrix} -1 \\ 2 \\ 3 \end{pmatrix}$, $\boldsymbol{b} = \begin{pmatrix} 2 \\ -1 \\ -2 \end{pmatrix}$ and $\boldsymbol{c} = \begin{pmatrix} -1 \\ -2 \\ 3 \end{pmatrix}$. Simplify:

a $\boldsymbol{a}+\boldsymbol{b}+\boldsymbol{c}$ *b* $3\boldsymbol{a}-2\boldsymbol{b}+4\boldsymbol{c}$ *c* $4\boldsymbol{a}-\boldsymbol{b}-2\boldsymbol{c}$.

4 If $\boldsymbol{a}$, $\boldsymbol{b}$ and $\boldsymbol{c}$ are defined as in question *3*, find $\boldsymbol{x}$ in component form from each of the following equations.

a $\boldsymbol{x}+\boldsymbol{a} = \boldsymbol{c}$ *b* $\boldsymbol{x}-\boldsymbol{a} = \boldsymbol{b}$ *c* $3\boldsymbol{x} = \boldsymbol{b}-\boldsymbol{a}$

5 A is the point (1, 2, 3), B is (2, 4, 6) and C is (0, 5, −1). Find the column vectors representing $\overrightarrow{OA}$, $\overrightarrow{OB}$, $\overrightarrow{OC}$, $\overrightarrow{AB}$, $\overrightarrow{BC}$, $\overrightarrow{AC}$ and $\overrightarrow{CA}$.
Show that O, A and B are collinear, and find the value of OA : OB.

6 O is the origin, A is (6, 4, 2), B is (8, 6, 4) and C is (2, 2, 2).
Show that OABC is a parallelogram.

7 $\overrightarrow{PQ}$, $\overrightarrow{QR}$ and $\overrightarrow{RS}$ represent $\begin{pmatrix} 5 \\ 4 \\ -2 \end{pmatrix}$, $\begin{pmatrix} -8 \\ 10 \\ -2 \end{pmatrix}$ and $\begin{pmatrix} 3 \\ -14 \\ 4 \end{pmatrix}$ respectively.

a Use $\overrightarrow{PQ}+\overrightarrow{QR}+\overrightarrow{RS}$ to show that P and S coincide.
b The midpoint of QR is M. Express the vector represented by $\overrightarrow{PM}$ in number triple form.

8 A is the point (2, 4, 6), B is (6, 6, 2) and C is (14, 10, −6). Show that A, B and C are collinear, and find the value of AB : BC.

9 Repeat question *8* for the points A(1, −2, 5), B(2, −4, 4) and C(−1, 2, 7).

10 Given the points P(2, 7, 8) and Q(−1, 1, −1) find in component form the vector represented by $\overrightarrow{PQ}$. R is the point on PQ such that $\overrightarrow{PR} = \frac{1}{3}\overrightarrow{PQ}$. Give the vector represented by $\overrightarrow{PR}$, and hence find the coordinates of R.

11 O is the origin, A is (x_1, y_1, z_1) and B is (x_2, y_2, z_2).
a State the coordinates of M and N, the midpoints of OA and OB respectively.
b Show that MN is parallel to AB.

Unit vectors

Since there has been no insistence on rectangular axes so far, the work in this chapter has been independent of distance and angle. In the next section we discuss distance and, in a later section, angle. To make this easier we use as a basis *mutually perpendicular* (or *orthogonal*) vectors of unit length. These are conventionally known as $\boldsymbol{i}$ (in the x-direction), $\boldsymbol{j}$ (in the y-direction) and $\boldsymbol{k}$ (in the z-direction), and are taken to form a *right-handed* set of axes.

This means that if you imagine you could move along the positive z-axis and look back, you would see the axes OX and OY in their usual positions. The axes OX, OY and OZ have the same relation to one another as the thumb, forefinger and middle finger of your *right* hand when these fingers are extended (see Figure 15(i)).

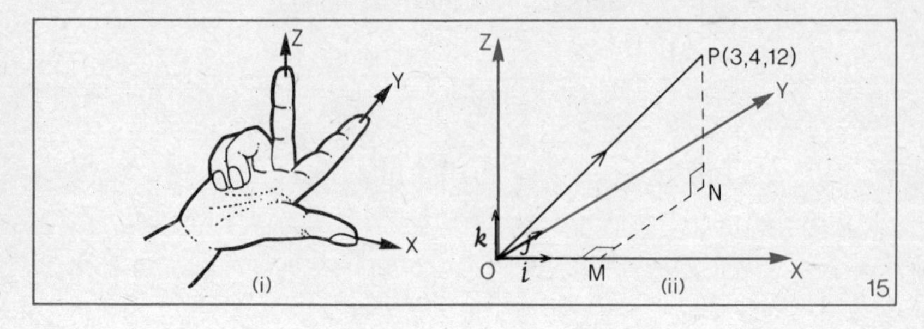

In component form $i = \begin{pmatrix}1\\0\\0\end{pmatrix}$, $j = \begin{pmatrix}0\\1\\0\end{pmatrix}$ and $k = \begin{pmatrix}0\\0\\1\end{pmatrix}$.

In Figure 15(ii), P is the point (3, 4, 12). If $\overrightarrow{OP}$ represents $\boldsymbol{p}$, then

$$\boldsymbol{p} = \begin{pmatrix}3\\4\\12\end{pmatrix} = 3\boldsymbol{i}+4\boldsymbol{j}+12\boldsymbol{k}.$$

We saw in Section 3 that there is only one way of expressing a vector in terms of three given non-coplanar vectors such as $\boldsymbol{i}$, $\boldsymbol{j}$ and $\boldsymbol{k}$. Thus if $p\boldsymbol{i}+q\boldsymbol{j}+r\boldsymbol{k} = 3\boldsymbol{i}-2\boldsymbol{j}+\boldsymbol{k}$, it follows that $p = 3$, $q = -2$ and $r = 1$.

Exercise 4

1 Express the following vectors in component form.

a $\boldsymbol{j}$ *b* $\boldsymbol{i}+\boldsymbol{j}$ *c* $\boldsymbol{j}+\boldsymbol{k}$ *d* $\boldsymbol{i}+\boldsymbol{j}+\boldsymbol{k}$

2 If $l\boldsymbol{i}+m\boldsymbol{j}+n\boldsymbol{k} = 3\boldsymbol{i}+2\boldsymbol{j}-\boldsymbol{k}$, write down the values of l, m and n.

3 a Express the vectors $\boldsymbol{a} = \begin{pmatrix}2\\3\\1\end{pmatrix}$, $\boldsymbol{b} = \begin{pmatrix}-1\\2\\0\end{pmatrix}$ and $\boldsymbol{c} = \begin{pmatrix}3\\-1\\4\end{pmatrix}$ in terms of $\boldsymbol{i}$, $\boldsymbol{j}$ and $\boldsymbol{k}$.

b Express the vectors $2\boldsymbol{a}+\boldsymbol{b}-\boldsymbol{c}$ and $-\boldsymbol{a}+2\boldsymbol{b}-4\boldsymbol{c}$ in terms of $\boldsymbol{i}$, $\boldsymbol{j}$ and $\boldsymbol{k}$ where $\boldsymbol{a}$, $\boldsymbol{b}$ and $\boldsymbol{c}$ are defined as in the first part of the question.

4 If $\boldsymbol{p} = 3\boldsymbol{i} - 2\boldsymbol{j} + \boldsymbol{k}$ and $\boldsymbol{q} = 2\boldsymbol{i} + \boldsymbol{j} - 3\boldsymbol{k}$ express in component form:

a $\boldsymbol{p}+\boldsymbol{q}$ *b* $\boldsymbol{p}-\boldsymbol{q}$ *c* $2\boldsymbol{p}+3\boldsymbol{q}$

5 Explain why the vectors $2\boldsymbol{i} - \boldsymbol{j} + 3\boldsymbol{k}$ and $-4\boldsymbol{i} + 2\boldsymbol{j} - 6\boldsymbol{k}$ are parallel.

6 In Figure 16, OABC, DEFG is a cuboid with edges along OX, OY and OZ. OA = 2, OC = 3 and OD = 4. Express in terms of $\boldsymbol{i}, \boldsymbol{j}$ and $\boldsymbol{k}$ the vectors represented by:

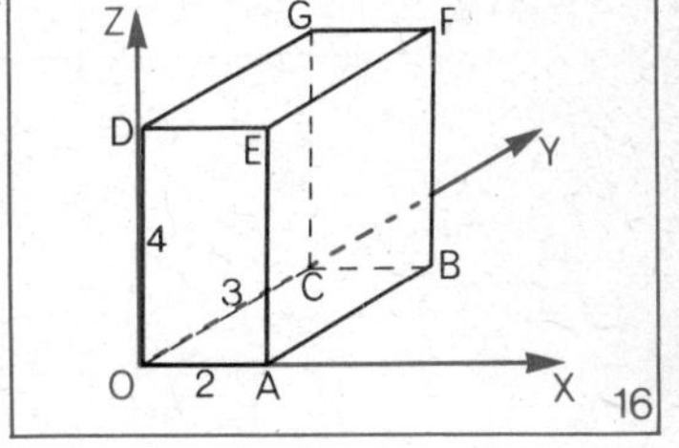

a $\overrightarrow{OB}$ *b* $\overrightarrow{OF}$ *c* $\overrightarrow{OG}$ *d* $\overrightarrow{CF}$ *e* $\overrightarrow{BG}$

f $\overrightarrow{OM}$, where M is the midpoint of GF.

4 *Magnitude of a vector; the distance formula*

In this section we take it for granted that the axes of reference are mutually perpendicular.

Magnitude of a vector

Example. In Figure 15 again, by Pythagoras' theorem,

$$\text{ON}^2 = \text{OM}^2 + \text{MN}^2 = 9 + 16 = 25$$
$$\text{OP}^2 = \text{ON}^2 + \text{NP}^2 = 25 + 144 = 169.$$

Hence $\text{OP} = \sqrt{169} = 13$.

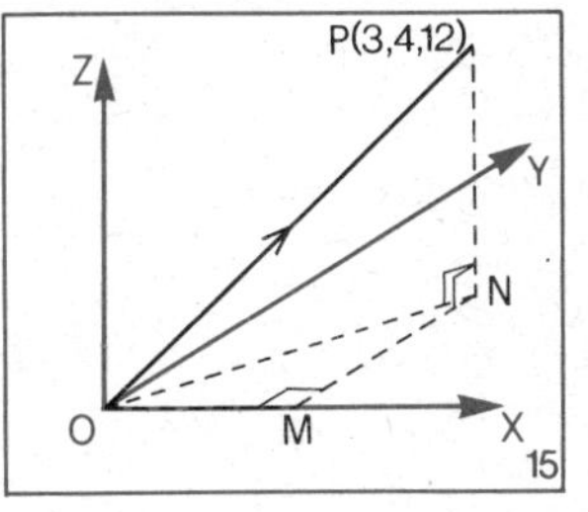

Using the method of this example we can find the magnitude of any directed line segment $\overrightarrow{OP}$, denoted by $|\overrightarrow{OP}|$. If the components of $\overrightarrow{OP}$ are l, m and n, then

$$\text{ON}^2 = l^2 + m^2 \text{ and } \text{OP}^2 = (l^2 + m^2) + n^2.$$

Hence $|\overrightarrow{OP}| = \sqrt{(l^2 + m^2 + n^2)}$.

Since all representatives of a vector have the same magnitude, it follows that if $\boldsymbol{p} = \begin{pmatrix} l \\ m \\ n \end{pmatrix}$ then the *length* or *magnitude* of $\boldsymbol{p}$, denoted by $|\boldsymbol{p}|$, can be expressed in terms of its components by

$$|\boldsymbol{p}| = \sqrt{(l^2+m^2+n^2)}.$$

Distance formula in three dimensions

If A is the point (x_1, y_1, z_1) and B is the point (x_2, y_2, z_2) then $\overrightarrow{AB}$ represents the vector $\begin{pmatrix} x_2-x_1 \\ y_2-y_1 \\ z_2-z_1 \end{pmatrix}$. (See page 119.)

Hence the distance between A and B is

$$|\overrightarrow{AB}| = \sqrt{[(x_2-x_1)^2+(y_2-y_1)^2+(z_2-z_1)^2]}.$$

Exercise 5

1 Calculate the distances between the following pairs of points.

a O(0, 0, 0), P(4, 2, 4) *b* O(0, 0, 0), Q$(1, 2\sqrt{2}, 3\sqrt{3})$

c A(4, 6, 2), B(3, 4, 4) *d* C(−2, 1, −1), D(5, −3, 3)

e E(−1, 2, 3), F(−1, −4, −5) *f* G$(2a, 4a, -3a)$, H$(-a, -8a, a)$

2 If $\boldsymbol{p} = \boldsymbol{i}+2\boldsymbol{j}-2\boldsymbol{k}$ and $\boldsymbol{q} = 3\boldsymbol{i}-2\boldsymbol{j}+6\boldsymbol{k}$, find:

a $|\boldsymbol{p}|$ *b* $|\boldsymbol{q}|$ *c* $|\boldsymbol{p}+\boldsymbol{q}|$ *d* $|\boldsymbol{q}-\boldsymbol{p}|$

3 If $\boldsymbol{f}_1 = 2\boldsymbol{i}-3\boldsymbol{j}+4\boldsymbol{k}$, $\boldsymbol{f}_2 = -\boldsymbol{i}+5\boldsymbol{j}$ and $\boldsymbol{f}_3 = 3\boldsymbol{i}+2\boldsymbol{j}+3\boldsymbol{k}$, find $\boldsymbol{f}_1+\boldsymbol{f}_2+\boldsymbol{f}_3$ and $|\boldsymbol{f}_1+\boldsymbol{f}_2+\boldsymbol{f}_3|$.

4 Show that P(3, 4, −1), Q(−9, −2, 3) and R(9, 8, 11) are vertices of an isosceles triangle.

5 P is the point (5, 7, −5), Q is (4, 7, −3) and R is (2, 7, −4). Show by the distance formula that △PQR is isosceles and right-angled.

6 Prove that A(1, 3, −1), B(3, 5, 0) and C(−1, 4, 1) are vertices of a right-angled isosceles triangle. Write down the coordinates of the fourth vertex of the square ABDC (note order of letters).

7 A is the point (2, 4, 6), B is (7, 5, 0), C is (6, 10, −6) and D is (1, 9, 0).

a Show that $\overrightarrow{AB} = \overrightarrow{DC}$. What type of figure is ABCD?
b Show that $|\overrightarrow{AB}| = |\overrightarrow{BC}|$. What type of figure is ABCD?

8 If $\overrightarrow{PQ}$, $\overrightarrow{QR}$ and $\overrightarrow{RS}$ represent the vectors $\begin{pmatrix} 2 \\ -1 \\ -3 \end{pmatrix}$, $\begin{pmatrix} 1 \\ -4 \\ 2 \end{pmatrix}$ and $\begin{pmatrix} -3 \\ 5 \\ 1 \end{pmatrix}$ respectively, show that S coincides with P, and that triangle PQR is right-angled.

9 Show that the points $(2, -3, 10)$, $(0, 5, -4)$ and $(4, 11, 2)$ lie on the surface of a sphere with centre $(2, 3, 4)$, and state the radius of the sphere.

10 If $(x-1)^2+(y-2)^2+(z-3)^2 = 25$, what is the distance between the point (x, y, z) and the point $(1, 2, 3)$? What can you say about all possible positions of the point (x, y, z)? Describe in words the locus defined by the set $\{(x, y, z):(x-1)^2+(y-2)^2+(z-3)^2 = 25\}$.

11 A sphere with centre $(3, 4, 5)$ has radius 4. Describe in set notation, as in question ***10***, the surface of this sphere.

12 $\triangle$PQR has vertices $P(4, 1, -1)$, $Q(2, 5, 3)$ and $R(5, -1, 1)$.

a Calculate PQ^2, QR^2, and RP^2.
b Use the cosine rule to calculate the size of $\angle$QPR.

5 *Position vectors; the section formula*

Dividing a line in the ratio m:n

A point P divides the line segment AB in the ratio $m:n$ if AP : PB = $m:n$. If P divides AB *internally*, $\overrightarrow{AB}$ and $\overrightarrow{PB}$ have the same direction and m and n have the same sign. If P divides AB *externally*, $\overrightarrow{AP}$ and $\overrightarrow{PB}$ have opposite directions and m and n have opposite signs.

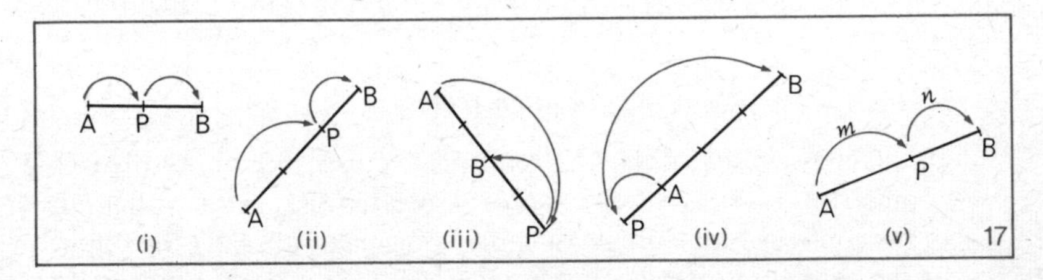

In Figure 17(i), AP : PB = 1 : 1, and AP : AB = 1 : 2.
In Figure 17(ii), AP : PB = 2 : 1, and AP : AB = 2 : 3.
In Figure 17(iii) AP : PB = 4 : −2 = 2 : −1, and AP:AB = 4:2 = 2:1.

In Figure 17(iv) AP:PB = $-1:4$, and AP:AB = $-1:3$.
In Figure 17(v) AP:PB = $m:n$, and AP:AB = $m:(m+n)$.

Exercise 6

1 Mark centimetre intervals on a straight line AB 6 cm long. Mark the points:

a P, so that AP = PB *b* Q, so that AQ:QB = 5:1
c R, so that AR:RB = 2:1 *d* S, so that AS:SB = 1:2.

2 Draw a line CD 2 cm long. Mark the points:

a X, so that CX:XD = 2:-1 *b* Y, so that CY:YD = 3:-2.

3 ABCD is a kite with its diagonals intersecting at O, so that DO = OB = 3 cm, AO = 6 cm and OC = 4 cm. Write down the values of the ratios:

a DO:OB *b* DO:DB *c* AO:OC *d* AO:AC *e* AC:CO.

4 A straight line AE is divided into four equal parts by the points B, C and D. Write down the values of the ratios:

a AB:BD *b* AC:CE *c* AB:AE *d* AC:AE
e AE:EC *f* BE:ED *g* BA:AD *h* DA:AC.

5 a Draw a straight line PQ, and on it mark points K and L that divide PQ internally and externally in the ratio 2:1.
b What are the values of: (*1*) PK:PQ (*2*) PQ:QL?

6 A(-3, 0), B(6, 0) and C(9, 0) are points on the *x*-axis. Calculate, in magnitude and sign, the following ratios.

a OB:BC *b* OC:CB *c* AB:BC *d* OA:OB *e* OB:BA.

Position vectors

In the last two questions of Exercise 2, we used vectors to prove a geometrical result. In each case we associated the vector with a *line* of the configuration. Often it is more convenient to relate the points of the configuration to some point O, called an origin, thus associating each *point* A, B, C, ..., with a *vector* $\boldsymbol{a}$, $\boldsymbol{b}$, $\boldsymbol{c}$, ..., such that $\boldsymbol{a}$, $\boldsymbol{b}$, $\boldsymbol{c}$, ... are represented by $\overrightarrow{OA}$, $\overrightarrow{OB}$, $\overrightarrow{OC}$, ... This idea was introduced in Book 5: $\boldsymbol{a}$ is called the *position vector* of A. When the point chosen as origin is otherwise of no particular interest, it may save complicating the diagram if it is not marked in; but this is not advisable until the idea of position vector is thoroughly grasped.

As we saw on page 119, if A is the point (x_1, y_1, z_1) then $\boldsymbol{a} = \begin{pmatrix} x_1 \\ y_1 \\ z_1 \end{pmatrix}$.

The *coordinates* of A are the *components* of its position vector.

The section formula in terms of vectors

In Figure 18, if $\boldsymbol{p}$ is the position vector of the point P which divides AB in the ratio $m:n$, then

$$\boldsymbol{p} = \frac{m\boldsymbol{b}+n\boldsymbol{a}}{m+n}.$$

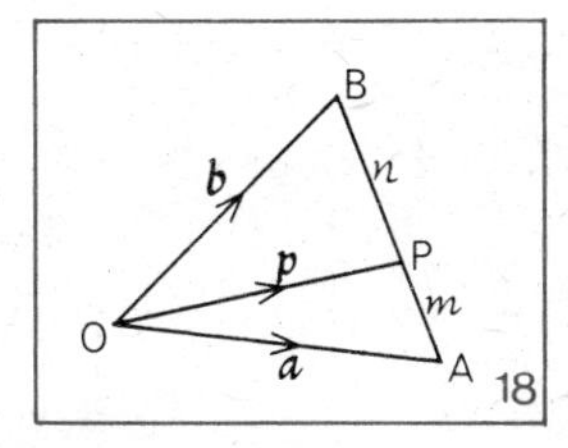

This is the *Section Formula* in terms of vectors.

Proof.

$$\text{AP}:\text{PB} = m:n \Rightarrow \text{AP}:\text{AB} = m:(m+n)$$

For all positions of P on AB, internal or external,

$$n\overrightarrow{\text{AP}} = m\overrightarrow{\text{PB}} \quad \textit{or} \quad \overrightarrow{\text{OP}} = \overrightarrow{\text{OA}} + \overrightarrow{\text{AP}}$$

$$\Leftrightarrow \quad n(\boldsymbol{p}-\boldsymbol{a}) = m(\boldsymbol{b}-\boldsymbol{p}) \qquad \Leftrightarrow \quad \overrightarrow{\text{OP}} = \overrightarrow{\text{OA}} + \frac{m}{m+n}\overrightarrow{\text{AB}}$$

$$\Leftrightarrow \quad n\boldsymbol{p}-n\boldsymbol{a} = m\boldsymbol{b}-m\boldsymbol{p}$$

$$\Leftrightarrow \quad (m+n)\boldsymbol{p} = m\boldsymbol{b}+n\boldsymbol{a} \qquad \Leftrightarrow \quad \boldsymbol{p} = \boldsymbol{a} + \frac{m}{m+n}(\boldsymbol{b}-\boldsymbol{a})$$

$$\Leftrightarrow \quad \boldsymbol{p} = \frac{m\boldsymbol{b}+n\boldsymbol{a}}{m+n} \qquad \Leftrightarrow \quad \boldsymbol{p} = \frac{m\boldsymbol{b}+n\boldsymbol{a}}{m+n}$$

The formula is more easily used in the external case if the numerically larger of m and n is taken as positive, e.g. $4:-1$, rather than $-4:1$.

The particular case where P is the midpoint of AB is worth noting. Here $m:n = 1:1$, and $\boldsymbol{p} = \frac{1}{2}(\boldsymbol{a}+\boldsymbol{b})$.

Note. The mathematical idea expressed by the section formula is the idea of the *weighted mean*: it has many applications. What is the average age of a roomful of people if one is aged 46 and the other twenty-nine are aged 16? Where on AB is the centre of gravity of a

1-kg mass at A and a 4-kg mass at B? What should be charged per litre for an 8 : 1 mixture of petrol at 10p per litre and oil at 20p per litre? An understanding of the idea behind all these makes the formula very easy to remember.

Example. Find the position vectors of P and Q which divide AB internally and externally in the ratio 5 : 2.

For P, $m : n = 5 : 2$.

Hence $\boldsymbol{p} = \dfrac{5\boldsymbol{b}+2\boldsymbol{a}}{5+2}$

$= \frac{1}{7}(5\boldsymbol{b}+2\boldsymbol{a})$.

For Q, $m : n = 5 : -2$.

Hence $\boldsymbol{q} = \dfrac{5\boldsymbol{b}-2\boldsymbol{a}}{5-2}$

$= \frac{1}{3}(5\boldsymbol{b}-2\boldsymbol{a})$.

Exercise 7

1 In Figure 19, $\boldsymbol{p}$ and $\boldsymbol{q}$ are the position vectors of P and Q. Show that $\overrightarrow{PQ}$ represents $\boldsymbol{q}-\boldsymbol{p}$.

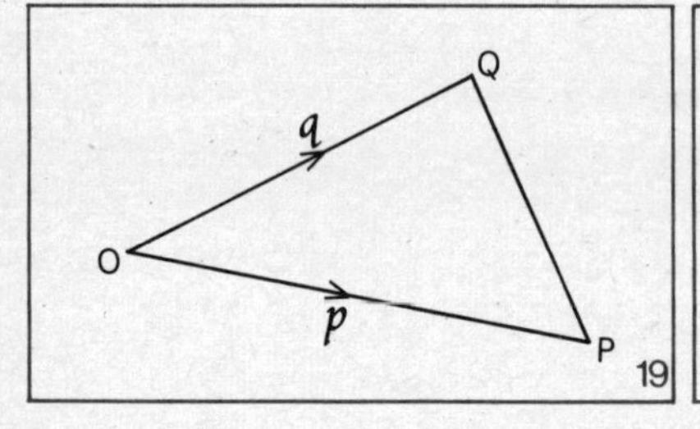

19

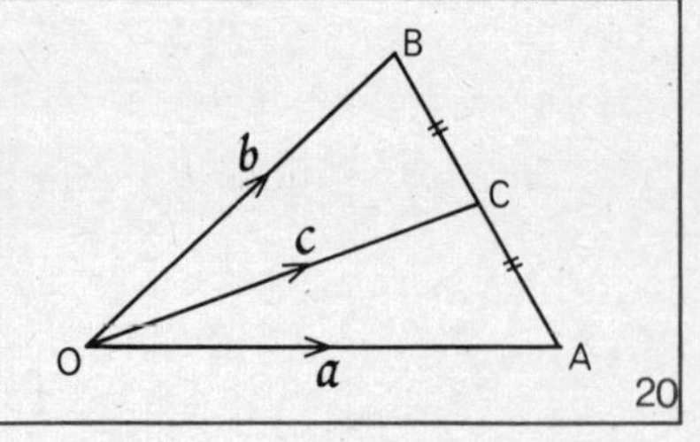

20

2 In Figure 20, C is the midpoint of AB. $\boldsymbol{a}$, $\boldsymbol{c}$ and $\boldsymbol{b}$ are the position vectors of A, C and B respectively. Copy and complete each of the following to show that $\boldsymbol{c} = \frac{1}{2}(\boldsymbol{a}+\boldsymbol{b})$.

a $\overrightarrow{AC} = \overrightarrow{CB}$

$\Leftrightarrow$ $\boldsymbol{c}-\boldsymbol{a} = \boldsymbol{b}-\boldsymbol{c}$

$\Leftrightarrow$ ....

b $\overrightarrow{OC} = \overrightarrow{OA}+\overrightarrow{AC}$

$\Leftrightarrow$ $\overrightarrow{OC} = \overrightarrow{OA}+\frac{1}{2}\overrightarrow{AB}$

$\Leftrightarrow$ $\boldsymbol{c} =$ ....

3 Prove the result in question **2** by using the formula $\boldsymbol{c} = \dfrac{m\boldsymbol{b}+n\boldsymbol{a}}{m+n}$.

4 Repeat the methods of questions **2** and ***3*** to find the position vector of the point C which divides AB internally in the ratio 2 : 1. (Choose one of the methods in question **2**.)

5 Use the formula $\boldsymbol{p} = \dfrac{m\boldsymbol{b}+n\boldsymbol{a}}{m+n}$ to find the position vectors of the following points in terms of $\boldsymbol{a}$ and $\boldsymbol{b}$.

a D, dividing AB in the ratio 3:2
b E, dividing AB in the ratio $3:-2$
c F and G, dividing AB internally and externally in the ratio 1:2
d H and K, dividing AB internally and externally in the ratio 5:3.

6 ABCD is a parallelogram. M is the midpoint of AB, and T divides DM in the ratio 2:1. If $\overrightarrow{AD}$ represents $\boldsymbol{u}$ and $\overrightarrow{AB}$ represents $\boldsymbol{v}$, find the vector represented by $\overrightarrow{AT}$ in terms of $\boldsymbol{u}$ and $\boldsymbol{v}$.

Deduce that A, T and C are collinear, and find AT:TC.

7 P, Q and R are the midpoints of BC, CA and AB respectively in $\triangle$ABC. $\boldsymbol{a}, \boldsymbol{b}, \ldots$, are the position vectors of A, B, ...

a Find the position vectors $\boldsymbol{p}$, $\boldsymbol{q}$ and $\boldsymbol{r}$ in terms of $\boldsymbol{a}$, $\boldsymbol{b}$ and $\boldsymbol{c}$.
b Prove that $\boldsymbol{p}+\boldsymbol{q}+\boldsymbol{r} = \boldsymbol{a}+\boldsymbol{b}+\boldsymbol{c}$.
c Show that $\overrightarrow{AP}+\overrightarrow{BQ}+\overrightarrow{CR} = \boldsymbol{0}$.

Exercise 7B

1 In $\triangle$OPR, Q is the midpoint of PR. Show, with reference to O as origin, that $\boldsymbol{p}+\boldsymbol{r} = 2\boldsymbol{q}$.

Deduce that if OABC is a quadrilateral and X and Y are the midpoints of AC and OB respectively, then the sum of the vectors represented by $\overrightarrow{OA}$, $\overrightarrow{OC}$, $\overrightarrow{BA}$ and $\overrightarrow{BC}$ is the vector represented by $4\overrightarrow{YX}$.

2 OA, OB and OC are three concurrent edges of a parallelepiped. The other vertices are O_1, A_1, B_1 and C_1 named so that OO_1, AA_1, BB_1 and CC_1 are space diagonals of the parallelepiped. Taking O as origin for position vectors, write down the position vectors of O_1, A_1, B_1 and C_1 in terms of $\boldsymbol{a}$, $\boldsymbol{b}$ and $\boldsymbol{c}$.

Find the position vectors of the midpoints of the four space diagonals. State a theorem about the four space diagonals.

3 ABCD is a quadrilateral. N, M, R and P are the midpoints of AB, BC, CD and DA respectively.

a Express in terms of $\boldsymbol{a}$, $\boldsymbol{b}$, $\boldsymbol{c}$ and $\boldsymbol{d}$ the position vectors of:
(*1*) N, M, R and P
(*2*) the midpoint S of NR and the midpoint T of PM.

b What can you say about S and T, and about NR and PM?
c What type of figure is NMRP?

4 ABCD is a tetrahedron. N, M, R, P, K and Q are the midpoints of AB, BC, CD, DA, AC and BD respectively.

a Express, in terms of $\boldsymbol{a}$, $\boldsymbol{b}$, $\boldsymbol{c}$ and $\boldsymbol{d}$, the position vectors of the midpoints S of NR, T of PM and V of KQ.
b What can you say about S, T and V? State a geometrical property of every tetrahedron.

5 *Vector proof that the medians of a triangle are concurrent*

a D, E and F are the midpoints of the sides BC, CA and AB of the triangle ABC. Find the position vectors of D, E and F in terms of $\boldsymbol{a}$, $\boldsymbol{b}$ and $\boldsymbol{c}$.
b Find the position vectors of G, H and K which divide AD, BE and CF respectively in the ratio 2:1.
c Give a reason why G lies on BE and CF. State a result about the medians of a triangle.

G is called the *centroid* of the triangle. (It is the centre of gravity for a uniform triangular plate, but not for a triangular wire frame.)

6 *a* In a tetrahedron ABCD, G_1 is the centroid of $\triangle$BCD. Show that, if AG_1 is divided at G in the ratio 3:1, then $\boldsymbol{g} = \frac{1}{4}(\boldsymbol{a}+\boldsymbol{b}+\boldsymbol{c}+\boldsymbol{d})$.
b G_2, G_3, G_4 are the centroids of triangles CDA, DAB, ABC respectively. H, K and L divide BG_2, CG_3 and DG_4 respectively in the ratio 3:1. Find $\boldsymbol{h}$, $\boldsymbol{k}$ and $\boldsymbol{l}$ in terms of $\boldsymbol{a}$, $\boldsymbol{b}$, $\boldsymbol{c}$ and $\boldsymbol{d}$.
c What conclusion can you draw about G, H, K and L? State a theorem about a tetrahedron corresponding to the theorem about medians of a triangle.

Note. Sometimes a careful choice of origin simplifies a problem by avoiding heavy algebra, e.g. by choosing G, the centroid of $\triangle$ABC, as origin we have $\frac{1}{3}(\boldsymbol{a}+\boldsymbol{b}+\boldsymbol{c}) = \boldsymbol{g} = \boldsymbol{0}$. Often, however, the choice of a particular point in a diagram may destroy the algebraic symmetry of the treatment (see questions **5** and ***6*** above). It is necessary to weigh these conflicting advantages. Questions ***4*** and ***6*** indicate the power of vectors in three-dimensional work.

The section formula in terms of coordinates

The section formula may also be expressed in terms of coordinates.

If $P(x_P, y_P, z_P)$ divides the straight line joining $A(x_1, y_1, z_1)$ and $B(x_2, y_2, z_2)$ in the ratio $m:n$, then the coordinates of P are

$$x_P = \frac{mx_2+nx_1}{m+n}, \quad y_P = \frac{my_2+ny_1}{m+n}, \quad z_P = \frac{mz_2+nz_1}{m+n}.$$

For the sake of completeness we repeat the proof of the section formula in terms of vectors on page 127.

Proof. As the coordinates of a point are the components of its position vector, then $\boldsymbol{a} = \begin{pmatrix} x_1 \\ y_1 \\ z_1 \end{pmatrix}$, $\boldsymbol{b} = \begin{pmatrix} x_2 \\ y_2 \\ z_2 \end{pmatrix}$ and $\boldsymbol{p} = \begin{pmatrix} x_P \\ y_P \\ z_P \end{pmatrix}$.

Since A, P and B are collinear

$$AP:PB = m:n$$

$$\Leftrightarrow \quad n\overrightarrow{AP} = m\overrightarrow{PB}$$

$$\Leftrightarrow n(\boldsymbol{p}-\boldsymbol{a}) = m(\boldsymbol{b}-\boldsymbol{p})$$

$$\Leftrightarrow n\boldsymbol{p}-n\boldsymbol{a} = m\boldsymbol{b}-m\boldsymbol{p}$$

$$\Leftrightarrow (m+n)\boldsymbol{p} = m\boldsymbol{b}+n\boldsymbol{a}$$

$$\Leftrightarrow \quad \boldsymbol{p} = \frac{m\boldsymbol{b}+n\boldsymbol{a}}{m+n}$$

(continued above, right)

i.e. $$\begin{pmatrix} x_P \\ y_P \\ z_P \end{pmatrix} = \frac{1}{m+n}\left[m\begin{pmatrix} x_2 \\ y_2 \\ z_2 \end{pmatrix} + n\begin{pmatrix} x_1 \\ y_1 \\ z_1 \end{pmatrix}\right]$$

from which $$x_P = \frac{mx_2+nx_1}{m+n}$$

$$y_P = \frac{my_2+ny_1}{m+n}$$

and $$z_P = \frac{mz_2+nz_1}{m+n}$$

In coordinate geometry these are called the *section formulae*. Figure 21 indicates how you may easily remember the order of multiplication.

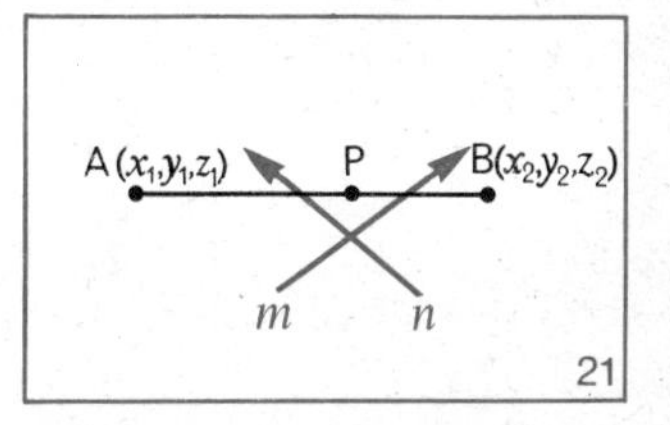

For coordinates in a plane exactly the same method of proof may be used. Thus in the x-y plane we put the z-coordinates equal to zero. Then, if the point $P(x_P, y_P)$ divides the line joining $A(x_1, y_1)$ and $B(x_2, y_2)$ in the ratio $m:n$,

$$x_P = \frac{mx_2+nx_1}{m+n} \quad \text{and} \quad y_P = \frac{my_2+ny_1}{m+n}.$$

Example. Find the coordinates of the points P and Q dividing the line joining A(1, 4, 6) and B(1, 0, 2) internally and externally in the ratio 3 : 1.

As in Figure 21, A(1, 4, 6) B(1, 0, 2), and A(1, 4, 6) B(1, 0, 2)

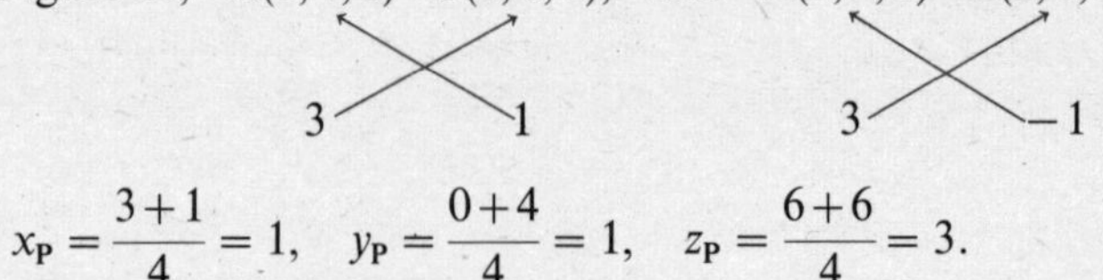

$$x_P = \frac{3+1}{4} = 1, \quad y_P = \frac{0+4}{4} = 1, \quad z_P = \frac{6+6}{4} = 3.$$

Hence P is the point (1, 1, 3).

$$x_Q = \frac{3-1}{2} = 1, \quad y_Q = \frac{0-4}{2} = -2, \quad z_Q = \frac{6-6}{2} = 0.$$

Hence Q is the point (1, −2, 0).

Exercise 8

1 P(x_P, y_P) is a point on AB. Find the coordinates of P if:

a A is (3, 2), B is (9, 5) and AP : PB = 2 : 1
b A is (−1, −3), B is (7, 5) and P is the midpoint of AB
c A is (−2, −3), B is (3, 7) and AP : PB = 3 : 2.

2 Q is a point on CD. Find the coordinates of Q if:

a C is (2, 1), D is (4, 7) and CQ : QD = 3 : −2
b C is (0, 0), D is (2, −4) and CQ : QD = 3 : −1
c C is (−1, −2), D is (4, 0) and CQ : QD = −2 : 1.

3 Find the coordinates of the points dividing the line joining R(−2, 4) and S(2, 0) internally and externally in the ratio 5 : 3.

4 P(x_P, y_P, z_P) lies on AB. Find the coordinates of P if:

a A is (1, 0, 2), B is (5, 4, 10) and AP : PB = 3 : 1
b A is (1, 1, 1), B is (3, −2, 5) and AP : PB = 3 : 2
c A is (−3, −2, −1), B is (0, −5, 2) and AP : PB = 4 : −3.

5 Find the coordinates of the points dividing the line joining E(5, 2, 1) and F(9, 10, 13) internally and externally in the ratio 1 : 3.

6 △ABC has vertices A(3, 6), B(0, 3) and C(6, 0). P divides AB in the ratio 1 : 2, Q divides CA in the ratio 1 : 2 and R divides BC in the ratio 4 : −1.

a Find the coordinates of P, Q and R.
b Show that P, Q and R are collinear, and calculate the ratio PQ : QR.

7 △ABC has vertices A($-1, -2$), B($4, 3$) and C($-6, 8$). P divides AB in the ratio 2 : 3, Q divides CA in the ratio 2 : 3 and R divides BC in the ratio 9 : -4.

a Find the coordinates of P, Q and R.
b Show that P, Q and R are collinear, and calculate the ratio PQ : PR.

8 P is the point ($1\frac{1}{2}, 2\frac{1}{2}, 1$), Q is ($1, 0, 0$) and R is ($2, 5, 2$). Show that P, Q and R are collinear, and find S such that $\overrightarrow{PS} = 3\overrightarrow{PQ}$.

9 The vertices of a triangle are P($-2, -2, -2$), Q($1, -2, 1$) and R($1, 0, -1$). S divides PQ in the ratio 2 : 1.

a Find the coordinates of S. *b* Show that ∠QSR is a right angle.

10 △ABC has vertices A($-3, 3$), B($0, 9$) and C($3, 6$). L divides BC in the ratio 2 : 1 and M divides CA in the ratio 1 : 2. N is the midpoint of AB.

a Calculate the coordinates of L, M and N.
b Show that BM and AL cut each other in the ratio 3 : 1 at K. (*Hint.* Calculate the coordinates of the points in which BM and AL are divided in the ratio 3 : 1.)
c Express the vectors represented by $\overrightarrow{CK}$ and $\overrightarrow{CN}$ in component form. What can you say about C, K and N?

11 The points A($-1, 5, 4$), B($2, -1, -2$) and C($3, p, q$) are collinear.

a From the formula $x_P = \dfrac{mx_2 + nx_1}{m+n}$ deduce that $\dfrac{m}{n} = \dfrac{x_P - x_1}{x_2 - x_P}$.
b Calculate the ratio in which B divides AC.
c Calculate the values of p and q.

---

6 *The scalar product of two vectors*

If $\boldsymbol{a}$ and $\boldsymbol{b}$ are two non-zero vectors, they can be represented by $\overrightarrow{OA}$ and $\overrightarrow{OB}$ as shown in Figure 22.

Let ∠AOB = θ, where $0 \leqslant \theta \leqslant \pi$.

The *scalar product* (or *dot product*, or *inner product*) of $\boldsymbol{a}$ and $\boldsymbol{b}$ is the real number defined by $\boldsymbol{a}\,.\,\boldsymbol{b} = |\boldsymbol{a}||\boldsymbol{b}|\cos\theta$. If either $\boldsymbol{a} = \boldsymbol{0}$ or $\boldsymbol{b} = \boldsymbol{0}$, the angle θ is not determined and we define $\boldsymbol{a}\,.\,\boldsymbol{b}$ to be 0.

Note that the scalar product of two vectors is *not* itself a vector.

Component form of scalar product

A convenient notation for the components of vectors $\boldsymbol{a}$, $\boldsymbol{b}$, etc., is $\boldsymbol{a} = \begin{pmatrix} a_1 \\ a_2 \\ a_3 \end{pmatrix}$, $\boldsymbol{b} = \begin{pmatrix} b_1 \\ b_2 \\ b_3 \end{pmatrix}$, and so on. Using the same letter for vector and components avoids the necessity of defining the components of each vector every time.

Let $\boldsymbol{a} = \begin{pmatrix} a_1 \\ a_2 \\ a_3 \end{pmatrix}$ and $\boldsymbol{b} = \begin{pmatrix} b_1 \\ b_2 \\ b_3 \end{pmatrix}$. If $\overrightarrow{\text{OA}}$ and $\overrightarrow{\text{OB}}$ in Figure 22 represent the vectors $\boldsymbol{a}$ and $\boldsymbol{b}$ respectively, then A is the point (a_1, a_2, a_3) and B is (b_1, b_2, b_3).

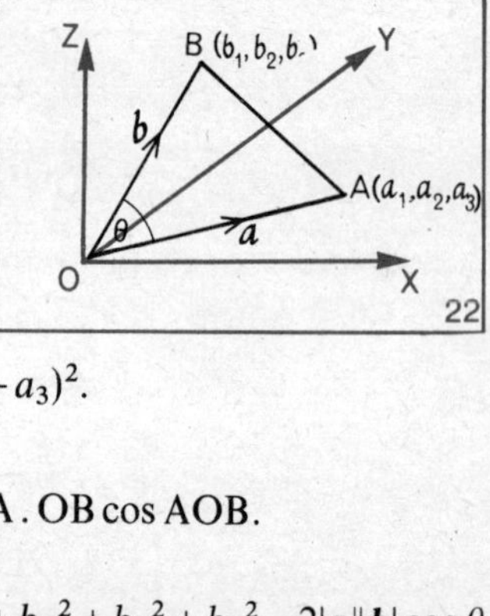

By the distance formula,

$$|\overrightarrow{\text{OA}}|^2 = a_1{}^2 + a_2{}^2 + a_3{}^2$$

$$|\overrightarrow{\text{OB}}|^2 = b_1{}^2 + b_2{}^2 + b_3{}^2$$

$$|\overrightarrow{\text{AB}}|^2 = |\boldsymbol{b} - \boldsymbol{a}|^2$$

$$= (b_1 - a_1)^2 + (b_2 - a_2)^2 + (b_3 - a_3)^2.$$

In $\triangle$AOB, by the cosine rule,

$$\text{AB}^2 = \text{OA}^2 + \text{OB}^2 - 2\text{OA}\,.\,\text{OB}\cos \text{AOB}.$$

So
$$(b_1 - a_1)^2 + (b_2 - a_2)^2 + (b_3 - a_3)^2 = a_1{}^2 + a_2{}^2 + a_3{}^2 + b_1{}^2 + b_2{}^2 + b_3{}^2 - 2|\boldsymbol{a}||\boldsymbol{b}|\cos\theta$$

$$\Leftrightarrow \quad 2|\boldsymbol{a}||\boldsymbol{b}|\cos\theta = 2a_1 b_1 + 2a_2 b_2 + 2a_3 b_3$$

$$\Leftrightarrow \quad |\boldsymbol{a}||\boldsymbol{b}|\cos\theta = a_1 b_1 + a_2 b_2 + a_3 b_3$$

$$\Leftrightarrow \quad \boldsymbol{a}\,.\,\boldsymbol{b} = a_1 b_1 + a_2 b_2 + a_3 b_3$$

It is clear that either of the equations

$$\boldsymbol{a}\,.\,\boldsymbol{b} = |\boldsymbol{a}||\boldsymbol{b}|\cos\theta$$

or

$$\boldsymbol{a}\,.\,\boldsymbol{b} = a_1 b_1 + a_2 b_2 + a_3 b_3$$

can be taken to define the scalar product of $\boldsymbol{a}$ and $\boldsymbol{b}$, and the other equation can be derived from this definition.

Note that the scalar product can be defined in terms of *components* of vectors and that these components may have to be calculated as on page 119.

In finding scalar products by components it may be helpful to write the components next to each other, for example

$$\begin{pmatrix}2\\3\\4\end{pmatrix}.\begin{pmatrix}-1\\2\\-3\end{pmatrix}=2\times(-1)+(3\times 2)+4\times(-3).$$

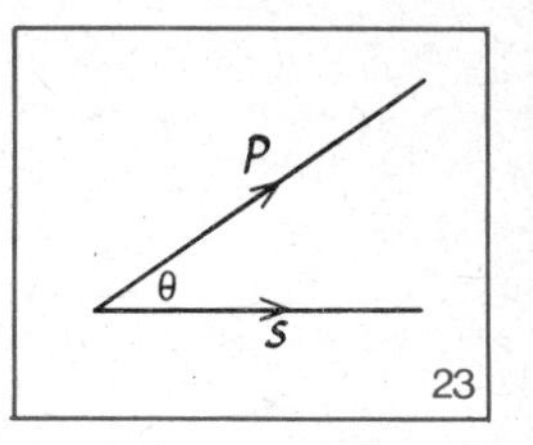

23

Note. Pupils who have some knowledge of physics may have met the concept of work. If a force P displaces a body through a distance s, the work done on the body is $Ps\cos\theta$ (see Figure 23). Since the force and the displacement have a magnitude and a direction assigned to them, it is reasonable for the physicist to represent them as vectors $\boldsymbol{P}$ and $\boldsymbol{s}$. Naturally he wants the product of $\boldsymbol{P}$ and $\boldsymbol{s}$ to be $Ps\cos\theta$. Hence the mathematician's definition of $\boldsymbol{a}\,.\,\boldsymbol{b}$ arises from physical considerations.

The sign of the number a.b ($a \neq 0$, $b \neq 0$)

From its definition, $\boldsymbol{a}\,.\,\boldsymbol{b}$ is a real number whose sign is determined by the size of θ as follows.

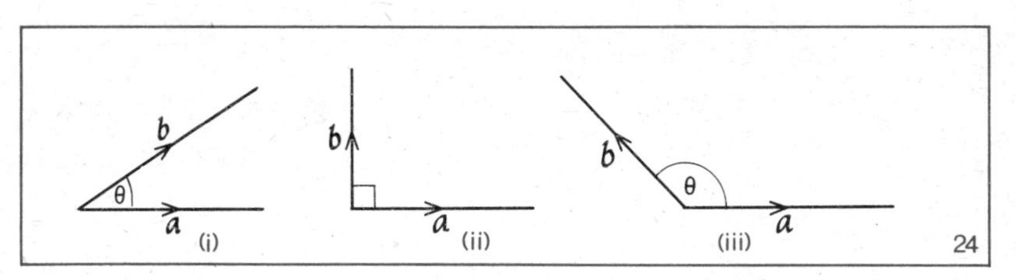

24

(i)	(ii)	(iii)
$0 \leqslant \theta < \frac{1}{2}\pi$	$\theta = \frac{1}{2}\pi$	$\frac{1}{2}\pi < \theta \leqslant \pi$
$\boldsymbol{a}\,.\,\boldsymbol{b} = \lvert\boldsymbol{a}\rvert\lvert\boldsymbol{b}\rvert\cos\theta > 0$	$\boldsymbol{a}\,.\,\boldsymbol{b} = \lvert\boldsymbol{a}\rvert\lvert\boldsymbol{b}\rvert\cos\theta = 0$	$\boldsymbol{a}\,.\,\boldsymbol{b} = \lvert\boldsymbol{a}\rvert\lvert\boldsymbol{b}\rvert\cos\theta < 0$

Note. Since $\cos\theta = \cos(-\theta)$ the direction in which we measure θ, from $\boldsymbol{a}$ to $\boldsymbol{b}$ or from $\boldsymbol{b}$ to $\boldsymbol{a}$, does not matter.

From (ii), $\theta = \frac{1}{2}\pi \;\Rightarrow\; \boldsymbol{a}\,.\,\boldsymbol{b} = 0 \;\Rightarrow\; a_1 b_1 + a_2 b_2 + a_3 b_3 = 0$.

Conversely, $\boldsymbol{a}\,.\,\boldsymbol{b} = 0 \;\Rightarrow\; \theta = \frac{1}{2}\pi$, i.e. $\boldsymbol{a}$ and $\boldsymbol{b}$ are perpendicular, *provided that* $\boldsymbol{a}$ *and* $\boldsymbol{b}$ *are non-zero vectors.*

Example. P is the point (1, 5, 8), Q is (−2, 1, 3) and R is (1, −6, 0). If $\overrightarrow{PQ}$ represents $\boldsymbol{u}$ and $\overrightarrow{QR}$ represents $\boldsymbol{v}$, calculate $\boldsymbol{u}.\boldsymbol{v}$.

$$\boldsymbol{u} = \begin{pmatrix}-2\\1\\3\end{pmatrix} - \begin{pmatrix}1\\5\\8\end{pmatrix} = \begin{pmatrix}-3\\-4\\-5\end{pmatrix} \text{ and } \boldsymbol{v} = \begin{pmatrix}1\\-6\\0\end{pmatrix} - \begin{pmatrix}-2\\1\\3\end{pmatrix} = \begin{pmatrix}3\\-7\\-3\end{pmatrix}$$

$$\boldsymbol{u}.\boldsymbol{v} = \begin{pmatrix}-3\\-4\\-5\end{pmatrix} . \begin{pmatrix}3\\-7\\-3\end{pmatrix} = (-3\times 3)+(-4)\times(-7)+(-5)\times(-3)$$

$$= -9+28+15 = 34$$

Exercise 9

1 Using $\boldsymbol{a}.\boldsymbol{b} = |\boldsymbol{a}||\boldsymbol{b}|\cos\theta$, calculate $\boldsymbol{a}.\boldsymbol{b}$ for each part of Figure 25.

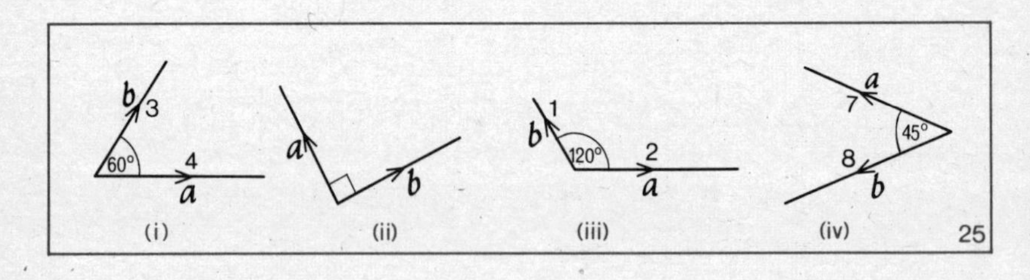

2 Using $\boldsymbol{a}.\boldsymbol{b} = a_1b_1 + a_2b_2 + a_3b_3$ and position vector notation, calculate $\boldsymbol{a}.\boldsymbol{b}$ in the cases where:

a A is (1, 2, 3), B is (−2, 4, −1) *b* A is (3, −1, 0), B is (2, 4, 6).

3 Using $\boldsymbol{a}.\boldsymbol{b} = a_1b_1 + a_2b_2 + a_3b_3$, find $\boldsymbol{a}.\boldsymbol{b}$ if:

a $\boldsymbol{a} = \boldsymbol{i}+2\boldsymbol{j}+5\boldsymbol{k}$, $\boldsymbol{b} = -\boldsymbol{i}+\boldsymbol{j}-2\boldsymbol{k}$
b $\boldsymbol{a} = 3\boldsymbol{i}-2\boldsymbol{j}+\boldsymbol{k}$, $\boldsymbol{b} = 2\boldsymbol{i}-2\boldsymbol{j}-\boldsymbol{k}$.

4 P is the point (1, 1, 1), Q is (−1, 1, 0) and R is (3, −2, −1). If $\overrightarrow{QP}$ represents $\boldsymbol{a}$ and $\overrightarrow{QR}$ represents $\boldsymbol{b}$, use components to calculate $\boldsymbol{a}.\boldsymbol{b}$.

5 Repeat question *4* for the points P(0, 2, 4), Q(1, 3, 5) and R(−2, 0, −4).

6 P is the point (1, 2, −3) and Q is (11, −3, 7). PQ is divided internally at R in the ratio 3 : 2. $\overrightarrow{PR}$ represents $\boldsymbol{a}$ and $\overrightarrow{RQ}$ represents $\boldsymbol{b}$. Calculate:

a the coordinates of R *b* the components of $\boldsymbol{a}$ and $\boldsymbol{b}$ *c* $\boldsymbol{a}.\boldsymbol{b}$

The angle between two non-zero vectors

From the definitions $\boldsymbol{a}\,.\,\boldsymbol{b} = a_1 b_1 + a_2 b_2 + a_3 b_3$
and $\boldsymbol{a}\,.\,\boldsymbol{b} = |\boldsymbol{a}||\boldsymbol{b}|\cos\theta$

we have $\cos\theta = \dfrac{\boldsymbol{a}\,.\,\boldsymbol{b}}{|\boldsymbol{a}||\boldsymbol{b}|} = \dfrac{a_1 b_1 + a_2 b_2 + a_3 b_3}{\sqrt{(a_1^2 + a_2^2 + a_3^2)(b_1^2 + b_2^2 + b_3^2)}}$

Example 1. $\boldsymbol{a} = \boldsymbol{i} + 2\boldsymbol{j} + 2\boldsymbol{k}$ and $\boldsymbol{b} = 2\boldsymbol{i} + 3\boldsymbol{j} - 6\boldsymbol{k}$. Calculate (i) $\boldsymbol{a}\,.\,\boldsymbol{b}$ (ii) the size of the angle between $\boldsymbol{a}$ and $\boldsymbol{b}$.

(i) $\boldsymbol{a} = \begin{pmatrix}1\\2\\2\end{pmatrix}$ and $\boldsymbol{b} = \begin{pmatrix}2\\3\\-6\end{pmatrix}$.

So $\boldsymbol{a}\,.\,\boldsymbol{b} = \begin{pmatrix}1\\2\\2\end{pmatrix}.\begin{pmatrix}2\\3\\-6\end{pmatrix} = 2 + 6 - 12 = -4.$

(ii) $\cos\theta = \dfrac{\boldsymbol{a}\,.\,\boldsymbol{b}}{|\boldsymbol{a}||\boldsymbol{b}|} = \dfrac{-4}{\sqrt{(1^2+2^2+2^2)(2^2+3^2+(-6)^2)}} = \dfrac{-4}{\sqrt{9}\sqrt{49}}$

$= \dfrac{-4}{21} = -0{\cdot}190.$

Hence, from tables, the required angle is $180° - 79° = 101°$.

Example 2. P is the point (4, 7, 0), Q is (6, 10, −6) and R is (1, 9, 0). Use the scalar product to show that $\angle\,\text{QPR} = \frac{1}{2}\pi$.

Let $\overrightarrow{\text{PQ}}$ represent $\boldsymbol{a}$, $\overrightarrow{\text{PR}}$ represent $\boldsymbol{b}$, and θ be $\angle\,\text{QPR}$.

Then $\boldsymbol{a} = \boldsymbol{q} - \boldsymbol{p} = \begin{pmatrix}6\\10\\-6\end{pmatrix} - \begin{pmatrix}4\\7\\0\end{pmatrix} = \begin{pmatrix}2\\3\\-6\end{pmatrix}$

and $\boldsymbol{b} = \boldsymbol{r} - \boldsymbol{p} = \begin{pmatrix}1\\9\\0\end{pmatrix} - \begin{pmatrix}4\\7\\0\end{pmatrix} = \begin{pmatrix}-3\\2\\0\end{pmatrix}$

Using $\boldsymbol{a}\,.\,\boldsymbol{b} = a_1 b_1 + a_2 b_2 + a_3 b_3$,

$\boldsymbol{a}\,.\,\boldsymbol{b} = \begin{pmatrix}2\\3\\-6\end{pmatrix}.\begin{pmatrix}-3\\2\\0\end{pmatrix} = -6 + 6 + 0 = 0,$

and $\cos\theta = \dfrac{\boldsymbol{a}\,.\,\boldsymbol{b}}{|\boldsymbol{a}||\boldsymbol{b}|} = \dfrac{0}{|\boldsymbol{a}||\boldsymbol{b}|} = 0.$

Hence $\theta = \frac{1}{2}\pi$. (Note that in this example it is not necessary to calculate the denominator.)

Exercise 10

1 Using $\cos\theta = \dfrac{a_1b_1 + a_2b_2 + a_3b_3}{|\boldsymbol{a}||\boldsymbol{b}|}$, calculate the cosines of the angles between the vectors:

a $\begin{pmatrix}2\\2\\0\end{pmatrix}$ and $\begin{pmatrix}2\\0\\2\end{pmatrix}$ *b* $\begin{pmatrix}3\\2\\1\end{pmatrix}$ and $\begin{pmatrix}-1\\3\\2\end{pmatrix}$ *c* $\begin{pmatrix}1\\2\\2\end{pmatrix}$ and $\begin{pmatrix}2\\3\\-6\end{pmatrix}$

2 Use the same formula to calculate the size of $\angle$AOB for each of the following, O being the origin:

a A(4, −1, 2), B(2, 4, −2) *b* A(1, 1, 0), B(0, −1, 1)

c A(3, 4, 0), B(0, 8, 6) *d* A(5, 0, 5), B(−5, 0, −5)

3 The vectors $\boldsymbol{a} = 2\boldsymbol{i} - 5\boldsymbol{j} + \boldsymbol{k}$ and $\boldsymbol{b} = x\boldsymbol{i} - 2\boldsymbol{j} + 4\boldsymbol{k}$ are perpendicular. Find x.

4 A is the point (3, −2), B is (6, −4), C(1, 5) and D(−1, 2). $\overrightarrow{AB}$ and $\overrightarrow{CD}$ represent vectors $\boldsymbol{u}$ and $\boldsymbol{v}$ respectively. Calculate:

a $\boldsymbol{u}.\boldsymbol{v}$ *b* the angle between $\boldsymbol{u}$ and $\boldsymbol{v}$.

5 Use the scalar product of two vectors to show that the triangle with vertices P(5, 7, −5), Q(4, 7, −3) and R(2, 7, −4) is right-angled at Q.

6 R is the point (4, 4, 1), S is (3, 2, 0) and T is (2, 0, 1). Calculate the sizes of the angles of triangle RST.

7 P divides AB in the ratio 2:1, where A is the point (1, 0, 1) and B is (4, 6, 10). Q is the midpoint of CD, where C is the point (5, −2, 8) and D is (9, 6, 6). Find the vectors represented by $\overrightarrow{AB}$, $\overrightarrow{CD}$ and $\overrightarrow{PQ}$ in component form, and prove that PQ is perpendicular to both AB and CD.

8 a Find the coordinates of the point R which divides the line joining P(−2, −1, 3) and Q(4, 2, 3) in the ratio 2:1.

b If S is the point (3, −1, 1), show that angle PRS is a right angle by means of : (*1*) scalar product (*2*) the converse of Pythagoras' theorem.

9 A is the point (4, 1, 0), B is (−2, 3, 2), C(0, 1, 4), D(2, −1, 0).

a Find the coordinates of P, Q, R, S, the midpoints of AB, BC, CD and DA respectively.

b Show that PQRS is a parallelogram, and calculate the sizes of its angles.

10a If $\boldsymbol{a} \cdot \boldsymbol{b} = 0$, state three ways in which this may come about.

b What important difference is there between the conclusion you can draw from $ab = 0$ $(a, b \in R)$, and $\boldsymbol{a} \cdot \boldsymbol{b} = 0$?

c What can you say about $\boldsymbol{p} \cdot \boldsymbol{q}$ if the vectors $\boldsymbol{p}$ and $\boldsymbol{q}$ are perpendicular?

11 Show that the angle between the vectors $\boldsymbol{i}-\boldsymbol{j}+2\boldsymbol{k}$ and $\boldsymbol{i}+\boldsymbol{k}$ is exactly $\frac{1}{6}\pi$.

12 Given that the angle between the vectors $2\boldsymbol{i}-\boldsymbol{j}+3\boldsymbol{k}$ and $\boldsymbol{i}+3\boldsymbol{j}-p\boldsymbol{k}$ is $\frac{1}{3}\pi$, find p.

---

7 *Properties of the scalar product*

By now it is clear that when we meet a new operation or a new type of mathematical combination it is of great interest to learn how it compares in properties with the familiar operations of addition and multiplication on real numbers. For example, vectors under vector addition and matrices under matrix addition behave exactly like numbers under addition; the closure, commutative, associative, identity element and inverse properties all apply. Consequently when only additive operations and properties are involved we can use exactly the same methods in all these cases for solving equations, etc. For matrices A and B, however, even if the products AB and BA are both defined, they are not necessarily equal, i.e. the commutative law does not apply to matrix multiplication.

Let us examine the properties possessed by the scalar product.

Commutative law for scalar products

It is required to prove that $\boldsymbol{a} \cdot \boldsymbol{b} = \boldsymbol{b} \cdot \boldsymbol{a}$.

Proof. Let $\boldsymbol{a} = \begin{pmatrix} a_1 \\ a_2 \\ a_3 \end{pmatrix}$ and $\boldsymbol{b} = \begin{pmatrix} b_1 \\ b_2 \\ b_3 \end{pmatrix}$.

Since multiplication of real numbers is commutative, $a_1 b_1 = b_1 a_1$.

$$\begin{aligned}\boldsymbol{a}.\boldsymbol{b} &= a_1 b_1 + a_2 b_2 + a_3 b_3 \text{ (definition of scalar product)}\\ &= b_1 a_1 + b_2 a_2 + b_3 a_3 \text{ (commutative law for multiplication of real numbers)}\\ &= \boldsymbol{b}.\boldsymbol{a} \text{ (definition of scalar product).}\end{aligned}$$

Distributive law for scalar product over vector addition

It is required to prove that $\boldsymbol{a}.(\boldsymbol{b}+\boldsymbol{c}) = \boldsymbol{a}.\boldsymbol{b}+\boldsymbol{a}.\boldsymbol{c}$.

Proof. In the usual notation $\boldsymbol{b}+\boldsymbol{c} = \begin{pmatrix} b_1+c_1 \\ b_2+c_2 \\ b_3+c_3 \end{pmatrix}$ and

$$\begin{aligned}\boldsymbol{a}.(\boldsymbol{b}+\boldsymbol{c}) &= a_1(b_1+c_1)+a_2(b_2+c_2)+a_3(b_3+c_3)\\ &\quad \text{(definition of scalar product)}\\ &= a_1 b_1 + a_1 c_1 + a_2 b_2 + a_2 c_2 + a_3 b_3 + a_3 c_3\\ &\quad \text{(distributive law for real numbers)}\\ &= (a_1 b_1 + a_2 b_2 + a_3 b_3)+(a_1 c_1 + a_2 c_2 + a_3 c_3)\\ &\quad \text{(commutative law for addition of real numbers)}\\ &= \boldsymbol{a}.\boldsymbol{b}+\boldsymbol{a}.\boldsymbol{c} \text{ (definition of scalar product).}\end{aligned}$$

The following table summarises the properties of *scalar product*:

Property	*Comment*	*Reason*
Closure	No	$\boldsymbol{a}.\boldsymbol{b}$ is *not* a vector.
Identity element	No	$\boldsymbol{a}.\boldsymbol{e} = \boldsymbol{a}$ is impossible; $\boldsymbol{a}.\boldsymbol{e}$ is not a vector.
Inverse	No	There is no identity element.
Associativity	No	$\boldsymbol{a}.(\boldsymbol{b}.\boldsymbol{c})$ and $(\boldsymbol{a}.\boldsymbol{b}).\boldsymbol{c}$ are both meaningless.
Commutativity	Yes	Proved above.
Distributivity	Yes	Proved above.

It is evident from this that we cannot treat $\boldsymbol{a}.\boldsymbol{b}$ as we would the product of two numbers. In particular, when dealing with a scalar product *there is no division by a vector* since there are no inverses. We find it convenient to call this method of combining two vectors a *scalar product* to distinguish it from other ways of combining vectors. It is called scalar because from its definition it is a real number. Physical quantities, such as mass, which have magnitude only, are often referred to as *scalar* quantities to distinguish them from *vector* quantities, such as velocity, which involve magnitude and direction.

Vectors and matrices

We have seen that column vectors like $\boldsymbol{a} = \begin{pmatrix} a_1 \\ a_2 \\ a_3 \end{pmatrix}$ and $\boldsymbol{b} = \begin{pmatrix} b_1 \\ b_2 \\ b_3 \end{pmatrix}$ may be added like matrices, and a vector may be treated in the same way as a matrix for multiplication by a scalar; but the scalar product is either

$$\boldsymbol{a}.\boldsymbol{b} = a_1 b_1 + a_2 b_2 + a_3 b_3 \quad \text{or} \quad \boldsymbol{b}.\boldsymbol{a} = b_1 a_1 + b_2 a_2 + b_3 a_3$$

$$= (a_1 \quad a_2 \quad a_3)\begin{pmatrix} b_1 \\ b_2 \\ b_3 \end{pmatrix} \qquad = (b_1 \quad b_2 \quad b_3)\begin{pmatrix} a_1 \\ a_2 \\ a_3 \end{pmatrix}.$$

Vectors have numerical magnitudes associated with them; matrices do not. Matrices with one row or column are often described as vectors, but great care has to be taken over the operations which can be carried out.

Exercise 11B

1 If the length of $\boldsymbol{a}$ is 6, what is $\boldsymbol{a}.\boldsymbol{a}$? Why can we write $\boldsymbol{a}.\boldsymbol{a} = |\boldsymbol{a}|^2$?

Note that this formula holds for all vectors; $\boldsymbol{p}.\boldsymbol{p} = |\boldsymbol{p}|^2$, $(\boldsymbol{a}+\boldsymbol{b}).(\boldsymbol{a}+\boldsymbol{b}) = |\boldsymbol{a}+\boldsymbol{b}|^2$, etc.

2 If $|\boldsymbol{a}| = 2$, $|\boldsymbol{b}| = 3$ and the angle between $\boldsymbol{a}$ and $\boldsymbol{b}$ is 60°, calculate $\boldsymbol{a}.(\boldsymbol{a}+\boldsymbol{b})$ and $\boldsymbol{b}.(\boldsymbol{a}+\boldsymbol{b})$.

3 If $\boldsymbol{a} = \begin{pmatrix} 1 \\ 2 \\ 1 \end{pmatrix}$, $\boldsymbol{b} = \begin{pmatrix} 1 \\ 3 \\ 2 \end{pmatrix}$ and $\boldsymbol{c} = \begin{pmatrix} 0 \\ -1 \\ 1 \end{pmatrix}$, evaluate $\boldsymbol{a}.(\boldsymbol{b}+\boldsymbol{c})$.

4 $\boldsymbol{a} = \begin{pmatrix} 1 \\ -1 \\ 1 \end{pmatrix}$, $\boldsymbol{b} = \begin{pmatrix} 1 \\ 2 \\ 1 \end{pmatrix}$ and $\boldsymbol{c} = \begin{pmatrix} 1 \\ 4 \\ x \end{pmatrix}$. Find x if $\boldsymbol{a}.(\boldsymbol{b}+\boldsymbol{c}) = \boldsymbol{a}.\boldsymbol{a}$.

5 $\boldsymbol{a} = 2\boldsymbol{u}+3\boldsymbol{v}$ where $\boldsymbol{u}$ and $\boldsymbol{v}$ are vectors of unit length inclined at $\frac{1}{3}\pi$ to each other. Calculate $\boldsymbol{u}.\boldsymbol{u}$, $\boldsymbol{v}.\boldsymbol{v}$ and $\boldsymbol{a}.\boldsymbol{a}$. What is the value of $|\boldsymbol{a}|$?

6 If $\boldsymbol{a}.\boldsymbol{b} = \boldsymbol{a}.\boldsymbol{c}$, $\boldsymbol{a} \neq 0$, which of the following *must* be true?

a $\boldsymbol{a}.(\boldsymbol{b}-\boldsymbol{c}) = 0$. *b* $\boldsymbol{b} = \boldsymbol{c}$.
c $\boldsymbol{a}$ is perpendicular to $\boldsymbol{b}-\boldsymbol{c}$. *d* $\boldsymbol{b} = \boldsymbol{c}$ or $\boldsymbol{a}$ is perpendicular to $\boldsymbol{b}-\boldsymbol{c}$.

7 $\boldsymbol{i}$, $\boldsymbol{j}$ and $\boldsymbol{k}$ are the mutually perpendicular vectors of unit length described in Section 3. Explain why $\boldsymbol{i}.\boldsymbol{i} = 1$ and $\boldsymbol{i}.\boldsymbol{j} = 0$. Copy and complete this table of scalar products.

.	i	j	k
i	1	0	
j	0		
k			

8 If $\boldsymbol{a} = a_1\boldsymbol{i}+a_2\boldsymbol{j}+a_3\boldsymbol{k}$ and $\boldsymbol{b} = b_1\boldsymbol{i}+b_2\boldsymbol{j}+b_3\boldsymbol{k}$, verify by using the properties in question 7 that $\boldsymbol{a}.\boldsymbol{b} = a_1b_1+a_2b_2+a_3b_3$.

9 The vectors $\boldsymbol{a} = 4\boldsymbol{i}-3\boldsymbol{j}+2\boldsymbol{k}$, $\boldsymbol{b} = 2\boldsymbol{i}-2\boldsymbol{j}+6\boldsymbol{k}$ and $\boldsymbol{c} = 3\boldsymbol{i}+4\boldsymbol{j}+5\boldsymbol{k}$ are position vectors of the vertices A, B and C of $\triangle$ABC. Use the scalar product to show that $\triangle$ABC is right-angled, and state the vertex of the right angle.

10 The vector $\boldsymbol{a} = 2\boldsymbol{i}+3\boldsymbol{j}-6\boldsymbol{k}$ makes angles α, β and γ with the positive directions of the x, y and z-axes respectively (i.e. with $\boldsymbol{i}$, $\boldsymbol{j}$ and $\boldsymbol{k}$). Use $\boldsymbol{a}.\boldsymbol{i}$, $\boldsymbol{a}.\boldsymbol{j}$ and $\boldsymbol{a}.\boldsymbol{k}$ to find $\cos\alpha$, $\cos\beta$ and $\cos\gamma$.

Note. $\cos\alpha$, $\cos\beta$ *and* $\cos\gamma$ *are called the direction cosines of* $\boldsymbol{a}$.

11 In an equilateral triangle ABC, $\overrightarrow{AB}$ represents $\boldsymbol{p}$ and $\overrightarrow{AC}$ represents $\boldsymbol{q}$. If each side is 2 units long, evaluate $\boldsymbol{p}.\boldsymbol{q}$.
What would be the answer if $\boldsymbol{q}$ was represented by $\overrightarrow{CA}$ instead of $\overrightarrow{AC}$? Illustrate with a sketch.

12 Expand $(\boldsymbol{a}+\boldsymbol{b}).(\boldsymbol{a}-\boldsymbol{b})$. Given that $|\boldsymbol{a}| = |\boldsymbol{b}|$, simplify your answer further. If also $\boldsymbol{a}+\boldsymbol{b}$ and $\boldsymbol{a}-\boldsymbol{b}$ are non-zero vectors what conclusion can you draw about them? Using $\overrightarrow{PQ}$ and $\overrightarrow{PR}$ to represent $\boldsymbol{a}$ and $\boldsymbol{b}$, illustrate with a sketch.

13 In $\triangle$PQR, $\overrightarrow{PQ}$ represents $\boldsymbol{u}$ and $\overrightarrow{QR}$ represents $\boldsymbol{v}$. If $(\boldsymbol{u}+\boldsymbol{v}).(\boldsymbol{u}+\boldsymbol{v}) = \boldsymbol{u}.\boldsymbol{u}+\boldsymbol{v}.\boldsymbol{v}$, show that $\angle$PQR is right.

14 Give a simple reason why $|\boldsymbol{a}+\boldsymbol{b}| \leqslant |\boldsymbol{a}|+|\boldsymbol{b}|$. When can the 'equals' sign be used?

15 Why is $\boldsymbol{a}.\boldsymbol{b} \leqslant |\boldsymbol{a}||\boldsymbol{b}|$? If equality exists, what must be true about $\boldsymbol{a}$ and $\boldsymbol{b}$?

16 Simplify $\boldsymbol{a}.(\boldsymbol{b}-\boldsymbol{c})+\boldsymbol{b}.(\boldsymbol{c}-\boldsymbol{a})+\boldsymbol{c}.(\boldsymbol{a}-\boldsymbol{b})$.

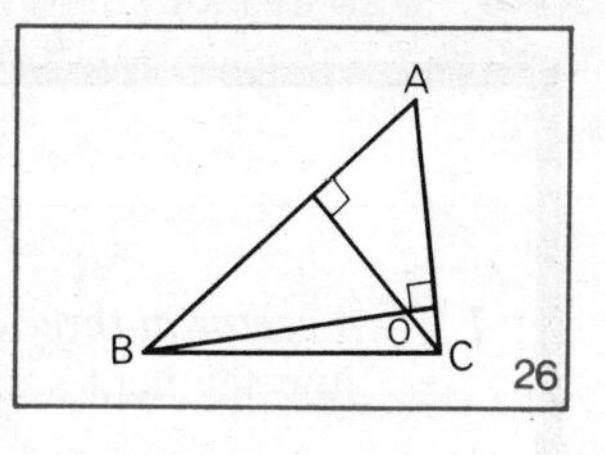

In $\triangle$ABC, in Figure 26, BO and OC are perpendicular to AC and AB respectively. If $\boldsymbol{a}$, $\boldsymbol{b}$, $\boldsymbol{c}$ denote the position vectors of A, B, C with respect to O as origin, show that $\boldsymbol{b}.(\boldsymbol{c}-\boldsymbol{a}) = 0$ and $\boldsymbol{c}.(\boldsymbol{a}-\boldsymbol{b}) = 0$. Hence show by vector algebra that $\boldsymbol{a}.(\boldsymbol{b}-\boldsymbol{c}) = 0$. What can you conclude about AO and BC? State your conclusion in the form of a theorem which is true for every triangle.

17 AC is a diameter of a circle, centre O, and B is a point on the circumference. $\overrightarrow{OA}$ represents $\boldsymbol{a}$ and $\overrightarrow{OB}$ represents $\boldsymbol{b}$. In terms of $\boldsymbol{a}$ and $\boldsymbol{b}$, what vectors do $\overrightarrow{OC}$, $\overrightarrow{BA}$ and $\overrightarrow{BC}$ represent?

By considering the dot product of the vectors represented by $\overrightarrow{BA}$ and $\overrightarrow{BC}$, prove that angle ABC is a right angle. Make a general statement, true for all positions of B on the circumference, distinct from A and C.

18 In a tetrahedron ABCD, AB = CD, AC = BD and AD = BC. Using the usual position vector notation:

a express the fact that AD = BC in vector terms
b find the dot product of the vector represented by the join of the midpoints of AB and CD with the vector represented by the join of the midpoints of AC and BD.
c What conclusions can you draw?

Summary

1 *A vector in three dimensions* can be represented by a directed line segment such as $\overrightarrow{AB}$, and can be defined by a number triple such as $\begin{pmatrix} 3 \\ -2 \\ 4 \end{pmatrix}$.

2 *Every vector in space* can be expressed uniquely in terms of three non-coplanar vectors $\boldsymbol{u}$, $\boldsymbol{v}$ and $\boldsymbol{w}$, which form a *basis* for the whole set of vectors. If $\boldsymbol{r} = l\boldsymbol{u}+m\boldsymbol{v}+n\boldsymbol{w}$, l, m and n are the components of $\boldsymbol{r}$, and $\boldsymbol{r} = \begin{pmatrix} l \\ m \\ n \end{pmatrix}$.

In Figure 1, $\overline{OC}$ represents $4\boldsymbol{u}+3\boldsymbol{v}+2\boldsymbol{w}$.

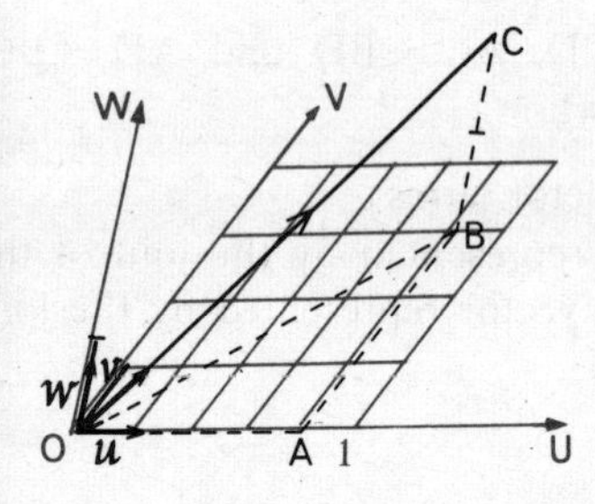

1

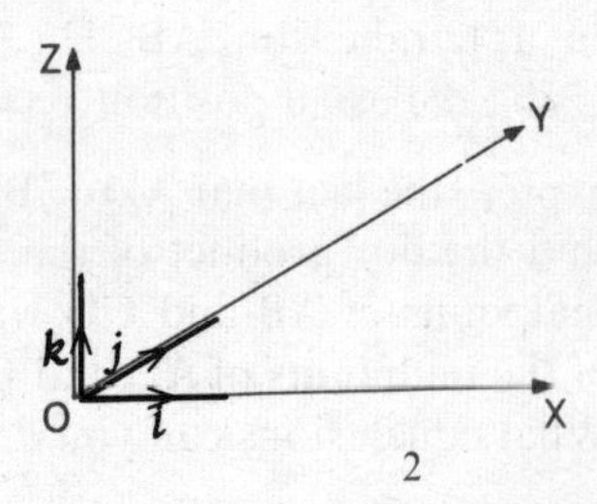

2

3 *Three mutually perpendicular vectors of unit length* (forming a right-handed system) are often used as a basis with rectangular axes. These are written $\boldsymbol{i}$, $\boldsymbol{j}$ and $\boldsymbol{k}$ (Figure 2). In this case $\boldsymbol{r} = l\boldsymbol{i}+m\boldsymbol{j}+n\boldsymbol{k}$.

4 a *The magnitude of a vector* $\boldsymbol{r}$ is $\sqrt{(l^2+m^2+n^2)}$ where $\boldsymbol{r} = l\boldsymbol{i}+m\boldsymbol{j}+n\boldsymbol{k}$.

b *The distance formula.* If A is the point (x_1, y_1, z_1) and B is (x_2, y_2, z_2) then $|\overrightarrow{AB}| = \sqrt{[(x_2-x_1)^2+(y_2-y_1)^2+(z_2-z_1)^2]}$.

5 *The position vector* of a point A is the vector represented by $\overrightarrow{OA}$ and is usually denoted by $\boldsymbol{a}$.

6 *The section formula.*

a *The position vector* of a point P dividing a line AB in the ratio $m:n$ is given by $\boldsymbol{p} = \dfrac{m\boldsymbol{b}+n\boldsymbol{a}}{m+n}$ (see Figure 3*a*).

The position vector of the midpoint of AB is $\frac{1}{2}(\boldsymbol{a}+\boldsymbol{b})$.

b *In coordinates*,, if $P(x_P, y_P, z_P)$ divides the line joining $A(x_1, y_1, z_1)$ and $B(x_2, y_2, z_2)$ in the ratio $m:n$, then

$$x_P = \frac{mx_2+nx_1}{m+n}, \quad y_P = \frac{my_2+ny_1}{m+n}, \quad z_P = \frac{mz_2+nz_1}{m+n} \text{ (Figure 3}b\text{).}$$

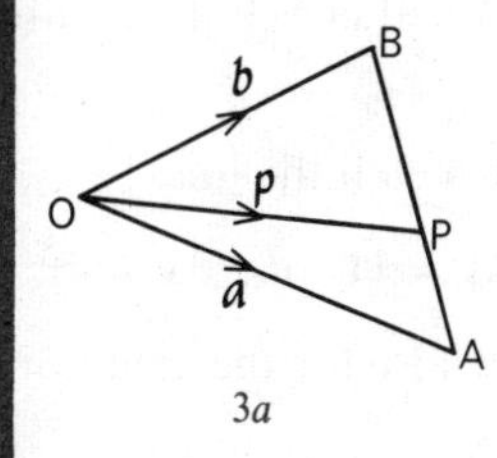

3*a*

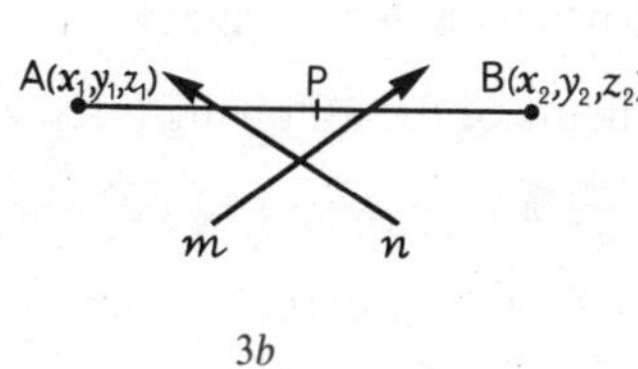

3*b*

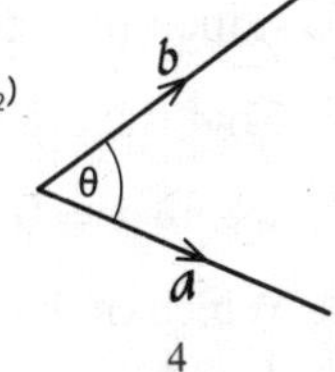

4

7 *The scalar product of two vectors* ***a*** *and* ***b***.

(i) $\boldsymbol{a}\,.\,\boldsymbol{b} = |\boldsymbol{a}||\boldsymbol{b}|\cos\theta \quad (\boldsymbol{a} \neq \boldsymbol{0}, \boldsymbol{b} \neq \boldsymbol{0})$; $\boldsymbol{a}\,.\,\boldsymbol{0} = \boldsymbol{0}\,.\,\boldsymbol{b} = 0$

(ii) $\boldsymbol{a}\,.\,\boldsymbol{b} = a_1b_1+a_2b_2+a_3b_3$

(iii) $\cos\theta = \dfrac{\boldsymbol{a}\,.\,\boldsymbol{b}}{|\boldsymbol{a}||\boldsymbol{b}|} = \dfrac{a_1b_1+a_2b_2+a_3b_3}{\sqrt{(a_1{}^2+a_2{}^2+a_3{}^2)(b_1{}^2+b_2{}^2+b_3{}^2)}}$

The scalar product is commutative and is distributive over vector addition.

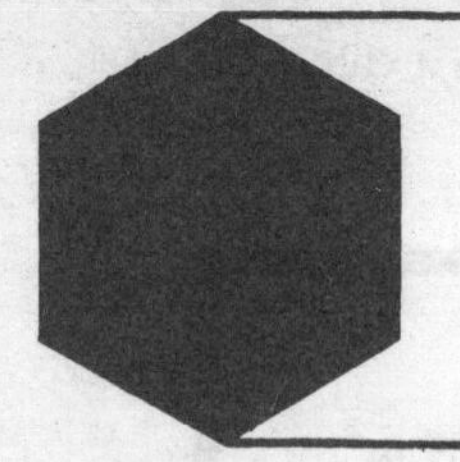

Revision Exercises

Revision Exercise on Chapter 1
The Gradient and Equations of a Straight Line

Revision Exercise 1

1 A is the point (3, 0), B is (8, 2), C(6, 5), D(1, 3). Use gradients to verify that ABCD is a parallelogram, and to find whether the diagonals slope up or down from left to right.

2 Find the gradients of the following lines, and sketch the lines:

a $x+y+1=0$ ***b*** $2y-3x=12$ ***c*** $2x+4y=11$ ***d*** $3x=4y$

3 Which of the following are true and which are false for the equation $2x-3y+1=0$?

a It is the equation of a straight line.
b $(-2, -1)$ lies on the line. ***c*** $(-5, 3)$ lies on the line.
d The gradient of the line is $\frac{3}{2}$. ***e*** The line cuts the y-axis at $(0, \frac{1}{3})$.

4 Find the equation of the image of the line with equation $y=2x+1$ in: ***a*** the y-axis ***b*** the x-axis.

5 The line with equation $2x-3y+6=0$ cuts the x-axis at A and the y-axis at B. Find the coordinates of A and B, and the length of AB.

6 A is the point (2, 1), B is (5, 2) and D is $(-3, 4)$. Parallelogram ABCD is completed. Find the coordinates of C.

AC is rotated anticlockwise through 90° to the position AE. Give the coordinates of E.

7 If the vertices of triangle ABC are A(2, 1), B$(-4, 3)$ and C(5, 2), find the equations of the medians BE and CF. Hence calculate the coordinates of the centroid of the triangle.

Verify that your answer is given by $[\frac{1}{3}(x_1+x_2+x_3), \frac{1}{3}(y_1+y_2+y_3)]$ when the vertices of the triangle are (x_1, y_1), (x_2, y_2) and (x_3, y_3).

8 Find the equations of the lines through the point $P(-1, 3)$, parallel and perpendicular to the line with equation $3x-4y = 8$.

9 Find the coordinates of the vertices and the lengths of the sides of triangle KLM, where KM, LM and KL have equations $2y-x = 11$, $y-4x+19 = 0$ and $5x+4y-29 = 0$ respectively.

10 Show that the diagonals of the quadrilateral with vertices $P(-6, -4)$, $Q(2, 2)$, $R(2, 12)$, $S(-6, 6)$ bisect each other at right angles. What shape is PQRS?

11 If k is a real number, then $x+2y-5+k(3x-2y-7) = 0$ is the equation of a straight line. Explain why.

Show that the point $(3, 1)$ lies on this straight line no matter what value is assigned to k.

Find the value of k:

a if the straight line passes through the origin
b if the straight line passes through the point $(-1, 2)$
c if the straight line is perpendicular to the line whose equation is $4x-3y = 2$.

12 Show that the straight lines $x-2y+7 = 0$, $2x+y-6 = 0$ and $x = k$ $(k \in R)$ always form a right-angled triangle except when $k = 1$. Calculate the area of the triangle when $k = 5$.

13 P is the point $(p\cos\alpha, p\sin\alpha)$, where $0 < \alpha < \frac{1}{2}\pi$. Show that the equation of the straight line through P, perpendicular to OP, may be written in the form $x\cos\alpha+y\sin\alpha = p$.

If this straight line has the same gradient as $\sqrt{3}x+y = 4$, find the value of α.

14 The line $2x+3y = 30$ cuts the x-axis at A and the y-axis at B. Find the coordinates of the point P which divides AB in the ratio 4:1.

Find also the equation of the line through P perpendicular to AB. If this line crosses the x-axis at S, calculate the area of triangle ASB.

15 Two sides of a parallelogram have equations $4x-9y = 27$ and $16x+3y = 147$, and one vertex is $(3, -8)$. Find the coordinates of the vertex in which the two given sides meet, and the equation of one of the other sides of the parallelogram.

Revision Exercise on Chapter 2
Vectors 2

Revision Exercise 2

1 P is the point (0, 2), Q is (3, 4), R(2, −2). Express the vectors represented by $\overrightarrow{PQ}$, $\overrightarrow{QR}$ and $\overrightarrow{RP}$ in number pair form, and verify that $\overrightarrow{PQ} = \overrightarrow{RQ} - \overrightarrow{RP}$.

2 A is the point (4, 3), and $\overrightarrow{AB}$ represents the vector $\begin{pmatrix} -10 \\ 5 \end{pmatrix}$. Find in component form the vector represented by $\overrightarrow{OB}$, and calculate $|\overrightarrow{OB}|$.

3 Show that A(2, 5, 0), B(5, 8, 3) and C(4, 7, 2) are collinear, and find the value of AB : BC.

4 *a* Show that the line joining A(2, 1, −1) and B(6, 3, −3) passes through the origin.

b Show that the line joining C(2, 2, 3) and D(4, 3, 2) is parallel to the line joining E(5, 3, −2) and F(9, 5, −4).

5 $\boldsymbol{u} = \begin{pmatrix} 0 \\ 2 \\ 1 \end{pmatrix}$, $\boldsymbol{v} = \begin{pmatrix} 2 \\ 1 \\ 0 \end{pmatrix}$ and $\boldsymbol{w} = \begin{pmatrix} -1 \\ 0 \\ 1 \end{pmatrix}$. If $a\boldsymbol{u}+b\boldsymbol{v}+c\boldsymbol{w} = \begin{pmatrix} 1 \\ 1 \\ 1 \end{pmatrix}$, find a, b, and c.

6 Prove that the triangle with vertices P(1, 0, 0), Q(1, 1, 1), R(0, 1, 1) is right-angled.

7 $\boldsymbol{p} = 3\boldsymbol{i}+4\boldsymbol{j}+12\boldsymbol{k}$, and $\boldsymbol{q} = 2\boldsymbol{i}+2\boldsymbol{j}+\boldsymbol{k}$. Calculate:

a $|\boldsymbol{p}|$ *b* $|\boldsymbol{q}|$ *c* $|\boldsymbol{p}-\boldsymbol{q}|$.

8 $\boldsymbol{a} = \begin{pmatrix} 1 \\ 2 \\ -2 \end{pmatrix}$, $\boldsymbol{b} = \begin{pmatrix} -4 \\ -3 \\ 1 \end{pmatrix}$, $\boldsymbol{c} = \begin{pmatrix} 5 \\ -2 \\ 0 \end{pmatrix}$.

a Find the components of the vector $\boldsymbol{a}-2\boldsymbol{b}+3\boldsymbol{c}$.

b Calculate the lengths of $\boldsymbol{a}+\boldsymbol{b}$ and $\boldsymbol{a}-\boldsymbol{b}$.

9 △ABC is right-angled at B. $\overrightarrow{AB}$ represents the vector $\boldsymbol{u} = \begin{pmatrix} u_1 \\ u_2 \\ u_3 \end{pmatrix}$, and $\overrightarrow{BC}$ represents the vector $\boldsymbol{v} = \begin{pmatrix} v_1 \\ v_2 \\ v_3 \end{pmatrix}$. Express the vector represented by $\overrightarrow{AC}$ in component form, and use Pythagoras' theorem to show that $u_1v_1 + u_2v_2 + u_3v_3 = 0$.

10 ABCD is a parallelogram whose vertices have position vectors $\boldsymbol{a}$, $\boldsymbol{b}$, $\boldsymbol{c}$, $\boldsymbol{d}$. Express the position vectors of the midpoints of AB, BC and CA, and also of the fourth vertex D, in terms of $\boldsymbol{a}$, $\boldsymbol{b}$, $\boldsymbol{c}$.

11 OABC and OPQR are rhombuses with P and R the midpoints of OA and OC. Express $\boldsymbol{b}$ and $\boldsymbol{q}$ in terms of $\boldsymbol{a}$ and $\boldsymbol{c}$. Hence show that O, Q and B are collinear, and that Q is the midpoint of OB.

12 The position vectors of A, B and C are $\boldsymbol{a}$, $\boldsymbol{b}$ and $3\boldsymbol{a}+\boldsymbol{b}$. Find the position vectors of D, E and F, the midpoints of BC, CA and AB of $\triangle$ABC.

Show that G and H, the points of trisection of AD and BE which are nearer to D and E, have the same position vector.

Express the position vector of the centroid of $\triangle$ABC in terms of $\boldsymbol{a}$ and $\boldsymbol{b}$.

13 E is the midpoint of side BC of parallelogram OABC, and F divides AC in the ratio 2:1.

a Express the position vectors of E and F in terms of $\boldsymbol{b}$ and $\boldsymbol{c}$.
b Show that OE passes through F, and that AC and OE trisect each other.

14 A, B, C and D are points in space. H, K, L, M are the midpoints of AB, BC, CD and DA. Prove that HKLM is a parallelogram.

15 Find the coordinates of the points dividing the line joining A(0, 1, 5) and B(0, −4, 5) internally and externally in the ratio 3:2.

16 $\triangle$ABC has vertices A(3, 0, 6), B(0, 3, −3) and C(1, 0, −4). P divides AB internally in the ratio 1:2, Q is the midpoint of AC and R divides BC externally in the ratio 2:1.

a Find the coordinates of P, Q and R.
b Show that P, Q, R are collinear, and find the value of PQ:QR.

17 O is the origin, A is the point (3, −6, 0) and B is (12, 6, −3). C and D divide OA and OB internally in the ratio 1:2.

a Find the coordinates of C and D.
b Show that CD is parallel to AB, and calculate the value of CD:AB.

18 P is the point (0, 2, 1) and Q is (0, −1, 2). Calculate the size of ∠POQ.

19 Find the angle between the vectors $\boldsymbol{a}+\boldsymbol{b}$ and $\boldsymbol{a}-\boldsymbol{b}$, where $\boldsymbol{a}=\begin{pmatrix}-1\\1\\2\end{pmatrix}$ and $\boldsymbol{b}=\begin{pmatrix}1\\-1\\1\end{pmatrix}$

20 A is the point (−1, 1, 2) and B is (−2, −1, 1). Calculate $|\boldsymbol{a}|$, $|\boldsymbol{b}|$ and the size of angle AOB. Hence show that △AOB is equilateral.

21 $\boldsymbol{a}=\begin{pmatrix}2\\-1\\-3\end{pmatrix}$ and $\boldsymbol{b}=\begin{pmatrix}k\\1\\-1\end{pmatrix}$. Find k if $\boldsymbol{a}$ is perpendicular to $\boldsymbol{b}$.

22 △ABC has vertices A(1, 4, 3), B(4, 1, 0) and C(6, 3, −2). CB is produced to D so that $\overrightarrow{BD}=\frac{1}{2}\overrightarrow{CB}$. Find the coordinates of D, and show that AD is an altitude of △ABC by means of:

a the converse of Pythagoras' theorem ***b*** scalar product.

23 △PQR has vertices P(0, 5, 4), Q(3, 0, 1) and R(4, 1, 0). S is a point on PR such that PS = $\frac{1}{4}$PR. Find the coordinates of S, and show that QS is perpendicular to QR by means of:

a the converse of Pythagoras' theorem ***b*** scalar product.

24 If vector $\begin{pmatrix}p\\q\\r\end{pmatrix}$ is perpendicular to both vectors $\begin{pmatrix}2\\-1\\1\end{pmatrix}$ and $\begin{pmatrix}1\\1\\-1\end{pmatrix}$, find p and express q in terms of r.

25 $\boldsymbol{a}=\begin{pmatrix}x\\y\\z\end{pmatrix}$, $\boldsymbol{b}=\begin{pmatrix}1\\0\\0\end{pmatrix}$ and $\boldsymbol{c}=\begin{pmatrix}0\\0\\1\end{pmatrix}$. If vector $\boldsymbol{a}$ is of unit length and makes an angle of 45° with vector $\boldsymbol{b}$, show that $x=1/\sqrt{2}$. If vector $\boldsymbol{a}$ is perpendicular to vector $\boldsymbol{c}$, show that $z=0$, and that there are two possible values of y. Show that the two vectors given by these values of x, y and z are mutually perpendicular.

26 ABCD is a parallelogram, with $\overrightarrow{AD}$ representing vector $\boldsymbol{u}$ and $\overrightarrow{AB}$ representing vector $\boldsymbol{v}$. Express the vectors represented by $\overrightarrow{AC}$ and $\overrightarrow{BD}$ in terms of $\boldsymbol{u}$ and $\boldsymbol{v}$. Show that if $|\boldsymbol{u}|=|\boldsymbol{v}|$, then $\boldsymbol{u}+\boldsymbol{v}$ is perpendicular to $\boldsymbol{u}-\boldsymbol{v}$. State and prove the converse.

Trigonometry

Note to the Teacher on Chapter 1

This chapter first revises and then extends the work on trigonometry. *Section* 1 summarises basic definitions and formulae from Books 5 and 6 which are particularly useful here and in later chapters.

Section 2 introduces certain necessary related angles, mainly for use in Section 4.

Section 3 defines the radian measure of an angle.

At the start of trigonometry it was agreed that to each angle there corresponds one value of each of x/r, y/r, and y/x where x, y, and r were defined (see Book 5, Trigonometry, Chapter 1). Thus we defined cosine, sine, and tangent as functions from angles to real numbers, and we measured angles in degrees. On this basis the size of a right angle, for example, is given as 90° (not 90). In general, the size of any angle can be given as a°, where a is some real number. Hence, at this earlier stage it was natural to ask for the solution of the equation $\sin x^\circ = 0{\cdot}5$, x° being an acute angle. The answer is $x = 30$.

The measurement of angles in degrees has its traditional application in numerical trigonometry. Tables of approximations to the values of the sine, cosine, and tangent functions have been compiled; and these are adequate for the calculation of lengths and angles, for solution of triangles, for areas of sectors of circles, and so on.

For the use of trigonometric functions in calculus (in Book 9) it is necessary to have the notation $\sin x$, etc., where x is a real number, and not an angle; for example, in the expression $\dfrac{\sin x}{x}$. The definition of $\sin x$ is approached via the *radian measure* of an angle, which is explained at the beginning of *Section* 3. From the relationship $180^\circ = \pi$ radians, the identities:

$$x^\circ = \frac{\pi x}{180} \text{ radians, } x \text{ radians} = \left(\frac{180x}{\pi}\right)^\circ, \sin(x \text{ radians}) = \sin\left(\frac{180x}{\pi}\right)^\circ$$

are easily obtained.

The notion of the *sine* of a *real number* is then defined by putting $\sin x = \sin(x \text{ radians})$ for every real number x, and thereafter the notation $\sin(x \text{ radians})$ is abandoned.

$\sin x$ is not the same as $\sin x^\circ$. '$\sin x$' denotes the sine of the real number x, which, by definition, is the sine of the angle whose *radian* measure is x. On the other hand, $\sin x^\circ$ is the sine of the angle whose size is x *degrees*, where x is a real number. The following identities emphasise the distinction:

$$\sin x = \sin\left(\frac{180x}{\pi}\right)^\circ, \text{ which is not the same as } \sin x^\circ$$

$$\sin x^\circ = \sin\left(\frac{\pi x}{180}\right), \text{ which is not the same as } \sin x.$$

Thus sin x and $\sin x^\circ$ express two different functions on the set of real numbers.

It is worth noting that '$180^\circ = \pi$ radians' is a true statement, but '$180^\circ = \pi$' is false, since π is a real number, not an angle.

Again, compare the two equations:

$$\text{(i) } \cos x = 0{\cdot}5, x \in R, \quad \text{and} \quad \text{(ii) } \cos x^\circ = 0{\cdot}5, x \in R.$$

The solution set of (i) is $\{x : x = 2n\pi \pm \frac{1}{3}\pi, n \in Z\}$, and of (ii) is $\{x : x = 360n \pm 60, n \in Z\}$.

The fundamental formula $\cos(\alpha + \beta) = \cos\alpha\cos\beta - \sin\alpha\sin\beta$ is proved in *Section* 4. The first method given is based on the distance formula and coordinates, with which pupils are familiar. Note that if angle AOB has radian measure α, then B is the point $(\cos\alpha, \sin\alpha)$; if angle AOB is equal to a°, then B is $(\cos a^\circ, \sin a^\circ)$. Hence the proof for $\cos(\alpha + \beta)$ is also valid for $\cos(a + b)^\circ$.

An alternative method of proof using the scalar product of two vectors is given in the text, and a further method using matrices is suggested in question ***12*** of Exercise 5.

The various 'addition' formulae and 'double angle' formulae are developed in *Sections* 4 and 5. The related Exercises have been compiled in the following way:

Exercises 4, 5, 7 and 8 provide practice in the use of the formulae.

Exercise 6B includes harder applications and manipulations of the addition formulae.

Exercise 9 introduces a variety of trigonometrical equations which bring together a number of formulae and techniques. Further equations involving sums and products of sines and cosines will be met in Book 9.

Exercise 10B consists of a selection of identities to be proved with the aid of the formulae.

1 The Addition Formulae

1 Revision

(i) The cosine, sine, and tangent functions

In Figure 1(i), P has Cartesian coordinates (x, y) and polar coordinates $(r, a°)$.

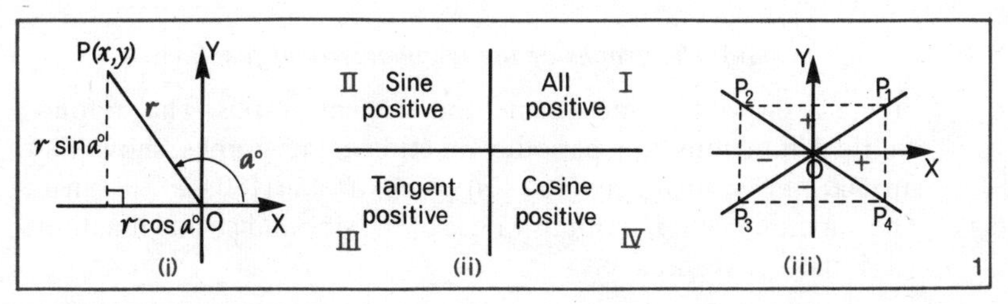

1

Then $\cos a° = \dfrac{x}{r}$ $\quad \sin a° = \dfrac{y}{r}$ $\quad \tan a° = \dfrac{y}{x}$

$\Leftrightarrow \quad x = r\cos a°$ $\quad \Leftrightarrow \quad y = r\sin a°$ $\quad \Leftrightarrow \quad y = x\tan a°$

From a study of the signs of x and y in each quadrant we obtain Figure 1(ii) which indicates the quadrants in which the functions are positive.

Using Figure 1(iii) we can express the trigonometrical functions of angles of all sizes in terms of functions of acute angles, e.g.

$\cos 150° = -\cos 30°$; $\quad \sin 300° = -\sin 60°$; $\quad \tan 200° = \tan 20°$

(ii) Special angles and triangles

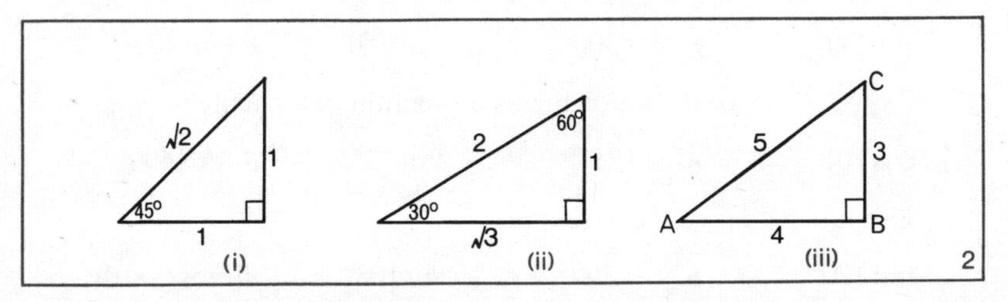

2

a°	30°	45°	60°
$\cos a^\circ$	$\frac{\sqrt{3}}{2}$	$\frac{1}{\sqrt{2}}$	$\frac{1}{2}$
$\sin a^\circ$	$\frac{1}{2}$	$\frac{1}{\sqrt{2}}$	$\frac{\sqrt{3}}{2}$
$\tan a^\circ$	$\frac{1}{\sqrt{3}}$	1	$\sqrt{3}$

Given $\cos A = \frac{4}{5}$ we can draw Figure 2(iii), from which BC = 3 and $\sin A = \frac{3}{5}$, $\tan A = \frac{3}{4}$.

(iii) Formulae connecting cos a°, sin a°, tan a°

If O is the point (0, 0) and P is the point $(r\cos a^\circ,\ r\sin a^\circ)$, then since $OP^2 = r^2$ we have $\cos^2 a^\circ + \sin^2 a^\circ = 1$.

Also from the definitions, $\tan a^\circ = \dfrac{\sin a^\circ}{\cos a^\circ}$.

(iv) The graphs of the trigonometrical functions

Figure 3 shows the sine, cosine and tangent graphs. The trigonometrical functions are periodic functions; the graphs show that $\sin(360+a)^\circ = \sin a^\circ$, $\cos(360+a)^\circ = \cos a^\circ$, $\tan(180+a)^\circ = \tan a^\circ$. The maximum and minimum values of the sine and cosine functions are 1 and -1 respectively.

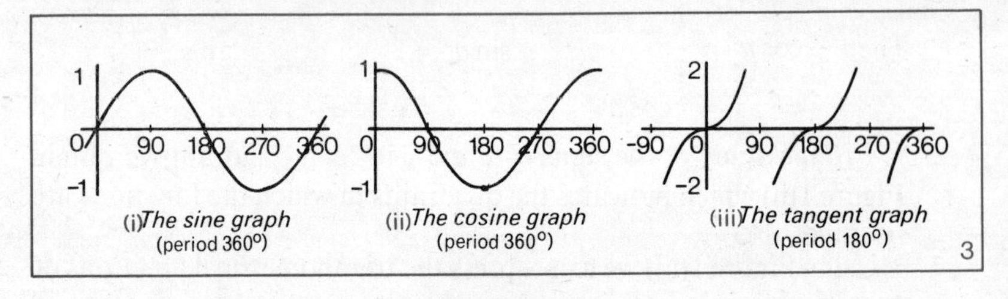

(i) *The sine graph* (period 360°) (ii) *The cosine graph* (period 360°) (iii) *The tangent graph* (period 180°) 3

Exercise 1

1 State the sign (+ or −) of each of the following:

a sin 73° *b* cos 158° *c* sin 285° *d* tan 141°

e sin 200° *f* cos 300° *g* tan 350° *h* sin 400°

2 Express each of the following as a function of an acute angle:

a cos 101° *b* sin 211° *c* tan 135° *d* cos 400°

3 Use tables to give the values of:

a tan 170° *b* cos 300° *c* sin 108° *d* cos (−10)°

4 Which pairs of the following are equal?

$\cos 57°$, $\cos 123°$, $\cos 237°$, $\cos 303°$, $\cos 0°$, $\cos 180°$, $\cos 360°$.

5 Solve the following equations, $x \in R$, $0 \leqslant x \leqslant 360$:

a $\sin x° = 0{\cdot}5$ *b* $\cos x° = 0{\cdot}5$ *c* $\tan x° = 1$

d $3\cos x° = -2$ *e* $4\sin x° = 3$ *f* $2\tan x° = -3$.

6 Evaluate $\sin(90+a)° + 2\cos(180-a)°$, when $a = 60$.

7 With the aid of the sine and cosine graphs, find x for which:

a $\sin x° = 1, 0 \leqslant x \leqslant 720$ *b* $\cos x° = 0, 0 \leqslant x \leqslant 720$

c $\sin x° = 0, -360 < x < 360$ *d* $\cos x° = -1, 0 \leqslant x \leqslant 360$

8 $\sin a° = \frac{4}{5}$, and $0 < a < 90$. Use the relation $\cos^2 a° + \sin^2 a° = 1$ to find the exact value of $\cos a°$. Hence find $\tan a°$.

9 $\cos b° = \frac{20}{29}$ and $0 < b < 180$. Find the exact value of $\sin b°$. Hence find $\tan b°$.

10a Sketch the graph with equation $y = \sin 2x°$, $0 \leqslant x \leqslant 360$.
b What is the period of the function f given by $f(x) = \sin 2x°$?

11 Repeat question ***10*** for the functions $g(x) = \sin 3x°$ and $h(x) = \cos 4x°$.

12 Sketch the graph with equation $y = \sin^2 x°$, $0 \leqslant x \leqslant 360$.

13 In terms of polar coordinates, P is the point $(4, 20°)$. Give the Cartesian coordinates of P, to two decimal places.

14 Repeat question ***13*** for the point $Q(5, 210°)$.

15 The circle having the origin as centre and passing through $R(10, 40°)$ cuts OX at A. Calculate the lengths of arc AR and chord AR, to 3 significant figures.

2 *The trigonometrical functions of certain related angles*

(i) $a°$ and $(180-a)°$

Under reflection in the y-axis,
$P(x, y) \rightarrow P'(-x, y)$.

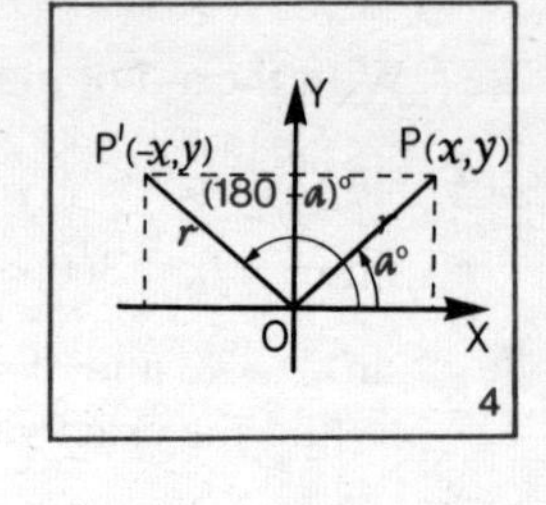

So $\cos(180-a)° = \dfrac{-x}{r} = -\cos a°$

$\sin(180-a)° = \dfrac{y}{r} = \sin a°$

$\tan(180-a)° = \dfrac{y}{-x} = -\tan a°$.

Note. Sin (angle) = sin (supplement); cos (angle) = − cos (supplement).

(ii) $a°$ and $(-a)°$

Under reflection in the x-axis,
$P(x, y) \rightarrow P'(x, -y)$.

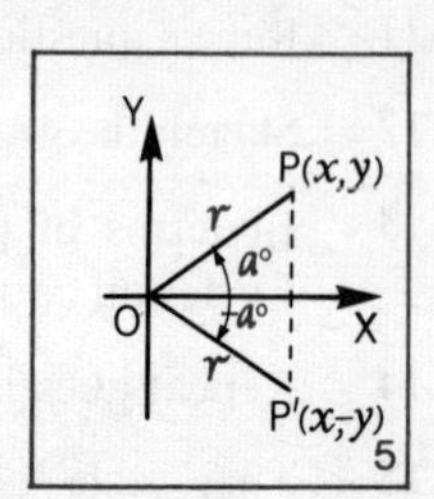

So $\cos(-a)° = \dfrac{x}{r} = \cos a°$

$\sin(-a)° = \dfrac{-y}{r} = -\sin a°$

$\tan(-a)° = \dfrac{-y}{x} = -\tan a°$.

(iii) $a°$ and $(90-a)°$

Under reflection in $y = x$, $P(x, y) \rightarrow P'(y, x)$.

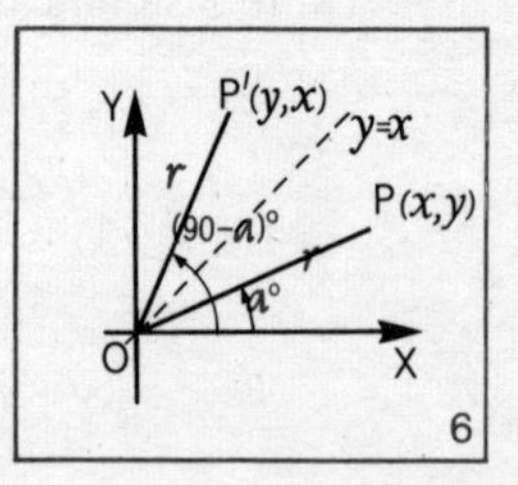

So $\cos(90-a)° = \dfrac{y}{r} = \sin a°$

$\sin(90-a)° = \dfrac{x}{r} = \cos a°$.

Note. Sin (angle) = cos (complement); cos (angle) = sin (complement).

Exercise 2

1 Simplify:

a $\sin(180-p)^\circ$ *b* $\cos(180-q)^\circ$ *c* $\tan(180-r)^\circ$

d $\cos(90-s)^\circ$ *e* $\sin(90-t)^\circ$ *f* $\cos(-u)^\circ$

g $\sin(180-b)^\circ$ *h* $\cos(90-c)^\circ$ *i* $\tan(-d)^\circ$

j $\cos(90-v)^\circ$ *k* $\sin(-k)^\circ$ *l* $\cos(180-z)^\circ$

2 Express each of the following as the sine or cosine of an acute angle:

a $\sin 120^\circ$ *b* $\cos 100^\circ$ *c* $\sin 95^\circ$ *d* $\cos 179^\circ$

3 Simplify:

a $\cos^2 a^\circ + \cos^2(90-a)^\circ$ *b* $\dfrac{\sin(180-a)^\circ}{\cos(180-a)^\circ}$ *c* $\dfrac{\cos(90-a)^\circ}{\sin(90-a)^\circ}$

d $\cos^2(180-a)^\circ + \sin^2(180-a)^\circ$ *e* $\sin^2 a^\circ + \cos^2(180-a)^\circ$

4 By means of reflection in the x-axis, show that:

$\cos(360-a)^\circ = \cos a^\circ$ and $\sin(360-a)^\circ = -\sin a^\circ$.

5 By means of a half turn about O, show that:

$\cos(180+a)^\circ = -\cos a^\circ$, $\sin(180+a)^\circ = -\sin a^\circ$, $\tan(180+a)^\circ = \tan a^\circ$.

6 By means of a quarter turn about O, simplify $\cos(90+a)^\circ$ and $\sin(90+a)^\circ$.

---

3 Measurement of angles

(i) Degree measure

We are already familiar with this system of measurement. The unit angle, called *one degree*, is $\frac{1}{360}$ revolution.

(ii) Radian measure

In Figure 7, O is the centre of both circles. OA′ and OB′ are radii of the larger circle cutting the smaller circle at A and B.

Sector A′OB′ can be obtained by dilatation of sector AOB, with O as the centre of dilatation.

Hence $\dfrac{\text{arc AB}}{\text{OA}} = \dfrac{\text{arc A}'\text{B}'}{\text{OA}'}$.

So the ratio $\dfrac{\text{arc AB}}{\text{OA}}$ is independent of the radius of the circle, and depends only on the size of $\angle$AOB.

The number given by $\dfrac{\text{arc AB}}{\text{OA}}$ gives a measure of $\angle$AOB; this number is called the *radian measure of* $\angle$AOB.

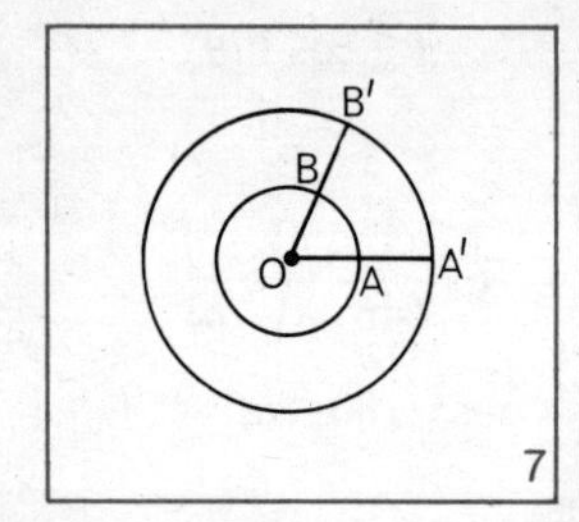

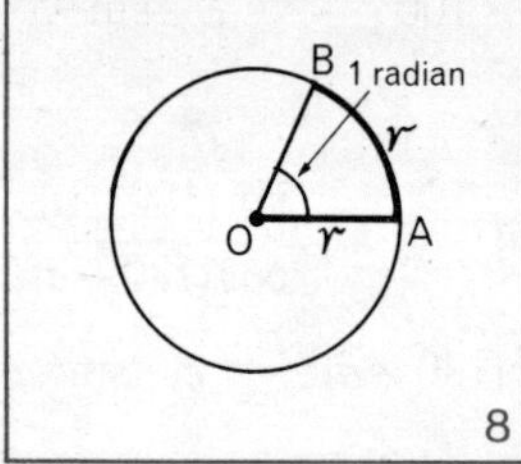

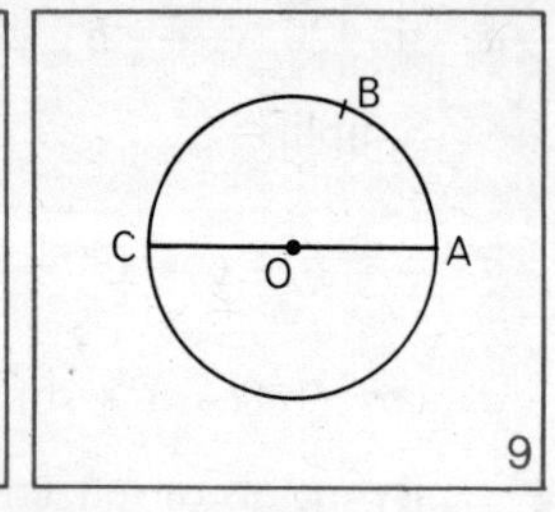

In Figure 8, arc AB = radius OA, so $\dfrac{\text{arc AB}}{\text{OA}} = 1$, i.e. the radian measure of $\angle$AOB is 1. We say that $\angle$AOB = 1 radian.

In Figure 9, $\dfrac{\text{arc ABC}}{\text{OA}} = \dfrac{\pi r}{r} = \pi$

i.e. $\angle$AOC = π radians

i.e. $180^\circ = \pi$ radians

Since $180^\circ = \pi$ radians,
we have $360^\circ = 2\pi$ radians
$90^\circ = \frac{1}{2}\pi$ radians
$60^\circ = \frac{1}{3}\pi$ radians, etc.

Also, $\sin(\pi \text{ radians}) = \sin 180^\circ = 0$;
$\cos(\frac{1}{3}\pi \text{ radians}) = \cos 60^\circ = 0{\cdot}5$.

We now define the sine of a *real number*. The radian measure of an angle is the ratio of two lengths and so is a real number.

For every real number x, $\sin x$ is defined to be $\sin(x \text{ radians})$; e.g. $\sin \pi = \sin(\pi \text{ radians}) = \sin 180^\circ = 0$.

Note. From now on we shall use only expressions like $\sin x$ and $\sin x^\circ$. These are not the same. Sin x is the sine of the real number x, which by definition is the sine of the angle whose radian measure is x. $\sin x^\circ$ is the sine of the angle whose size is x *degrees*, x being a real number. Thus $\sin x$ and $\sin x^\circ$ define two distinct functions on the set of real numbers.

Example 1. $\sin \frac{1}{6}\pi = \sin 30^\circ = 0{\cdot}5$.

Example 2. Express 1 radian in degree measure.

$$\pi \text{ radians} = 180^\circ$$
$$\Leftrightarrow \quad 1 \text{ radian} = 180 \times \frac{1}{\pi}^\circ$$
$$\doteqdot \frac{180}{3{\cdot}14}^\circ$$
$$\doteqdot 57^\circ$$

Example 3. Convert 29° to radian measure.

$$180^\circ = \pi \text{ radians}$$
$$\Leftrightarrow \quad 29^\circ = \pi \times \frac{29}{180} \text{ radians}$$
$$\doteqdot 3{\cdot}14 \times \frac{29}{180} \text{ radians}$$
$$\doteqdot 0{\cdot}5 \text{ radian}$$

Note. From radians–degrees conversion tables,

$$1 \text{ radian} = 57{\cdot}25^\circ \quad \text{and} \quad 29^\circ = 0{\cdot}506 \text{ radian.}$$

Exercise 3

1 Find the radian measure which is equivalent to:

0°, 30°, 45°, 60°, 90°, 180°, 360°.

2 Find the degree measure which is equivalent to:

$\frac{1}{4}\pi$ radians, $\frac{1}{2}\pi$ radians, $\frac{1}{5}\pi$ radians, $\frac{4}{3}\pi$ radians.

3 Give the values of:

a $\sin \frac{1}{6}\pi$ *b* $\tan \frac{1}{4}\pi$ *c* $\sin \pi$ *d* $\cos 2\pi$ *e* $\cos \frac{5}{3}\pi$

4 Use the relation π radians $= 180^\circ$ to convert:

a 2 radians to degrees *b* 20° to radian measure.

5 Use conversion tables to complete the following:

Angle in degrees	4°	66°	308°		
Angle in radians				0·65	0·123

6 Simplify:

a $\sin(\pi - \theta)$ *b* $\cos(\pi - \theta)$ *c* $\tan(\pi - \theta)$ *d* $\sin(\frac{1}{2}\pi - \theta)$

e $\cos(\frac{1}{2}\pi - \theta)$ *f* $\sin(\pi + \theta)$ *g* $\tan(\pi + \theta)$ *h* $\sin(2\pi - \theta)$

7 Find the solution sets of the following, for $0 \leqslant x \leqslant 2\pi$, $x \in R$:

a $\tan x = 1$ *b* $\cos x = \frac{1}{2}$ *c* $\sin x = \frac{1}{\sqrt{2}}$

d $\cos x = -2$ *e* $\sin x + 1 = 0$ *f* $\sqrt{3}\tan x = 1$

8 A radar scanner rotates at 30 revolutions per minute. Express its speed of rotation in: ***a*** degrees per second ***b*** radians per second.

9 A circle, centre O, has two perpendicular radii OA and OB r units long. Prove that $\frac{\text{arc AB}}{\text{OA}} \doteq \frac{\pi}{2}$.

10 An arc DE of a circle, centre O, is $\frac{1}{6}$ of the circumference. Calculate the size of $\angle$DOE in radians.

11 The angle of a sector of a circle has radian measure 2. Show that area of sector: area of circle $= 1:\pi$.

4 *The addition formulae*

(i) Formulae for cos $(\alpha+\beta)$ and cos $(\alpha-\beta)$

In Figure 10, the circle has unit radius, so A is the point (1,0).

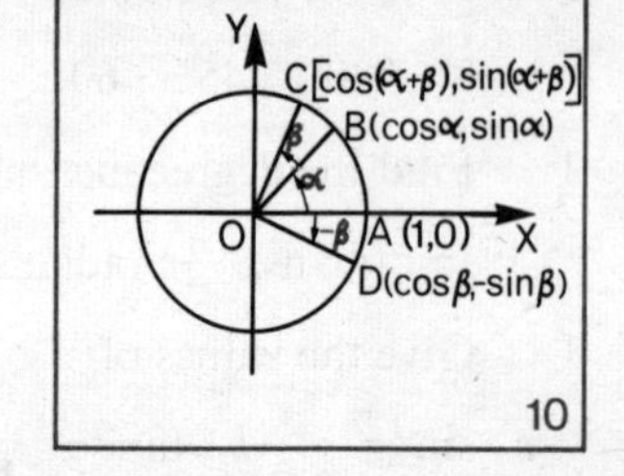

10

Let $\angle \text{AOB} = \alpha$ radians
$\angle \text{BOC} = \beta$ radians
So $\angle \text{AOC} = (\alpha+\beta)$ radians
Let $\angle \text{AOD} = -\beta$ radians

Using $(r\cos a^\circ, r\sin a^\circ)$ for the coordinates of a point, B is $(\cos\alpha, \sin\alpha)$, C is $[\cos(\alpha+\beta), \sin(\alpha+\beta)]$ and D is $[\cos(-\beta), \sin(-\beta)]$, i.e. $(\cos\beta, -\sin\beta)$.

$$\angle \text{AOC} = \angle \text{DOB} \Rightarrow \text{AC} = \text{DB} \Rightarrow \text{AC}^2 = \text{DB}^2.$$

Using the distance formula '$\text{PQ}^2 = (x_2-x_1)^2+(y_2-y_1)^2$',

$$[\cos(\alpha+\beta)-1]^2+[\sin(\alpha+\beta)-0]^2 = (\cos\alpha-\cos\beta)^2+(\sin\alpha+\sin\beta)^2$$

i.e. $\cos^2(\alpha+\beta)-2\cos(\alpha+\beta)+1+\sin^2(\alpha+\beta) =$
$\cos^2\alpha-2\cos\alpha\cos\beta+\cos^2\beta+\sin^2\alpha+2\sin\alpha\sin\beta+\sin^2\beta.$

Hence $2-2\cos(\alpha+\beta) = 2-2\cos\alpha\cos\beta+2\sin\alpha\sin\beta$,
since $\cos^2 a+\sin^2 a = 1$.

Therefore $\cos(\alpha+\beta) = \cos\alpha\cos\beta - \sin\alpha\sin\beta$.

One diagram is sufficient, as the statements are independent of the sizes of α and β which can be positive or negative.

$$\begin{aligned}\cos(\alpha-\beta) &= \cos[\alpha+(-\beta)]\\ &= \cos\alpha\cos(-\beta)-\sin\alpha\sin(-\beta)\\ &= \cos\alpha\cos\beta+\sin\alpha\sin\beta\end{aligned}$$

Therefore $\cos(\alpha-\beta) = \cos\alpha\cos\beta+\sin\alpha\sin\beta$

Note. There are corresponding formulae for angles measured in degrees.

Alternative proof for the formula for cos $(\alpha-\beta)$

Note. The above proof makes use of the *Distance Formula* and certain *properties of the circle.* The proof given below depends on the *scalar product* of *vectors* (see Geometry, Chapter 2).

Let $\overrightarrow{OA}$ and $\overrightarrow{OB}$ represent unit vectors. where $\angle XOA = \alpha$ radians and $\angle XOB = \beta$ radians as shown in Figure 11. $\angle BOA = (\alpha-\beta)$ radians. A is the point $(\cos\alpha, \sin\alpha)$, and B is $(\cos\beta, \sin\beta)$. The position vector of A is $\begin{pmatrix}\cos\alpha\\ \sin\alpha\end{pmatrix}$, and of B is $\begin{pmatrix}\cos\beta\\ \sin\beta\end{pmatrix}$.

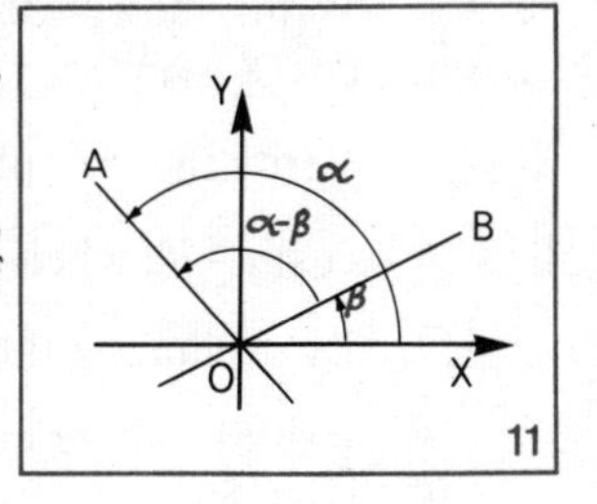

11

$$\overrightarrow{OA}.\overrightarrow{OB} = \begin{pmatrix}\cos\alpha\\ \sin\alpha\end{pmatrix}.\begin{pmatrix}\cos\beta\\ \sin\beta\end{pmatrix}$$

$$\Leftrightarrow \quad |\overrightarrow{OA}||\overrightarrow{OB}|\cos BOA = \cos\alpha\cos\beta+\sin\alpha\sin\beta$$
$$\Leftrightarrow \quad \cos(\alpha-\beta) = \cos\alpha\cos\beta+\sin\alpha\sin\beta$$

Example 1. $\cos(p-q)^\circ = \cos p^\circ\cos q^\circ + \sin p^\circ \sin q^\circ$.

Example 2. Use the 30°, 60°, 90° triangle to verify that $\cos(a-b)^\circ = \cos a^\circ\cos b^\circ+\sin a^\circ\sin b^\circ$ when $a = 90$, $b = 30$.

$$\begin{aligned}\text{Left side} &= \cos 60^\circ = \tfrac{1}{2}\\ \text{Right side} &= \cos 90^\circ\cos 30^\circ+\sin 90^\circ\sin 30^\circ\\ &= (0\times\tfrac{\sqrt{3}}{2})+(1\times\tfrac{1}{2})\\ &= \tfrac{1}{2}.\end{aligned}$$

Hence $\cos(a-b)^\circ = \cos a^\circ\cos b^\circ+\sin a^\circ\sin b^\circ$, for $a = 90$, $b = 30$.

Example 3. $\cos(\frac{1}{4}\pi-\theta)\cos(\frac{1}{4}\pi+\theta)-\sin(\frac{1}{4}\pi-\theta)\sin(\frac{1}{4}\pi+\theta)$
$= \cos[(\frac{1}{4}\pi-\theta)+(\frac{1}{4}\pi+\theta)]$, using
$\cos\alpha\cos\beta-\sin\alpha\sin\beta = \cos(\alpha+\beta)$
$= \cos\frac{1}{2}\pi$
$= 0$

Example 4. Find the exact value of cos 15°.

$$\cos 15^\circ = \cos(60-45)^\circ = \cos 60^\circ\cos 45^\circ+\sin 60^\circ\sin 45^\circ$$
$$= \left(\frac{1}{2}\times\frac{1}{\sqrt{2}}\right)+\left(\frac{\sqrt{3}}{2}\times\frac{1}{\sqrt{2}}\right)$$
$$= \frac{1+\sqrt{3}}{2\sqrt{2}}$$

Exercise 4

1 Write down formulae for $\cos(x-y)$, $\cos(A-B)$, $\cos(p+q)$, $\cos(X+Y)$.

2 Verify the formula for:

a $\cos(\alpha-\beta)$, when $\alpha=\beta=\frac{1}{4}\pi$ *b* $\cos(\alpha+\beta)$, when $\alpha=\beta=0$

3 By expanding the left side, show that:

a $\cos(90-a)^\circ = \sin a^\circ$ *b* $\cos(90+a)^\circ = -\sin a^\circ$
c $\cos(180-a)^\circ = -\cos a^\circ$

4 Prove that: *a* $\cos(\alpha+\beta)+\cos(\alpha-\beta) = 2\cos\alpha\cos\beta$
b $\cos(\alpha-\beta)-\cos(\alpha+\beta) = 2\sin\alpha\sin\beta$

5 Use the addition formulae to simplify:

a $\cos M\cos N+\sin M\sin N$ *b* $\cos 2\alpha\cos\alpha-\sin 2\alpha\sin\alpha$
c $\cos 100^\circ\cos 10^\circ+\sin 100^\circ\sin 10^\circ$ *d* $\cos 40^\circ\cos 5^\circ-\sin 40^\circ\sin 5^\circ$

6 Given that $\sin A = \frac{3}{5}$ and $\sin B = \frac{7}{25}$, and that A and B are acute angles, show that $\cos(A+B) = \frac{3}{5}$.

7 Given that $\tan X = \frac{12}{5}$ and $\tan Y = \frac{4}{3}$, find values of $\cos(X-Y)$ and $\cos(X+Y)$, assuming that X and Y are acute angles.

8 By expressing 15° as 45° − 30°, show that $\cos 15^\circ = \dfrac{1+\sqrt{3}}{2\sqrt{2}}$

9 By expressing 75° as $45^\circ+30^\circ$, and using the formula for $\cos(\alpha+\beta)^\circ$, find the exact value of $\cos 75^\circ$.

10 Prove that $\dfrac{\cos(\alpha+\beta)}{\cos\alpha\cos\beta} = 1-\tan\alpha\tan\beta$.

11 Prove that $\cos(270+a)^\circ = \sin a^\circ$, by:

a using an addition formula *b* using a quarter turn about O.

12 Prove that $\cos\theta+\cos(\theta+\frac{2}{3}\pi)+\cos(\theta+\frac{4}{3}\pi) = 0$.

(ii) Formulae for sin (α+β) and sin (α−β)

$$\begin{aligned}\sin(\alpha+\beta) &= \cos\left[\tfrac{1}{2}\pi-(\alpha+\beta)\right]\\ &= \cos\left[(\tfrac{1}{2}\pi-\alpha)-\beta\right]\\ &= \cos(\tfrac{1}{2}\pi-\alpha)\cos\beta+\sin(\tfrac{1}{2}\pi-\alpha)\sin\beta\\ &= \sin\alpha\cos\beta+\cos\alpha\sin\beta\end{aligned}$$

Therefore $\sin(\alpha+\beta) = \sin\alpha\cos\beta+\cos\alpha\sin\beta$

$$\begin{aligned}\sin(\alpha-\beta) &= \sin\left[\alpha+(-\beta)\right]\\ &= \sin\alpha\cos(-\beta)+\cos\alpha\sin(-\beta)\\ &= \sin\alpha\cos\beta-\cos\alpha\sin\beta\end{aligned}$$

Therefore $\sin(\alpha-\beta) = \sin\alpha\cos\beta-\cos\alpha\sin\beta$

Note. There are corresponding formulae for angles measured in degrees.

Exercise 5

1 Write down formulae for $\sin(x+y)$, $\sin(A+B)$, $\sin(p-q)$, $\sin(X-Y)$.

2 Verify the formula for:

a $\sin(\alpha+\beta)$, when $\alpha = \frac{1}{2}\pi$, $\beta = \frac{1}{4}\pi$ *b* $\sin(\alpha-\beta)$, when $\alpha = \frac{1}{3}\pi$, $\beta = \frac{1}{6}\pi$

3 By expanding the left side, show that:

a $\sin(90-a)^\circ = \cos a^\circ$ *b* $\sin(90+a)^\circ = \cos a^\circ$
c $\sin(180+a)^\circ = -\sin a^\circ$

4 Prove that: *a* $\sin(\alpha+\beta)+\sin(\alpha-\beta) = 2\sin\alpha\cos\beta$
b $\sin(\alpha+\beta)-\sin(\alpha-\beta) = 2\cos\alpha\sin\beta$

5 Use the addition formulae to simplify:

a $\sin M \cos N + \cos M \sin N$ *b* $\sin 2\alpha \cos \alpha - \cos 2\alpha \sin \alpha$
c $\sin 100^\circ \cos 10^\circ - \cos 100^\circ \sin 10^\circ$ *d* $\sin 123^\circ \cos 57^\circ + \cos 123^\circ \sin 57^\circ$

6 Given that $\sin A = \frac{5}{13}$ and $\cos B = \frac{3}{5}$, and A and B are acute angles, show that $\sin(A+B) = \frac{63}{65}$.

7 Given that $\tan P = \frac{3}{4}$ and $\tan Q = \frac{7}{24}$, find values of $\sin(P+Q)$ and $\sin(P-Q)$, assuming that P and Q are acute angles.

8 Use addition formulae to show that:

a $\sin 75^\circ = \dfrac{\sqrt{3}+1}{2\sqrt{2}}$ *b* $\sin 15^\circ = \dfrac{\sqrt{3}-1}{2\sqrt{2}}$

9 Prove that $\dfrac{\sin(\alpha+\beta)}{\cos\alpha\cos\beta} = \tan\alpha + \tan\beta$.

10 Prove that $\sin\theta + \sin(\theta + \frac{2}{3}\pi) + \sin(\theta + \frac{4}{3}\pi) = 0$.

11 The point A has Cartesian coordinates (x, y) and polar coordinates (r, θ). Under a rotation α about O, $A \to A_1(x_1, y_1)$.

a Express: *(1)* x and y in terms of r and θ
(2) x_1 and y_1 in terms of r, θ and α.

b Show that $x_1 = x\cos\alpha - y\sin\alpha$, and find y_1 in a similar form.

c Find the 2×2 matrix M_α such that $\begin{pmatrix} x_1 \\ y_1 \end{pmatrix} = M_\alpha \begin{pmatrix} x \\ y \end{pmatrix}$.

d Use this matrix equation to find the images of $P(a, b)$ under anti-clockwise rotations about the origin of: *(1)* $\frac{1}{2}\pi$ *(2)* π *(3)* $\frac{1}{4}\pi$ *(4)* $\frac{1}{6}\pi$ radians.

12 In this question assume that the matrix $M_\alpha = \begin{pmatrix} \cos\alpha & -\sin\alpha \\ \sin\alpha & \cos\alpha \end{pmatrix}$ maps $A(x, y)$ to $A_1(x_1, y_1)$ under a rotation α about O.

a Write down similar matrices which map:
(1) $A_1(x_1, y_1)$ to $A_2(x_2, y_2)$ under a rotation β about O (i.e. matrix M_β).
(2) $A(x, y)$ to $A_2(x_2, y_2)$ under a rotation $(\alpha+\beta)$ about O (i.e. matrix $M_{\alpha+\beta}$).

b Explain why $M_\beta M_\alpha \begin{pmatrix} x \\ y \end{pmatrix} = M_{\alpha+\beta} \begin{pmatrix} x \\ y \end{pmatrix}$.

c Use the matrix equation in ***b*** to derive formulae for $\cos(\alpha+\beta)$ and $\sin(\alpha+\beta)$.

Exercise 6B

1 Expand:

a $\cos(3a+2b)$ *b* $\sin(x+2y)$ *c* $\sin(3a-2b)$ *d* $\cos(2x-y)$

2 Simplify:

a $\cos 200^\circ \cos 115^\circ - \sin 200^\circ \sin 115^\circ$
b $\cos(\frac{1}{2}\pi+\alpha)\cos\alpha+\sin(\frac{1}{2}\pi+\alpha)\sin\alpha$

3 If $\cos\alpha = \frac{1}{\sqrt{5}}$ and $\sin\beta = \frac{3}{5}$, where α and β are acute angles, show that $\cos(\alpha-\beta) = \frac{2}{5}\sqrt{5}$.

4 Given that $\tan\alpha = \frac{3}{4}$ and $\tan\beta = \frac{1}{7}$, and that α and β are acute angles, show that $\cos(\alpha+\beta) = \dfrac{1}{\sqrt{2}}$.

5 Prove that $(\cos x+\cos y)^2+(\sin x-\sin y)^2 = 2[1+\cos(x+y)]$.

6 Simplify $\dfrac{\cos 2a}{\sin a}+\dfrac{\sin 2a}{\cos a}$.

7 If $2\cos(x+\frac{1}{4}\pi) = \cos(x-\frac{1}{4}\pi)$, show that $\tan x = \frac{1}{3}$.

8 If $\sin(x+30)^\circ = \sin x^\circ$, show that $\tan x^\circ = 2+\sqrt{3}$.

9 Given $2\cos(30+t)^\circ = \cos(30-t)^\circ$, show that $\tan t^\circ = 1/\sqrt{3}$, and hence find t for $0 < t < 360$.

10 Given $\sin(x+45)^\circ+2\cos(x+45)^\circ = 0$, find an equation in $\tan x^\circ$, and hence find x for $0 < x < 360$.

11 If $\dfrac{F}{\sin(\theta-\alpha)} = \dfrac{W}{\sin\theta}$, show that $\tan\theta = \dfrac{W\sin\alpha}{W\cos\alpha - F}$.

12 Prove the following identities:

a $\sin(x+y)\sin(x-y) = \sin^2 x - \sin^2 y$
b $\cos(x+y)\cos(x-y) = \cos^2 y - \sin^2 x$.

13 $\cos(a-b)^\circ = \frac{1}{2}\sqrt{3}$ and $\cos a^\circ \cos b^\circ = \frac{1}{2}$, where a° and b° are acute angles. Find an expression for $\sin a^\circ \sin b^\circ$, and hence calculate the value of $\cos(a+b)^\circ$. Use tables to calculate a and b.

14 If $\cos\beta\sin(\theta-\alpha)=\cos\alpha\sin(\beta-\theta)$, prove $\tan\theta=\dfrac{\sin(\alpha+\beta)}{2\cos\alpha\cos\beta}$.

15 The equation $W\cos(\alpha-45)^\circ+F\cos(\alpha+45)^\circ=0$ occurs in mechanics. Show that $F=\dfrac{W(\tan\alpha^\circ+1)}{\tan\alpha^\circ-1}$.

(iii) Formulae for tan $(\alpha+\beta)$ and tan $(\alpha-\beta)$

$$\begin{aligned}
\tan(\alpha+\beta) &= \frac{\sin(\alpha+\beta)}{\cos(\alpha+\beta)} = \frac{\sin\alpha\cos\beta+\cos\alpha\sin\beta}{\cos\alpha\cos\beta-\sin\alpha\sin\beta} \\
&= \frac{\dfrac{\sin\alpha\cos\beta}{\cos\alpha\cos\beta}+\dfrac{\cos\alpha\sin\beta}{\cos\alpha\cos\beta}}{\dfrac{\cos\alpha\cos\beta}{\cos\alpha\cos\beta}-\dfrac{\sin\alpha\sin\beta}{\cos\alpha\cos\beta}} = \frac{\tan\alpha+\tan\beta}{1-\tan\alpha\tan\beta}
\end{aligned}$$

$$\begin{aligned}
\tan(\alpha-\beta) &= \frac{\sin(\alpha-\beta)}{\cos(\alpha-\beta)} = \frac{\sin\alpha\cos\beta-\cos\alpha\sin\beta}{\cos\alpha\cos\beta+\sin\alpha\sin\beta} \\
&= \frac{\dfrac{\sin\alpha\cos\beta}{\cos\alpha\cos\beta}-\dfrac{\cos\alpha\sin\beta}{\cos\alpha\cos\beta}}{\dfrac{\cos\alpha\cos\beta}{\cos\alpha\cos\beta}+\dfrac{\sin\alpha\sin\beta}{\cos\alpha\cos\beta}} = \frac{\tan\alpha-\tan\beta}{1+\tan\alpha\tan\beta}
\end{aligned}$$

So $\tan(\alpha+\beta)=\dfrac{\tan\alpha+\tan\beta}{1-\tan\alpha\tan\beta}$ and $\tan(\alpha-\beta)=\dfrac{\tan\alpha-\tan\beta}{1+\tan\alpha\tan\beta}$

Note. There are corresponding formulae for angles measured in degrees.

Example. Show that $\tan 15^\circ = 2-\sqrt{3}$.

$$\tan 15^\circ = \tan(45-30)^\circ = \frac{\tan 45^\circ-\tan 30^\circ}{1+\tan 45^\circ\tan 30^\circ} = \frac{1-\frac{1}{\sqrt{3}}}{1+\frac{1}{\sqrt{3}}}$$

$$= \frac{\sqrt{3}-1}{\sqrt{3}+1} = \frac{\sqrt{3}-1}{\sqrt{3}+1}\times\frac{\sqrt{3}-1}{\sqrt{3}-1} = \frac{3-2\sqrt{3}+1}{3-1} = 2-\sqrt{3}$$

Exercise 7

1 Expand: $\tan(x+y)$, $\tan(\mathrm{A}+\mathrm{B})$, $\tan(p-q)$, $\tan(\mathrm{M}-\mathrm{N})$.

2 Verify the formula for:

a $\tan(\alpha+\beta)$, when $\alpha=\beta=\frac{1}{3}\pi$ ***b*** $\tan(\alpha-\beta)$ when $\alpha=\beta=\frac{1}{4}\pi$.

3 Expand: $\tan(x+2y)$, $\tan(3a+2b)$, $\tan(2x-y)$, $\tan(3p-2q)$, $\tan(\frac{1}{4}\pi - x)$.

4 If $\tan\alpha = \frac{1}{2}$ and $\tan\beta = \frac{1}{3}$, show that $\tan(\alpha+\beta) = 1$, and find the value of $\tan(\alpha-\beta)$.

5 Given $\sin A = \frac{3}{5}$ and $\cos B = \frac{12}{13}$, show that $\tan(A-B) = \frac{16}{63}$. (A and B are acute angles.)

6 Given $\cos P = \frac{5}{13}$ and $\sin Q = \frac{4}{5}$, find the value of $\tan(P+Q)$. (P and Q are acute angles.)

7 Given $\sin x = \frac{3}{5}$ and $\tan y = \frac{1}{7}$, prove without using tables that $x+y = \frac{1}{4}\pi$. (x and y are acute angles.)

8 Use $\tan(\alpha-\beta) = \tan[\alpha+(-\beta)]$ to deduce the formula for $\tan(\alpha-\beta)$ from the formula for $\tan(\alpha+\beta)$.

9 Prove that $\tan(\frac{1}{4}\pi+\alpha) = \dfrac{\cos\alpha+\sin\alpha}{\cos\alpha-\sin\alpha}$.

10 By expressing 15° as 60° − 45°, show that $\tan 15° = 2-\sqrt{3}$.

11 Show that $\tan 75° = 2+\sqrt{3}$.

12 Use the formula for $\tan(\alpha+\beta)$ to show that, if $\alpha+\beta = \frac{1}{4}\pi$, then $(1+\tan\alpha)(1+\tan\beta) = 2$.

13 Prove that, if $2x+y = \frac{1}{4}\pi$, then $\tan 2x = \dfrac{1-\tan y}{1+\tan y}$.

14 Given $\tan(\frac{1}{4}\pi+\theta) = 4\tan(\frac{1}{4}\pi-\theta)$, show that, if $\tan\theta = t$, then $t = \frac{1}{3}$ or 3.

15 If $\tan x = \dfrac{1}{a-1}$ and $\tan y = \dfrac{1}{a+1}$, prove that $\tan(x-y) = \dfrac{2}{a^2}$.

16 $A+B+C = \pi \Rightarrow A+B = \pi - C \Rightarrow \tan(A+B) = \tan(\pi - C)$. Use this relation to deduce that $\tan A + \tan B + \tan C = \tan A \tan B \tan C$. Show that, if $\tan A = 2$ and $\tan B = \frac{4}{3}$, then $C = A$.

5 *Formulae and applications involving 2α*

(i) $\sin 2\alpha = \sin(\alpha+\alpha) = \sin\alpha\cos\alpha + \cos\alpha\sin\alpha = 2\sin\alpha\cos\alpha$

$$\sin 2\alpha = 2\sin\alpha\cos\alpha$$

(ii) $\cos 2\alpha = \cos(\alpha+\alpha) = \cos\alpha\cos\alpha - \sin\alpha\sin\alpha = \cos^2\alpha - \sin^2\alpha$

$\cos 2\alpha = \cos^2\alpha - \sin^2\alpha = \cos^2\alpha - (1-\cos^2\alpha) = 2\cos^2\alpha - 1$

$\cos 2\alpha = \cos^2\alpha - \sin^2\alpha = (1-\sin^2\alpha) - \sin^2\alpha = 1 - 2\sin^2\alpha$

$\cos 2\alpha = \cos^2\alpha - \sin^2\alpha = 2\cos^2\alpha - 1 = 1 - 2\sin^2\alpha$

Note that in cosine formulae, cosines come first:

$$\begin{aligned} \cos(\alpha+\beta) &= \cos\alpha\cos\beta - \sin\alpha\sin\beta \\ \cos 2\alpha &= \cos^2\alpha \quad\quad - \sin^2\alpha \\ &= 2\cos^2\alpha \quad\quad - 1 \\ &= 1 \quad\quad\quad\quad - 2\sin^2\alpha \end{aligned}$$

(iii) $\cos 2\alpha = 2\cos^2\alpha - 1$ and $\cos 2\alpha = 1 - 2\sin^2\alpha$

$\Leftrightarrow \cos^2\alpha = \frac{1}{2}(1+\cos 2\alpha)$ $\quad\Leftrightarrow \sin^2\alpha = \frac{1}{2}(1-\cos 2\alpha)$

(iv) $\tan 2\alpha = \tan(\alpha+\alpha) = \dfrac{\tan\alpha+\tan\alpha}{1-\tan\alpha\tan\alpha} = \dfrac{2\tan\alpha}{1-\tan^2\alpha}$

$$\tan 2\alpha = \frac{2\tan\alpha}{1-\tan^2\alpha}$$

Note. There are corresponding formulae for angles measured in degrees.

Exercise 8

1 Write down formulae for sin 2A, cos 2A and tan 2A.

2 Write down formulae for $\sin\theta$, $\cos\theta$ and $\tan\theta$ in terms of $\frac{1}{2}\theta$.

3 Express $\sin 4\alpha$ in terms of $\sin 2\alpha$ and $\cos 2\alpha$.

4 Express $\cos 4\alpha$ in terms of: ***a*** $\cos 2\alpha$ ***b*** $\sin 2\alpha$.

5 Write down formulae for sin 8A, cos 6B and tan C in terms of 4A, 3B and $\frac{1}{2}$C respectively.

6 Given $\sin A = \frac{3}{5}$, and $0 < A < \frac{1}{2}\pi$, calculate sin 2A, cos 2A and tan 2A.

7 Given $\tan B = \frac{1}{2}$, and $0 < B < \frac{1}{2}\pi$, calculate tan 2B, cos 2B and sin 2B.

8 Express each of the following in terms of a single sine, cosine or tangent.

a $2\sin p\cos p$ *b* $2\cos^2 n-1$ *c* $1-2\cos^2 x$ *d* $2\sin 35°\cos 35°$

e $1-2\sin^2 y$ *f* $\cos^2 5°-\sin^2 5°$ *g* $\dfrac{2\tan k}{1-\tan^2 k}$ *h* $\dfrac{2\tan 50°}{1-\tan^2 50°}$

9 Use the formulae to simplify the following, and calculate their values:

a $2\sin 15°\cos 15°$ *b* $2\cos^2 30°-1$ *c* $\cos^2\frac{1}{6}\pi-\sin^2\frac{1}{6}\pi$ *d* $\sin\frac{1}{4}\pi\cos\frac{1}{4}\pi$

10 Which of the following are true and which are false?

a $\cos 2x = \cos^2 x+\sin^2 x$ *b* $\sin x = 2\sin\frac{1}{2}x\cos\frac{1}{2}x$

c $\tan 4x = \dfrac{2\tan 2x}{1-\tan^2 2x}$ *d* $\cos(x+y) = \cos x+\cos y$

e $\sin(x-y) = \sin x-\sin y$ *f* $\tan(x+y) = \tan x+\tan y$

11 If $\tan x = \frac{1}{2}$ and $\tan y = \frac{1}{3}$, calculate:

a $\tan 2x$ *b* $\tan 2y$ *c* $\tan(2x+y)$ *d* $\tan(x+2y)$

12 By expressing 3θ as $2\theta+\theta$ prove that:

a $\cos 3\theta = 4\cos^3\theta-3\cos\theta$ *b* $\sin 3\theta = 3\sin\theta-4\sin^3\theta$

Equations

Example. Solve the equation $\cos 2x°+\sin x° = 0$, $0 \leqslant x \leqslant 360$, $x \in R$.

$$\cos 2x°+\sin x° = 0$$
$$\Leftrightarrow \quad 1-2\sin^2 x°+\sin x° = 0$$
$$\Leftrightarrow \quad 2\sin^2 x°-\sin x°-1 = 0$$
$$\Leftrightarrow \quad (\sin x°-1)(2\sin x°+1) = 0$$
$$\Leftrightarrow \quad \sin x° = 1 \text{ or } -\tfrac{1}{2}$$
$$\Rightarrow \quad x = 90, 210, 330, \text{ so the solution set is } \{90, 210, 330\}.$$

Exercise 9

Solve the following equations, for $0 \leqslant x \leqslant 360$, $x \in R$:

1 $\sin 2x^\circ + \sin x^\circ = 0$

2 $\sin 2x^\circ - \cos x^\circ = 0$

3 $\cos 2x^\circ - \cos x^\circ = 0$

4 $\cos 2x^\circ - \sin x^\circ = 0$

5 $\cos 2x^\circ - 3\cos x^\circ + 2 = 0$

6 $\cos 2x^\circ - 3\sin x^\circ - 1 = 0$

7 $\cos 2x^\circ - 4\sin x^\circ + 5 = 0$

8 $\cos 2x^\circ - \sin x^\circ - 1 = 0$

9 $\cos 2x^\circ + 5\cos x^\circ - 2 = 0$

10 $\cos 2x^\circ + 3\cos x^\circ + 2 = 0$

11 $\cos 2x^\circ + \cos x^\circ = 0$

12 $5\cos 2x^\circ - \cos x^\circ + 2 = 0$

13 $3\sin 2x^\circ + 5\cos x^\circ = 0$

14 $6\cos 2x^\circ - 5\cos x^\circ + 4 = 0$

15 $4\cos 2x^\circ - 2\sin x^\circ - 1 = 0$

16 $5\cos 2x^\circ + 7\sin x^\circ + 7 = 0$

Solve the following equations, for $0 \leqslant \theta \leqslant 2\pi$, $\theta \in R$:

17 $\sin 2\theta - \sin\theta = 0$

18 $\sin 2\theta + \cos\theta = 0$

19 $\cos 2\theta + \cos\theta = 0$

20 $\cos 2\theta + \sin\theta = 0$

Identities

Example. Prove the identities:

(i) $2\cos\theta(\cos\tfrac{1}{2}\theta + \sin\tfrac{1}{2}\theta)^2 = 2\cos\theta + \sin 2\theta$ (ii) $\dfrac{1-\cos 2\theta}{\sin 2\theta} = \tan\theta$

(i) *Left side*

$$
\begin{aligned}
&= 2\cos\theta(\cos^2\tfrac{1}{2}\theta + 2\cos\tfrac{1}{2}\theta\sin\tfrac{1}{2}\theta + \sin^2\tfrac{1}{2}\theta) \\
&= 2\cos\theta(1+\sin\theta) \\
&= 2\cos\theta + 2\cos\theta\sin\theta \\
&= 2\cos\theta + \sin 2\theta \\
&= \textit{right side}
\end{aligned}
$$

(ii) *Left side*

$$
\begin{aligned}
&= \frac{1-(1-2\sin^2\theta)}{2\sin\theta\cos\theta} \\
&= \frac{2\sin^2\theta}{2\sin\theta\cos\theta} \\
&= \frac{\sin\theta}{\cos\theta} \\
&= \tan\theta \\
&= \textit{right side}
\end{aligned}
$$

Exercise 10B

Prove the following identities:

1 $(\sin\theta+\cos\theta)^2 = 1+\sin 2\theta$

2 $(\cos\theta-\sin\theta)^2 = 1-\sin 2\theta$

3 $(\cos\theta+\sin\theta)(\cos\theta-\sin\theta) = \cos 2\theta$

4 $\cos^4\theta-\sin^4\theta = \cos 2\theta$

5 $(2\cos\theta-1)(2\cos\theta+1) = 2\cos 2\theta+1$

6 $(\cos\frac{1}{2}\theta-\sin\frac{1}{2}\theta)^2 = 1-\sin\theta$

7 $\dfrac{\sin 2\theta}{1+\cos 2\theta} = \tan\theta$

8 $\dfrac{1-\cos 2\theta}{1+\cos 2\theta} = \tan^2\theta$

9 $\dfrac{2\tan\theta}{1+\tan^2\theta} = \sin 2\theta$

10 $\dfrac{1-\tan^2\theta}{1+\tan^2\theta} = \cos 2\theta$

11 Using $\cos 4\theta = 2\cos^2 2\theta-1$ and $\cos 2\theta = 2\cos^2\theta-1$, prove that $\cos 4\theta = 8\cos^4\theta-8\cos^2\theta+1$.

12 In a similar way to question ***11***, express $\cos 4\theta$ in terms of powers of $\sin\theta$.

13a Using $\cos^2\theta = \frac{1}{2}(1+\cos 2\theta)$, show that $\cos^4\theta = \frac{1}{4}+\frac{1}{2}\cos 2\theta+\frac{1}{4}\cos^2 2\theta$.

b Hence show that $\cos^4\theta = \frac{3}{8}+\frac{1}{2}\cos 2\theta+\frac{1}{8}\cos 4\theta$.

14a Using $\sin^2\theta = \frac{1}{2}(1-\cos 2\theta)$, express $\sin^4\theta$ in the form $a+b\cos 2\theta+c\cos^2 2\theta$.

b Hence express $\sin^4\theta$ in the form $d+e\cos 2\theta+f\cos 4\theta$.

15 Express $\sin^2\theta\cos^2\theta$ in the form $a+b\cos 4\theta$.

Summary

1 *Related angles*

$$\left.\begin{aligned}\cos(180-a)^\circ &= -\cos a^\circ\\ \sin(180-a)^\circ &= \sin a^\circ\end{aligned}\right\} \qquad \left.\begin{aligned}\cos(-a)^\circ &= \cos a^\circ\\ \sin(-a)^\circ &= -\sin a^\circ\end{aligned}\right\}$$

$$\left.\begin{aligned}\cos(90-a)^\circ &= \sin a^\circ\\ \sin(90-a)^\circ &= \cos a^\circ\end{aligned}\right\}$$

2 *Radian measure*

$$\text{Radian measure of } \angle \text{AOB at centre of circle} = \frac{\text{arc AB}}{\text{radius OA}}.$$

$$180^\circ = \pi \text{ radians}; \quad 1 \text{ radian} \doteqdot 57^\circ; \quad \sin\pi = \sin 180^\circ$$

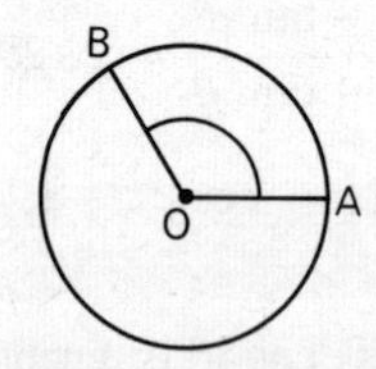

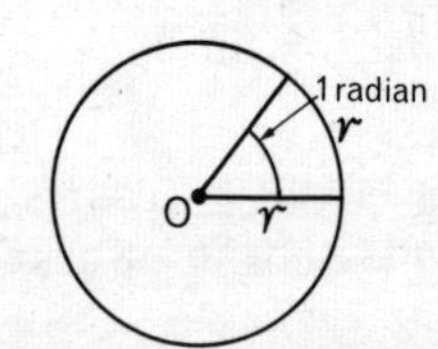

3 *Addition formulae*

$$\begin{aligned}\cos(\alpha+\beta) &= \cos\alpha\cos\beta-\sin\alpha\sin\beta\\ \cos(\alpha-\beta) &= \cos\alpha\cos\beta+\sin\alpha\sin\beta\\ \sin(\alpha+\beta) &= \sin\alpha\cos\beta+\cos\alpha\sin\beta\\ \sin(\alpha-\beta) &= \sin\alpha\cos\beta-\cos\alpha\sin\beta\\ \tan(\alpha+\beta) &= \frac{\tan\alpha+\tan\beta}{1-\tan\alpha\tan\beta}\\ \tan(\alpha-\beta) &= \frac{\tan\alpha-\tan\beta}{1+\tan\alpha\tan\beta}\end{aligned}$$

4 *'Double angle' formulae*

$$\begin{aligned}\sin 2\alpha &= 2\sin\alpha\cos\alpha\\ \cos 2\alpha &= \cos^2\alpha-\sin^2\alpha\\ &= 2\cos^2\alpha-1 \qquad\qquad \cos^2\alpha = \tfrac{1}{2}(1+\cos 2\alpha)\\ &= 1-2\sin^2\alpha \qquad\qquad \sin^2\alpha = \tfrac{1}{2}(1-\cos 2\alpha)\end{aligned}$$

$$\tan 2\alpha = \frac{2\tan\alpha}{1-\tan^2\alpha}$$

Revision Exercises

Revision Exercises on Chapter 1
The Addition Formulae

Revision Exercise 1A

1 Find the values of: ***a*** $\sin 104°$ ***b*** $\cos 104°$ ***c*** $\tan 200°$ ***d*** $\cos 333°$.

2 If $0 < a < 90$ and $\tan a° = \frac{5}{\sqrt{11}}$, find the exact value of $\sin a°$.

3. If $90 < b < 180$ and $\sin b° = \frac{\sqrt{7}}{4}$, find the exact value of $\cos b°$.

4 What are the periods of the functions defined by:

a $f(x) = \sin x°$ ***b*** $g(x) = \sin 4x°$ ***c*** $h(x) = \cos \frac{1}{3}x°$

5 If the value of $\cos u°$ is v, express the following in terms of v.

a $\cos(180-u)°$ ***b*** $\cos(180+u)°$ ***c*** $\cos(360-u)°$ ***d*** $\cos(-u)°$

6 If the value of $\sin m°$ is n, express the following in terms of n:

a $\sin(180-m)°$ ***b*** $\cos(90-m)°$ ***c*** $\sin(180+m)°$ ***d*** $\sin(360+m)°$

7 Find the solution sets of the following for $0 < x < 360$:

a $\tan x° = 0{\cdot}070$ ***b*** $\cos x° = -0{\cdot}438$ ***c*** $\sin x° = 1{\cdot}322$

d $\tan x° = 1{\cdot}322$ ***e*** $\cos x° = 0{\cdot}714$ ***f*** $\sin x° = -0{\cdot}667$

8 Copy and complete this table:

Angle in degrees	180°	45°	225°			
Angle in radians				2π	$\pi/5$	$3\pi/2$

9 An arc of a circle is 1·5 cm long and subtends an angle of 3 radians at the centre of the circle. Calculate the radius of the circle.

10 An arc AB of a circle with centre O subtends an angle of 1·5 radians at the centre. Find, to 2 significant figures, the ratio length of arc AB: circumference of circle.

11 Find the solution sets of the following for $0 < x < 2\pi$:

a $\tan x = -1$ *b* $2\sin x = 1$ *c* $2\cos x + 1 = 0$

12 Without using tables, simplify:

a $\sin 63^\circ \cos 27^\circ + \cos 63^\circ \sin 27^\circ$ *b* $\cos 70^\circ \cos 65^\circ - \sin 70^\circ \sin 65^\circ$

13 If $\sin A = \frac{4}{5}$ and $\sin B = \frac{7}{25}$, where A and B are acute angles, prove that $\cos(A+B) = \frac{44}{125}$ and $\sin(A-B) = \frac{3}{5}$.

14 Given $\sin\alpha = \frac{3}{5}$ and $\tan\beta = \frac{5}{12}$, where α and β are acute angles, calculate: *a* $\cos(\alpha+\beta)$ *b* $\sin(\alpha-\beta)$.

15 If $\tan x = 3$ and $\tan y = \frac{1}{7}$, calculate the values of $\tan 2x$ and $\tan(2x-y)$. Deduce that $2x - y = \frac{3}{4}\pi$, assuming that $0 < x < \frac{1}{2}\pi$ and $0 < y < \frac{1}{2}\pi$.

16 Solve these equations, $0 \leqslant x \leqslant 360$, $x \in R$:

a $3\cos 2x^\circ - \cos x^\circ + 2 = 0$ *b* $5\cos 2x^\circ - 12\sin x^\circ + 11 = 0$

17 Prove that $\dfrac{\sin(\alpha-\beta)}{\cos\alpha\cos\beta} = \tan\alpha - \tan\beta$.

Revision Exercise 1B

1 *a* P is the point (3, 4) and Q is (15, 8). Use the formula for $\sin(\alpha-\beta)$ to find sin POQ as a fraction, where O is the origin.

b R is the point (4, 3) and S is (−5, 12). Find cos ROS.

2 Using $\frac{1}{12}\pi = \frac{1}{3}\pi - \frac{1}{4}\pi$, show that $\tan\frac{1}{12}\pi = 2 - \sqrt{3}$.

3 Given $2\cos(x+y) = \cos(x-y)$, prove that $\tan x \tan y = \frac{1}{3}$.

4 Show that solutions of the equation $2\sin 2\pi t - \sin\pi t = 0$ are given by the pair of equations $\sin\pi t = 0$, $\cos\pi t = \frac{1}{4}$.

5 If $\dfrac{\cos(30+t)^\circ}{\cos(30-t)^\circ} = n$, show that $\tan t^\circ = \sqrt{3}\left(\dfrac{1-n}{1+n}\right)$. Hence find t in the case where $n = 2$, giving all solutions between 0 and 360.

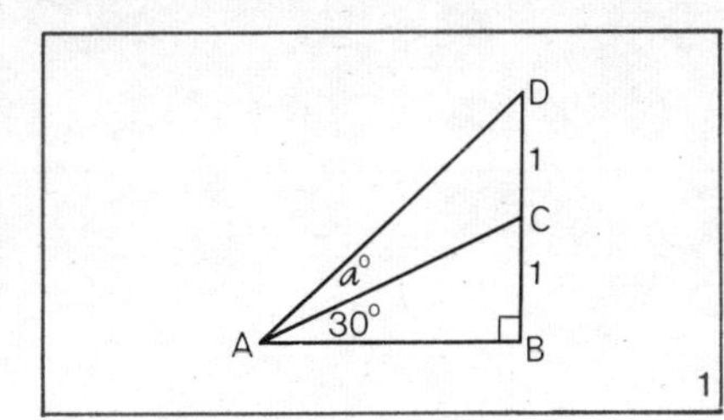

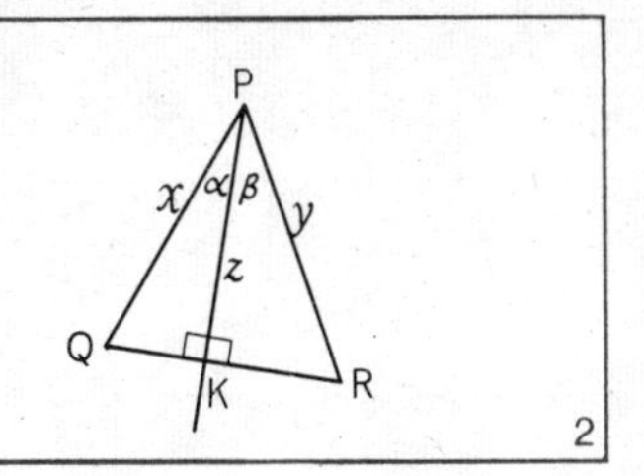

6 In Figure 1, BC = CD = 1 unit. Use the formula for $\sin(a+b)°$ to show that $\sqrt{3}\sin a° + \cos a° = 4/\sqrt{7}$.

7 Noting in Figure 2 that $\triangle PQR = \triangle PQK + \triangle PKR$, use the formula $\triangle = \frac{1}{2}bc\sin A$ to show that $\sin(\alpha+\beta) = \sin\alpha\cos\beta + \cos\alpha\sin\beta$.

8 Without using tables, find the value of $\cos 56° + \sin 56° \tan 28°$.

9 Prove the identities:

a $\sin(A+B)\sin(A-B) = \sin^2 A - \sin^2 B$ ***b*** $\dfrac{1+\cos 2x}{1-\cos 2x} = \dfrac{1}{\tan^2 x}$

10 If $p = \sin x + \sin y$ and $q = \cos x + \cos y$, show that $\frac{1}{2}(p^2+q^2) = 1 + \cos(x-y)$.

11 If $x = a\tan\theta$ and $y = a\tan 2\theta$, prove that $y(a^2-x^2) = 2a^2x$.

12 If $\sin(\alpha+\beta)\cos\theta = 2\sin\alpha\cos(\beta-\theta)$, prove $\tan\theta = \dfrac{\sin(\beta-\alpha)}{2\sin\alpha\sin\beta}$.

13 Prove that $\dfrac{2\sin x - \sin 2x}{2\sin x + \sin 2x} = \tan^2 \frac{1}{2}x$.

14 Show that $\cos 2\theta + 4\cos\theta$ can be expressed in the form $2(\cos\theta+1)^2 - 3$. Deduce that $-3 \leqslant \cos 2\theta + 4\cos\theta \leqslant 5$.

15 Use the method of question ***14*** to find the maximum and minimum values of: ***a*** $\cos 2\theta - 4\cos\theta$ ***b*** $\cos 2\theta - 2\cos\theta$ ***c*** $\cos 2\theta + \cos\theta$.

Calculus

Note to the Teacher on Chapter 1

Calculus involves a number of basic difficulties which arise from topological properties of the real numbers. Little progress can be made without the use of properties of limits, results on continuous functions, the mean-value theorem, the intermediate-value theorem, suprema and infina, etc. Even although it is impossible to consider these at the beginning we must, because of the needs of mathematics and its applications, give some idea of the power of calculus at a reasonably early stage. As in the past, our objective is simply to present certain elementary techniques and their applications in the differential (and later integral) calculus, but making more use than many previous books of the notion of *function.* We stress from the beginning that calculus is essentially concerned with *real functions*, i.e. mappings from R (the real numbers), or a subset of R, to R.

From the beginner's point of view the first difficulty in the differential calculus is that of *limit.* Although one may have met the idea of a type of limit in dealing with the sum to infinity of a geometric series, this does not really help with the 'neighbourhood limit' required in differential calculus. We introduce the necessary limit idea in *Section* 2 by means of the idea of speed at an instant, approached as usual through average speed over an interval of time. Note that the ideas and notation of *function* are used for clarity of exposition of the processes involved. Thus, associated with the equation $s = t^2$ is the real function defined by $f(x) = x^2$. Since the x is a dummy variable, any other letter can be used equally well, and here we have used t (since time is involved). From *Section* 2 the main fact that emerges is that the question of rate of change at a point is approached through the limit of the average change over an interval. In *Section* 3 we jump straight to a definition of rate of change of function $f: x \rightarrow f(x)$ at $x = a$ and introduce the derivative notation $f'(a)$. Practice in forming the difference ratio

$$\frac{f(a+h)-f(a)}{h} \ (h \neq 0), \text{ and } f'(a) = \lim_{h \to 0} \frac{f(a+h)-f(a)}{h}$$

will help to lay a strong foundation for the differential calculus. Clearly some facility with algebraic manipulation will be necessary. At this stage some practical problems involving rates of change have been included and these should indicate that the idea of a rate of change can arise in a variety of forms.

As a natural development from the evaluation of the real number $f'(a)$ the derived function f' is introduced in *Section* 4, and this is followed, in *Section* 5, by the determination of the derived function when f is defined by $f(x) = ax^n$ (a a constant and n a rational number), or by a sum of such expressions. In reaching this stage we have required and assumed implicitly certain results on limits such as the following:

(a) If $\lim_{h\to 0} f(h) = A$ and $\lim_{h\to 0} g(h) = B$, then

(i) $\lim_{h\to 0} [f(h) \pm g(h)] = A \pm B$

(ii) $\lim_{h\to 0} [cf(h)] = cA$ (c a constant)

(iii) $\lim_{h\to 0} f(h) . g(h) = A . B$

(iv) $\lim_{h\to 0} \dfrac{1}{f(h)} = \dfrac{1}{A}$ (provided $A \neq 0$)

(b) If f is differentiable (and so continuous), $\lim_{h\to 0} f(x+h) = f(x)$;

e.g. if n is a positive integer,

(i) $\lim_{h\to 0} h^n = 0$ (ii) $\lim_{h\to 0} (x+h)^n = x^n$.

It may be noted that $\lim_{h\to 0} \dfrac{f(x+h)-f(x)}{h} = L$ means:

'$\left|\dfrac{f(x+h)-f(x)}{h} - L\right|$ can be made as small as we please by taking $|h|(\neq 0)$ small enough.'

Here $h = 0$ is omitted, and for this reason the neighbourhood of x is said to be 'punctured'. The pupil need not of course be concerned with the finer points of the definition at this stage. Looking ahead to the chain rule it is important to build up a feeling for the pair of functions f, f'; we can think of: $f(\ \ ) = (\ \ )^n, f'(\ \ ) = n(\ \ )^{n-1}$ with the idea that we have something to insert in the brackets, e.g. $f(x) = x^n, f'(x) = nx^{n-1}; f(g(x)) = (g(x))^n, f'(g(x)) = n(g(x))^{n-1} . g'(x)$, and so on. Again the pupils will not be aware of this type of thinking, but sowing the seeds will make the chain rule more acceptable.

In *Section* 6 the derivative is related to the gradient of the tangent at a point on a graph, and the differential notation $\frac{df}{dx}$ is introduced; in particular, from $y=f(x)$ we write $\frac{dy}{dx}=f'(x)$. If we are given a curve, e.g. $y = x^3 - 6x$, it is obviously unnecessary to introduce the function f for which $f(x) = x^3 - 6x$ in finding the tangent at the point $(1, -5)$; we write $\frac{dy}{dx} = 3x^2 - 6$ and so the gradient at $x = 1$ has value -3, etc. It is also important to have the notation $\frac{d}{dx}(x^3 - 6x)$ available in order to avoid pedantic use of function notation.

The remainder of the chapter is concerned with applying differentiation to the study of functions and their graphs, to determining intervals of increase and of decrease, stationary values, maxima and minima, and the interesting and difficult problem of sketching graphs themselves. Finally, in the last two sections, there are collections of practical problems involving the determination of maxima and minima and discussion of velocity, acceleration and other rates of change. Many of these illustrate the important concept of mathematical model building, in which, with the introduction of suitable variables, a practical problem is represented by a mathematical problem; the solution of the mathematical problem can then be used to obtain information about the practical problem. An effort has been made to stress that maxima and minima need not occur at stationary points, but that nevertheless calculus methods together with graphical considerations are important in obtaining maximum and minimum values of functions.

1 An Introduction: The Differential Calculus

1 Introduction

(i) How fast will a chemical reaction go? How fast can a population of bacilli be expected to grow? How fast does heat travel along a poker? What force is needed to accelerate a car from rest to 50 km/h in 15 seconds? What are the dimensions of the open box with largest volume that can be made with a given area of cardboard?
(ii) How can the formulae for the area of a circle and the volume of a sphere be obtained? What speed does a rocket need if it is to escape from the earth's gravitational field? What is the best shape for a girder if it is to be strong but light? If a liquid cools at a rate proportional to the temperature, how long will it take for the temperature to reach half its initial value? What is the mass of the atmosphere?

These are a few of the many problems that can be investigated by means of the Calculus. The problems in the first group are concerned mainly with the idea of *rate of change*, which is the basis of the *differential calculus*; the systematic study of such problems started around the beginning of the seventeenth century. The problems in the second group are concerned mainly with the idea of *summation*, which is the basis of the *integral calculus*; some fairly elementary but basic ideas of this type were developed more than 2000 years ago by Greek mathematicians such as Archimedes.

Although a number of seventeenth-century mathematicians contributed to the development of the ideas and methods of the subject, the main honours for this work are shared by the English mathematician Newton and the German mathematician Leibniz. However, the subject was put on a firm mathematical basis only during the nineteenth century.

2 *Rate of change; the idea of a limit*

Speed; rate of change of distance with respect to time

We are all familiar with statements such as 'The car was travelling at 20 metres per second when it passed the police car'.

What does the phrase '20 m/s' (at the time in question) mean? If the car had been travelling at a *steady* speed of 20 m/s it would have covered 20 metres in one second, 40 metres in two seconds, and so on; in general, the distance s metres covered in each period of t seconds would be given by $s = 20t$, so that $20 = \frac{s}{t}$, i.e. speed $= \frac{\text{distance travelled}}{\text{time taken}}$.

If the speed is *not steady*, the situation is more complicated. As a particular case, suppose that a car starts at a point O and is at a distance s metres from O along its path at a time t seconds later; and suppose that the distance s metres at time t seconds is given by the formula $s = t^2$.

We note that $\frac{s}{t} = t$, so that $\frac{s}{t}$ is not constant, and the car is not travelling at a steady speed. But we feel, from our experience, that the car will have some speed at each instant, as shown on the speedometer, for example.

We might ask 'What is the speed in m/s at $t = 1$?', i.e. the speed at the end of 1 second from its starting time $t = 0$. The answer to this question clearly depends on the formula $s = t^2$ as well as on $t = 1$.

This equation associates a non-negative real number s with each real number t. Hence the formula $s = t^2$ defines a function whose domain is R and whose range is the set of non-negative real numbers. This function is the function $f: t \to t^2$ $(t \in R)$, and it may be described as the function associated with the equation $s = t^2$. The speed which we are investigating is concerned with this function at $t = 1$.

To find a meaning for the speed at $t = 1$ when $s = t^2$
Using $f: t \to t^2$ to denote the function associated with the formula $s = t^2$ we have $f(t) = t^2$, which gives the distance travelled in time t from $t = 0$.

The distance travelled from $t = 0$ to $t = 1$ is $f(1) = 1^2 = 1$ m.
The distance travelled from $t = 0$ to $t = 2$ is $f(2) = 2^2 = 4$ m.
The *average speed* over thc time interval $t = 1$ to $t = 2$

$$= \frac{\text{change in distance}}{\text{change in time}} = \frac{f(2)-f(1)}{2-1} = \frac{4-1}{1} = 3 \text{ m/s.}$$

In the same way, the average speed from $t = 1$ to $t = 1{\cdot}5$

$$= \frac{f(1{\cdot}5)-f(1)}{1{\cdot}5-1} = \frac{1{\cdot}5^2-1^2}{0{\cdot}5} = \frac{2{\cdot}25-1}{0{\cdot}5} = \frac{1{\cdot}25}{0{\cdot}5} = 2{\cdot}5 \text{ m/s.}$$

To try to find an answer to the question 'What is the speed at $t = 1$?' we make a table of average speeds in the short intervals from $t = 1$ to $t = 1+h$ for smaller and smaller positive values of h.

To do this we calculate $\dfrac{f(1+h)-f(1)}{(1+h)-1}$ for each h, obtaining:

h	0·05	0·04	0·03	0·02	0·01
$\dfrac{f(1+h)-f(1)}{h}$	2·05	2·04	2·03	2·02	2·01

A glance at the second row shows that the average speed $\dfrac{f(1+h)-f(1)}{h}$ over the interval from $t = 1$ to $t = 1+h$ seems to be very close to 2(m/s) for small positive h. It can be shown that the same is true for small negative h.

In fact the ratio can be made as close as we wish to 2 by taking $h(\neq 0)$ sufficiently small. Neither the above table, nor any we could make, is sufficiently detailed to prove this, but we can prove it as follows.

Since $f(t) = t^2$, and assuming $h \neq 0$,

$$\frac{f(1+h)-f(1)}{h} = \frac{(1+h)^2-1^2}{h} = \frac{(1+2h+h^2)-1}{h} = \frac{h(2+h)}{h} = 2+h.$$

Hence $\dfrac{f(1+h)-f(1)}{h}$ can be made as close as we wish to 2 by taking h sufficiently small. For example, if we wish the ratio to be even nearer to 2 than 2·001, $h = 0{\cdot}0001$ would do.

We say that $\dfrac{f(1+h)-f(1)}{h}$ has *limit* 2 as h tends to 0, and we write $\displaystyle\lim_{h\to 0}\frac{f(1+h)-f(1)}{h} = 2.$

This limit evidently provides an answer to our question 'What is the speed of the car at $t = 1$?' In other words, our feeling that it made sense to talk about 'speed at an instant' is justified if we define this speed by a *limit*. The speed at $t = 1$ is 2 m/s.

(In this context, the arrow in '$h \to 0$' has no connection with mapping. '$h \to 0$' means 'h tends to zero', which is a way of stating 'taking h sufficiently small'.)

Example. Find the speed of the car above at the end of 10 seconds from the start.

The speed at $t = 10$ is given by $\displaystyle\lim_{h\to 0}\frac{f(10+h)-f(10)}{h}$.

$$\frac{f(10+h)-f(10)}{h} = \frac{(10+h)^2-10^2}{h} = \frac{100+20h+h^2-100}{h}$$

$$= \frac{h(20+h)}{h} = 20+h,\ h \neq 0.$$

$$\lim_{h\to 0}\frac{f(10+h)-f(10)}{h} = \lim_{h\to 0}(20+h) = 20.$$

So the speed after 10 seconds is 20 m/s.

Exercise 1

1 From the above example in which $f(t) = t^2$, calculate the *average speed* over the intervals from:

a $t = 2$ to $t = 2{\cdot}5$, by evaluating $\dfrac{f(2{\cdot}5)-f(2)}{2{\cdot}5-2}$.

b $t = 2$ to $t = 2{\cdot}1$.

c $t = 2$ to $t = 2+h$. Deduce the speed at $t = 2$ by evaluating $\displaystyle\lim_{h\to 0}\frac{f(2+h)-f(2)}{h}$.

2 For the same car, calculate the *average speed* over the intervals from:

a $t = 4$ to $t = 4{\cdot}5$ *b* $t = 4$ to $t = 4{\cdot}1$

c $t = 4$ to $t = 4+h$. Deduce the speed at $t = 4$.

3 Suppose that the formula connecting s and t is $s = 2t^2$, so that $f(t) = 2t^2$.

a Find the average rate of change of s with respect to t from $t = 1$ to $t = 1{\cdot}2$ (i.e. the average speed over the interval from $t = 1$ to $t = 1{\cdot}2$) by evaluating $\dfrac{f(1{\cdot}2)-f(1)}{1{\cdot}2-1}$.

b Find the average rate of change of s with respect to t from $t = 1$ to $t = 1+h$ by evaluating $\dfrac{f(1+h)-f(1)}{h}$. Deduce the speed at $t = 1$.

c Calculate the speed at $t = 3$ by evaluating $\lim\limits_{h\to 0}\dfrac{f(3+h)-f(3)}{h}$.

4 The distance s metres travelled in t seconds by a falling object is given approximately by the formula $s = 5t^2$, so that $f(t) = 5t^2$.

a Calculate the speed after falling 4 seconds from rest by evaluating $\lim\limits_{h\to 0}\dfrac{f(4+h)-f(4)}{h}$.

b Calculate the speed of the object after 8 seconds.

3 Rate of change of value of function $f{:}x \to f(x)$ at $x=a$; derivative of f at $x=a$

In Section 2 we saw that to deal with speed at a given instant we first had to consider average speed over a time interval. This involved a function of time t and a ratio $\dfrac{\text{change in value of function}}{\text{change in value of } t}$

This idea can be extended as follows. Suppose that we have a function $f{:}x \to f(x)$; in the interval $a \leqslant x \leqslant a+h$, the value of f changes from $f(a)$ at $x = a$ to $f(a+h)$ at $x = a+h$, and the *average rate of change* in value of f with respect to x in the interval from

$x = a$ to $x = a+h$ $(h \neq 0)$ is

$$\frac{\text{change in value of function}}{\text{change in variable}}, \text{ i.e. } \frac{f(a+h)-f(a)}{(a+h)-a}, \text{ i.e. } \frac{f(a+h)-f(a)}{h}.$$

Figure 1 shows the graph of f and the changes in variable and in value of function.

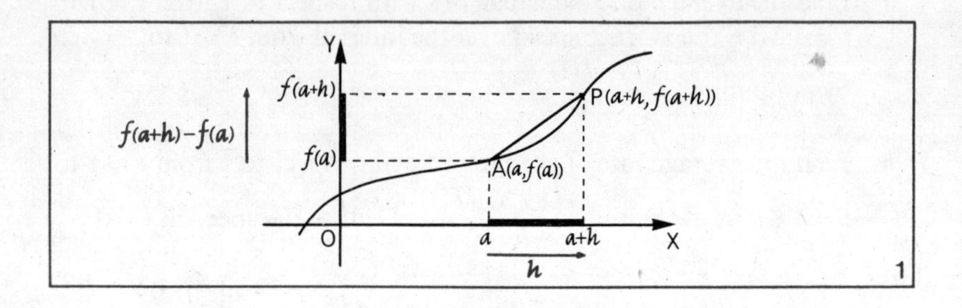

By the limiting process introduced in Section 2 we can now form $\lim_{h\to 0}\frac{f(a+h)-f(a)}{h}$. *This limit is called the rate of change of f at $x = a$. The limit ('derived from $f(x)$') is denoted by $f'(a)$, and is often called the derivative of f at $x = a$.*

In the examples in Section 2 the rate of change was the speed of an object at a particular instant.

In Section 6 we will find a geometrical meaning for the derivative.

Example. Find the derivative of the function f defined by $f(x) = x^2+5$ at $x = 3$.

The derivative of f at $x = 3$ is $f'(3) = \lim_{h\to 0}\frac{f(3+h)-f(3)}{h}$.

$$\frac{f(3+h)-f(3)}{h} = \frac{(3+h)^2+5-(3^2+5)}{h} = \frac{9+6h+h^2+5-9-5}{h}$$

$$= \frac{h(6+h)}{h} = 6+h,\ h \neq 0.$$

$$f'(3) = \lim_{h\to 0}\frac{f(3+h)-f(3)}{h} = \lim_{h\to 0}(6+h) = 6.$$

Exercise 2

In questions *1–6* find the derivative of each function f at the element x stated:

1 $f(x) = x^2$, at $x = 1$

2 $f(x) = 2x$, at $x = 5$

3 $f(x) = 2x+1$, at $x = 4$

4 $f(x) = x^2+1$, at $x = 3$

5 $f(x) = 4x^2$, at $x = 2$

6 $f(x) = x^2+x$, at $x = 7$

7 If $g(x) = x^2+4x$, find $g'(2)$.

8 If $h(x) = 2x^2+1$, find $h'(1)$.

9 The area $A\,\text{cm}^2$ of a square of side $x\,\text{cm}$ is given by $A = x^2$, which defines a function f given by $f(x) = x^2$. Find the rate of change of area with respect to x when the side is 5 cm long, by finding $f'(5)$.

10 The circumference C cm of a circle of radius r is given by $C = 2\pi r$, so $f(r) = 2\pi r$. Find the rate of change of circumference with respect to r when the radius is 3 cm.

4 *The derived function, or derivative, of f; differentiation*

Let a function f have a derivative for each element of domain D. If $a, b, \ldots \in D$, then

$$f'(a) = \lim_{h\to 0}\frac{f(a+h)-f(a)}{h}, \quad f'(b) = \lim_{h\to 0}\frac{f(b+h)-f(b)}{h}, \quad \ldots,$$

and *for each element of D we obtain the corresponding value of f'*. This gives us a new function f', with domain D, called the *derived function* of f, or the *derivative* of f.

Hence the derived function f' is defined by the formula

$$f'(x) = \lim_{h\to 0}\frac{f(x+h)-f(x)}{h}.$$

The process of obtaining f' from f is called *differentiation*, and we say that we *differentiate* f to obtain f'. In using the above formula we are finding the derivative *from first principles*.

Example 1. Find the derivative of the function f defined by $f(x) = x^3$, from first principles.

$$\frac{f(x+h)-f(x)}{h} = \frac{(x+h)^3-x^3}{h} = \frac{x^3+3x^2h+3xh^2+h^3-x^3}{h}$$

$$= \frac{h(3x^2+3xh+h^2)}{h} = 3x^2+3xh+h^2,\ h \neq 0.$$

$$f'(x) = \lim_{h\to 0}\frac{f(x+h)-f(x)}{h} = \lim_{h\to 0}(3x^2+3xh+h^2) = 3x^2.$$

Example 2. Given $f(x) = \frac{1}{x^2}$, find $f'(x)$ from first principles.

$$\frac{f(x+h)-f(x)}{h} = \frac{\frac{1}{(x+h)^2}-\frac{1}{x^2}}{h} = \frac{x^2-(x^2+2xh+h^2)}{hx^2(x+h)^2} = \frac{-2x-h}{x^2(x+h)^2},$$

$$h \neq 0.$$

$$f'(x) = \lim_{h\to 0}\frac{f(x+h)-f(x)}{h} = \lim_{h\to 0}\left(\frac{-2x-h}{x^2(x+h)^2}\right) = \frac{-2x}{x^2 . x^2} = -\frac{2}{x^3}.$$

(In this example, we require $x \neq 0$. The same restriction applies in questions ***9–12*** of Exercise 3.)

Exercise 3

Use the formula $f'(x) = \lim_{h\to 0}\frac{f(x+h)-f(x)}{h}$ to verify the value of $f'(x)$ for each $f(x)$ in questions **1–12**.

1 $f(x) = x;\ f'(x) = 1$

2 $f(x) = 3x;\ f'(x) = 3$

3 $f(x) = x^2;\ f'(x) = 2x$

4 $f(x) = 5x^2;\ f'(x) = 10x$

5 $f(x) = 3x^2+1;\ f'(x) = 6x$

6 $f(x) = 3;\ f'(x) = 0$

7 $f(x) = 2x-5;\ f'(x) = 2$

8 $f(x) = 2x^3;\ f'(x) = 6x^2$

9 $f(x) = \frac{1}{x};\ f'(x) = -\frac{1}{x^2}$

10 $f(x) = \frac{2}{x^2};\ f'(x) = -\frac{4}{x^3}$

11 $f(x) = x+\frac{1}{x};\ f'(x) = 1-\frac{1}{x^2}$

12 $f(x) = 1+\frac{1}{x^2};\ f'(x) = -\frac{2}{x^3}$

5 Some particular derived functions

(i) The derivative of a constant function

If $f(x) = c$, where c is a constant, then

$$f'(x) = \lim_{h\to 0}\frac{f(x+h)-f(x)}{h} = \lim_{h\to 0}\frac{c-c}{h} = 0.$$

The derivative of a constant function is zero.

(ii) The derivative of a positive power of x, i.e. x^n (n a positive integer)

From Example 1 and Exercise 3 in Section 4, we have:

$f(x)$	x	x^2	x^3
$f'(x)$	1	$2x$	$3x^2$

If you look closely at the pattern developing, you should be able to write down the derivatives of x^4, x^5, ...

Using the method of the worked example in Section 4 it can be shown that the derivative of x^n (n a positive integer) is nx^{n-1}.

(You would have to satisfy yourself that $(x+h)^n = x^n + nx^{n-1}h + h^2(.....)$, where $(.....)$ is some expression in x and h, and so $\dfrac{(x+h)^n - x^n}{h} = nx^{n-1} + h(.....)$. Then proceed as usual.)

If $f(x) = x^n$, then $f'(x) = nx^{n-1}$, where n is a positive integer.

(iii) The derivative of ax^n (n a positive integer)

If $f(x) = ax^2$, where a is a constant, then

$$\frac{f(x+h)-f(x)}{h} = \frac{a(x+h)^2 - ax^2}{h} = \frac{ax^2 + 2axh + ah^2 - ax^2}{h} =$$

$$= \frac{ah(2x+h)}{h} = a(2x+h)$$

$$f'(x) = \lim_{h\to 0}\frac{f(x+h)-f(x)}{h} = \lim_{h\to 0} a(2x+h) = 2ax.$$

In general, if $f(x) = cg(x)$, then $f'(x) = cg'(x)$, where c is a constant.

Also, if $f(x) = g(x) + h(x)$, then $f'(x) = g'(x) + h'(x)$. This can be extended to the sum of any finite number of terms.

Example 1. Given $f(x) = x^3 - 2x + 6$, find $f'(x)$ and $f'(-1)$.

$$\begin{aligned} f(x) &= x^3 - 2x + 6 \\ \Rightarrow \quad f'(x) &= 3x^2 - 2 \\ \Rightarrow \quad f'(-1) &= 3 - 2 = 1. \end{aligned}$$

Example 2. Given $f(x) = (x^2 - 3)^2$, find $f'(x)$ and the rate of change of f at $x = 2$.

At present we can differentiate only terms of the form ax^n, so we must first expand $(x^2 - 3)^2$.

$$\begin{aligned} f(x) &= (x^2 - 3)^2 = x^4 - 6x^2 + 9 \\ \Rightarrow \quad f'(x) &= 4x^3 - 12x \\ \Rightarrow \quad f'(2) &= 32 - 24 = 8. \end{aligned}$$

The rate of change of f at $x = 2$ is 8.

Exercise 4

Find the derivatives of the following (e.g. $f(x) = x^4 \Rightarrow f'(x) = 4x^3$).

1 x^6 2 x^8 *3* $4x^3$ 4 $\frac{1}{2}x^2$ 5 $\frac{1}{4}x^4$

6 $-2x^5$ 7 ax^3 *8* $5x$ 9 5 *10* 1

11 $x^2 + 2x + 3$ *12* $x^3 - 7x^2 + 2$ *13* $4x^4 - x^2 + 9$

14 $\frac{1}{3}x^3 + \frac{1}{2}x^2 + 1$ *15* $6 - 4x^5 + 2x^9$ *16* $ax^2 + bx + c$

17 $(x+1)^2$ *18* $(x-2)^2$ *19* $(3x+4)^2$

20 $(x+3)(2x-1)$ *21* $(5x+1)(5x-1)$ 22 $(2-x)(3-x)$

23 $(x^2+2)^2$ 24 $(x^3-1)^2$ 25 $(x^4+3)^2$

26 $(x+1)(x-2)^2$ 27 $(x+3)(x-3)(2x+5)$ 28 $(x+4)^3$

29 Given $f(x) = 3 + x - x^2$, find the values of $f'(0), f'(\frac{1}{2}), f'(1), f'(-10)$.

30a Given the function $f: x \rightarrow (x^3 - 2)^2$, find a formula for the derived function f'.

b Find the rate of change of f at $x = -1$ and at $x = 2$.

31 Given $f(x) = x^2 - 4x + 1$, find x for which:

a $f'(x) = 0$ *b* $f'(x) = 2$ *c* $f'(x) > 0$.

32 Given $f(x) = \frac{1}{3}x^3 + \frac{1}{2}x^2 - 6x$, find x for which:

a $f'(x) = 0$ *b* $f'(x) = -4$ *c* $f'(x) < 0$.

33 For each of the following, find f' and sketch the graphs of f and f'.

a $f: x \to x$ *b* $f: x \to x^2$ *c* $f: x \to x^3$ *d* $f: x \to x^4$.

34 For a marble moving in a groove the distance s cm from one end at time t seconds is given by $s = 5t - t^2$ so $f(t) = 5t - t^2$.

a Find the speed of the marble at $t = 2$.
b Find t when the speed of the marble is zero.

35 A ball thrown vertically upwards with an initial speed of 30 m/s moves according to the equation $h = 30t - 5t^2$, where h is the height in metres above the starting point after t seconds. *a* Find its speed after 1·5 seconds. *b* When does the ball stop rising?

(iv) The derivatives of negative and rational powers of x

From Example 2 and Exercise 3 in Section 4, we have:

$f(x)$	$\frac{1}{x}$, i.e. x^{-1}	$\frac{1}{x^2}$, i.e. x^{-2}
$f'(x)$	$-\frac{1}{x^2}$, i.e. $-x^{-2}$	$-\frac{2}{x^3}$, i.e. $-2x^{-3}$

If you look closely at the pattern developing you should be able to write down the derivatives of $x^{-3}, x^{-4}, \ldots$

Using the method of the worked example in Section 4 it can be shown that the derivative of x^{-n} ($x \neq 0$, and n a positive integer) is $-nx^{-n-1}$, or $-\frac{n}{x^{n+1}}$.

It is in fact true that for all rational powers of x:

If $f(x) = x^n$, then $f'(x) = nx^{n-1}$, where n is a rational number.

Example 1. Given $f(x) = \frac{1}{3x^2}$, find $f'(x)$

$$f(x) = \frac{1}{3x^2} = \frac{1}{3} \times \frac{1}{x^2} = \frac{1}{3}x^{-2}$$

$$\Rightarrow \quad f'(x) = \frac{1}{3}(-2x^{-3}) = -\frac{2}{3x^3}$$

Example 2. Given $f(x) = \left(2x + \dfrac{1}{2x}\right)^2$, find $f'(x)$.

$$f(x) = \left(2x + \frac{1}{2x}\right)^2 = 4x^2 + 2 + \frac{1}{4x^2} = 4x^2 + 2 + \tfrac{1}{4}x^{-2}$$

$$\Rightarrow \quad f'(x) = 8x - \tfrac{2}{4}x^{-3} = 8x - \frac{1}{2x^3}.$$

Example 3. Given $f(x) = \dfrac{x+1}{\sqrt{x}}$, find $f'(x)$.

$$f(x) = \frac{x+1}{\sqrt{x}} = \frac{x}{x^{1/2}} + \frac{1}{x^{1/2}} = x^{1/2} + x^{-1/2}$$

$$\Rightarrow \quad f'(x) = \tfrac{1}{2}x^{-1/2} - \tfrac{1}{2}x^{-3/2} = \frac{1}{2x^{1/2}} - \frac{1}{2x^{3/2}}.$$

Exercise 5

In this Exercise express all answers with positive indices, as in the Worked Examples.

Find the derivative of each of the following:

1	$x^{3/2}$	2	$x^{4/3}$	3	$x^{5/2}$	4	$x^{1/2}$	5	$x^{1/3}$
6	x^{-1}	7	x^{-2}	8	x^{-6}	9	$2x^{-3}$	10	$\tfrac{1}{2}x^{-4}$

Express each of the following in the form ax^n, and then find its derivative. $\left(\textit{Reminders.}\ \sqrt[n]{x^m} = x^{m/n};\ \dfrac{1}{x^n} = x^{-n}\right)$.

11	$\sqrt{x}$	12	$\sqrt[3]{x}$	13	$\sqrt[3]{x^2}$	14	$\dfrac{1}{x^2}$	15	$\dfrac{1}{x^4}$
16	$\dfrac{1}{x}$	17	$\dfrac{2}{x}$	18	$\dfrac{1}{\sqrt{x}}$	19	$\dfrac{1}{\sqrt[3]{x}}$	20	$\dfrac{2}{x^3}$
21	$\dfrac{1}{2x^{1/2}}$	22	$\dfrac{2}{3x^2}$	23	$\dfrac{4}{3x^3}$	24	$\dfrac{1}{5x^4}$	25	$\dfrac{1}{2\sqrt{x}}$

Express each of the following as a sum of terms of the form ax^n, and then find its derivative:

26	$x + \dfrac{1}{x}$	27	$\sqrt{x} + \dfrac{1}{\sqrt{x}}$	28	$2x - \dfrac{2}{x}$

29 $2x^2 - \frac{1}{4x^2}$ **30** $x^2 + 6 - \frac{1}{x^2}$ **31** $\frac{x}{5} + \frac{5}{x}$

32 $2x^4 + \frac{1}{2x^3}$ **33** $8x^{3/4} - \frac{6}{x^{2/3}}$ **34** $x^2(1+\sqrt{x})$

35 $\sqrt{x}(1-\sqrt{x})$ **36** $\left(x+\frac{1}{x}\right)^2$ **37** $\left(x^2-\frac{1}{x^2}\right)^2$

38 $\left(5x+\frac{5}{x}\right)^2$ **39** $\left(\sqrt{x}-\frac{1}{\sqrt{x}}\right)^2$ **40** $\left(x+\frac{1}{x}\right)\left(x-\frac{1}{x}\right)$

41 $\frac{x-4}{x}$ **42** $\frac{x^2+2x-1}{x}$ **43** $\frac{2x^3-3x^2+4}{x^3}$

44 $\frac{x-1}{x^{1/2}}$ **45** $\frac{2+4x}{x^2}$ **46** $\frac{x+1}{2x^{1/3}}$

47 $\frac{(1+x)^2}{2x}$ **48** $\frac{(1-x)(2-x)}{x}$ **49** $\frac{x^2-4x}{x\sqrt{x}}$

50 Given $f(x) = 4x^{3/2}$, find the values of:

a $f'(0)$ *b* $f'(1)$ *c* $f'(4)$ *d* $f'(\frac{1}{9})$

51 Given $f(x) = \frac{1}{x^2}$, find the values of:

a $f'(1)$ *b* $f'(-1)$ *c* $f'(2)$ *d* $f'(-\frac{1}{2})$

52 If $f(x) = \left(x+1+\frac{1}{x}\right)\left(x+1-\frac{1}{x}\right)$, find $f'(x)$.

53 If $f(x) = \left(2x^3+\frac{1}{x}\right)^2$, find $f'(x)$.

54 Calculate the rate of change of the function $f: x \rightarrow \sqrt[3]{x} + \frac{1}{\sqrt[3]{x}}$ at $x = 8$.

55 Given $f(x) = x^{1/2}\left(x+\frac{1}{x}\right)\left(x-\frac{1}{x}\right)$, show that $f'(x) = \frac{5x^4+3}{2x^{5/2}}$.

6 *A geometrical interpretation of the derivative*

In this section we change the emphasis from the function itself to the Cartesian graph of the function.

From the graph of the function f, shown in Figure 2, we can obtain a geometrical meaning for $\lim_{h\to 0} \dfrac{f(x+h)-f(x)}{h}$.

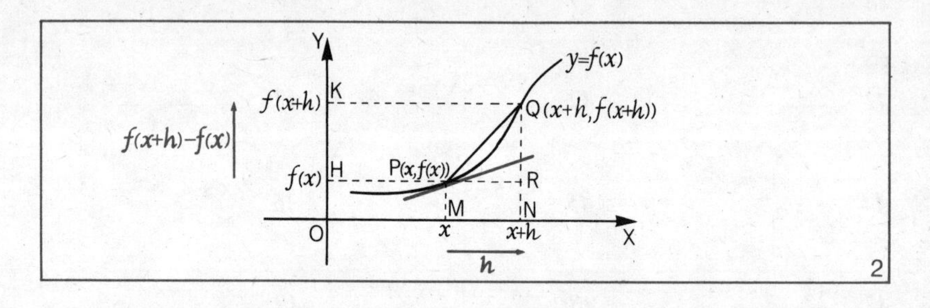

$$\frac{f(x+h)-f(x)}{h} = \frac{\text{HK}}{\text{MN}} = \frac{\text{RQ}}{\text{PR}} = \text{gradient of chord PQ.}$$

$$\text{Hence } \lim_{h\to 0} \frac{f(x+h)-f(x)}{h} = \lim_{h\to 0} (\text{gradient of chord PQ})$$

$$= \text{gradient of tangent at P,}$$

since by taking h sufficiently small the direction of the chord through P can be made as close as we please to the direction of the tangent at P.

$$f'(x) = \lim_{h\to 0} \frac{f(x+h)-f(x)}{h} = \text{gradient of tangent at point } (x, f(x))$$

on the graph of f.

Another notation for the derivative

In many applications of calculus it is important to stress that we are dealing with a change in x, and instead of h the symbol Δx ('delta x') is often used, as shown in Figure 3.

Δx = change, or *increment*, in x. The corresponding increment in the value of f is denoted by Δf. So $\Delta f = f(x+\Delta x)-f(x)$.

Then $f'(x) = \lim\limits_{\Delta x \to 0} \dfrac{f(x+\Delta x)-f(x)}{\Delta x} = \lim\limits_{\Delta x \to 0} \dfrac{\Delta f}{\Delta x}$, which is denoted by $\dfrac{df}{dx}$ ('df by dx'), notation which was introduced by Leibniz.

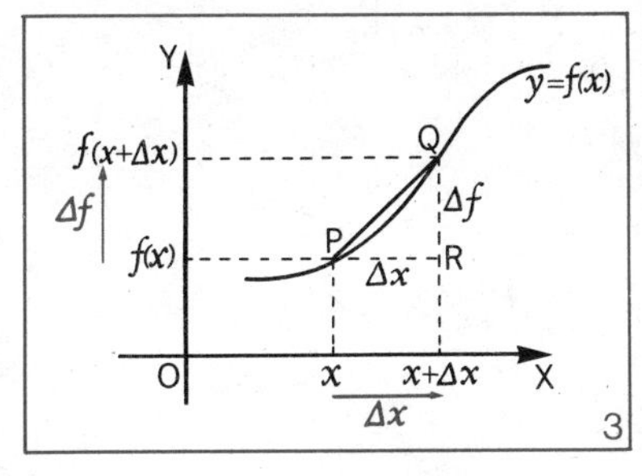

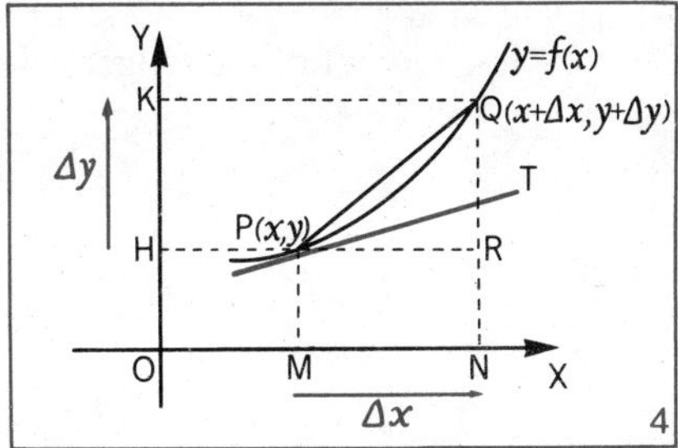

As another illustration, consider the curve $y = f(x)$ shown in Figure 4. If $P(x, y)$ is any point on the curve, we can denote the coordinates of a neighbouring point Q on the curve by $(x+\Delta x, y+\Delta y)$. Then $y+\Delta y = f(x+\Delta x)$.

Then $f'(x) = \lim\limits_{\Delta x \to 0} \dfrac{f(x+\Delta x)-f(x)}{\Delta x} = \lim\limits_{\Delta x \to 0} \dfrac{(y+\Delta y)-y}{\Delta x} = \lim\limits_{\Delta x \to 0} \dfrac{\Delta y}{\Delta x}$

$\dfrac{\Delta y}{\Delta x} = \dfrac{HK}{MN} = \dfrac{RQ}{PR}$ = gradient of chord PQ.

So $\lim\limits_{\Delta x \to 0} \dfrac{\Delta y}{\Delta x} = \lim\limits_{\Delta x \to 0}$ (gradient of chord PQ)

= gradient of tangent at point $P(x, y)$ on curve $y = f(x)$.

Denoting $\lim\limits_{\Delta x \to 0} \dfrac{\Delta y}{\Delta x}$ by $\dfrac{dy}{dx}$ ('dy by dx'), we have

$$\frac{dy}{dx} = f'(x) = \text{gradient of tangent to curve } y = f(x) \text{ at point } (x, y).$$

Example 1. Find the gradient, and hence the equation, of the tangent to the parabola $y = 3x^2+4x-5$ at the point $(1, 2)$.

$$y = 3x^2+4x-5$$

$$\Rightarrow \quad \frac{dy}{dx} = 6x+4 = 10 \text{ at } x = 1.$$

So the gradient of the tangent at the point (1, 2) is 10.
The equation of the tangent is $y-2 = 10(x-1)$

$$\Leftrightarrow \quad y = 10x-8$$

Example 2. Find the points on the curve $y = \sqrt[3]{x}$ at which the tangents are perpendicular to the line with equation $12x+y = 1$.

$$12x+y = 1 \quad \Leftrightarrow \quad y = -12x+1$$

The gradient of the tangents is $\frac{1}{12}$. $(m_1 m_2 = -1)$.

$$y = x^{1/3}$$

$$\Rightarrow \quad \frac{dy}{dx} = \frac{1}{3}x^{-2/3} = \frac{1}{3x^{2/3}}$$

$$\frac{1}{3x^{2/3}} = \frac{1}{12}$$

$$\Leftrightarrow \quad x^{2/3} = 4$$

$$\Leftrightarrow \quad x = 4^{3/2} = \pm 2^3 = \pm 8$$

The points are (8, 2) and (−8, −2).

Exercise 6

Find the gradient, and hence the equation, of the tangent to each of the following curves at the given point:

1 $y = x^2$, at (3, 9)

2 $y = 5x$, at (−1, −5)

3 $y = \sqrt{x}$, at (4, 2)

4 $y = \frac{1}{x}$ at (1, 1)

Find the equations of the tangents to the curves:

5 $y = x^3$, at $x = -2$

6 $y = x^2-3x+2$, at $x = 1$

7 $y = 1-x^2$, at $x = 0$

8 $y = x^3+3x-5$, at $x = 2$

9 $y = 2x^4$, at $x = -1$

10 $y = (2x-3)(x-1)$, at $x = 3$

11 $y = 10\sqrt{x}$, at $x = 25$

12 $y = \sqrt[4]{x}$, at $x = 16$

13 $y = \frac{4}{x^2}$, at $x = 1$

14 $y = x+\frac{1}{x}$, at $x = \frac{1}{2}$

15 Find the equation of the tangent to the curve $y = \frac{1}{4}x^2$ at the point given by $x = 2$. Show that if the tangent cuts the axes at P and Q the midpoint of PQ is $(\frac{1}{2}, -\frac{1}{2})$.

16 Find the equations of the tangents to the curve $y = 2x^2$ at the points given by $x = 1$ and $x = -1$, and then find the point of intersection of these tangents.

17 Show that there is one tangent to the curve $y = x^2+5$ which has gradient 4, and find its equation.

18 Show that the tangents to the curve $y = x^3$ at the points where $x = 1$ and $x = -1$ are parallel. Find the coordinates of the points where these tangents cut the x- and y-axes.

19 Find the equations of the tangents to the parabola $y = x^2+2$ at the points A and B with x-coordinates -1 and 2 respectively. Show that the tangents intersect at a point on the x-axis.

20 The tangent to the parabola $y = (x-2)^2$ at the point $(4, 4)$ cuts the x-axis at P and the y-axis at Q. Find the length of PQ.

21 Find the coordinates of the point on the curve $y = x^2+4x+6$ at which the tangent has gradient 12.

22 Find the points on the curve $y = x^2(x-3)$ at which the gradient of the tangent is 9, and find also the equations of the tangents at these points.

23 The tangent at a point P on the curve $4y = x^2+4x-16$ is parallel to the line $6x-2y = 5$. Find the coordinates of P.

24 Find the equation of the tangent to the curve $y = x-\dfrac{1}{x^2}$ at the point where the curve crosses the x-axis.

25 Show that the gradient of the tangent at each point of the curve $y = \dfrac{1}{x}$ is negative. Find the equations of the tangent at $x = \frac{1}{2}$ and $x = -\frac{1}{2}$.

26 Prove that the gradient of the tangent to the curve with equation $y = x^3-6x^2+12x+1$ is never negative. Find the point on the curve where the gradient is zero.

27 A curve has equation $y = 8\sqrt{x}$. Find the equation of the tangent to the curve at the point where $x = 4$, and show that this tangent cuts the y-axis at $(0, 8)$.

28 Find the equation of the tangent to the curve $y = (x-3)\sqrt{x}$ at the point where $x = 1$.

29 Find the point on the curve $y = 2x^2 - 3x + 1$ at which the tangent makes an angle of 45° with OX.

30 Find the point on the curve $y = \frac{1}{2}x^2 + \frac{8}{x}$ at which the tangent is parallel to the x-axis.

31 Find the equation of the tangent to the curve $y = x^3 + \frac{2}{x}$ at the point where $x = 2$.

32 A curve has equation $y = x^2 + ax + b$, where a and b are constants. If the line $y = 2x$ is a tangent to the curve at $(2, 4)$, find a and b.

33 The curve $y = \left(a + \frac{b}{x}\right)\sqrt{x}$ passes through the point $(4, 8)$, at which the gradient of the tangent is 2. Find the constants a and b.

---

7 *Increasing and decreasing functions*

Figure 5 shows the graph of the function f defined by $f(x) = x^2 - 1$.

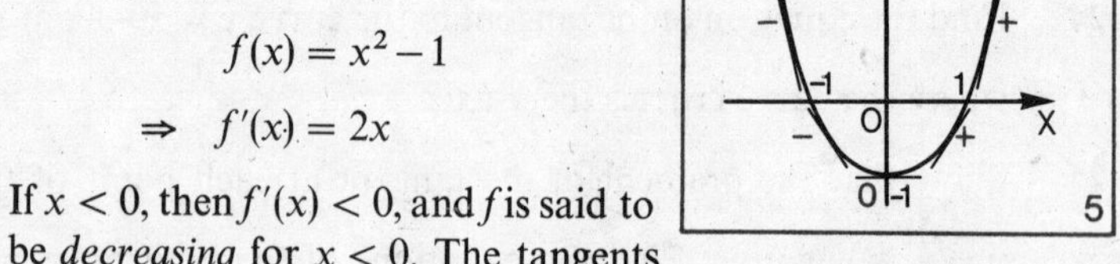

$$f(x) = x^2 - 1$$

$$\Rightarrow \quad f'(x) = 2x$$

(i) If $x < 0$, then $f'(x) < 0$, and f is said to be *decreasing* for $x < 0$. The tangents at all such points have *negative gradients*, as indicated in Figure 5.

(ii) If $x > 0$, then $f'(x) > 0$, and f is said to be *increasing* for $x > 0$. The tangents at all such points have *positive gradients*, as shown in the figure.

(iii) If $x = 0$, then $f'(x) = 0$, and the tangent at $(0, -1)$ is parallel to the x-axis. In this case, f is *neither increasing nor decreasing*, and is said to have a *stationary value* $f(0) = -1$.

To find the intervals for which a function f is increasing or decreasing we first find $f'(x)$ and then study the solutions of the inequations $f'(x) > 0$ and $f'(x) < 0$.

Example. Find the intervals for which the function f given by $f(x) = 2+x^2-\frac{1}{3}x^3$ is (i) increasing (ii) decreasing.

$$f(x) = 2+x^2-\tfrac{1}{3}x^3$$
$$\Rightarrow \quad f'(x) = 2x-x^2 = x(2-x)$$

(i) $f'(x) > 0$
$\Leftrightarrow x(2-x) > 0$
$\Leftrightarrow 0 < x < 2$

(ii) $f'(x) < 0$
$\Leftrightarrow x(2-x) < 0$
$\Leftrightarrow x < 0$ or $x > 2$

(Figure 6)

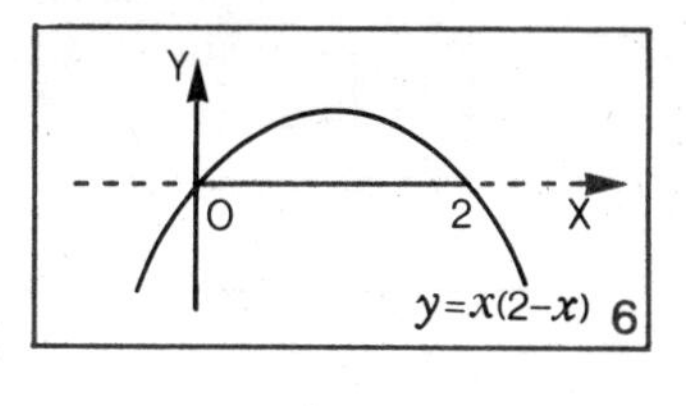

Exercise 7

For each function defined in questions ***1–13***, find the intervals in which it is increasing and those in which it is decreasing.

1 $f(x) = x^2$ ***2*** $f(x) = x^2-2x$ ***3*** $f(x) = x^2-8x+10$

4 $f(x) = x-x^2$ ***5*** $f(x) = x^2+6x-6$ ***6*** $f(x) = 4-2x+x^2$

7 $f(x) = x^3$ ***8*** $f(x) = x^3-6x^2+5$ ***9*** $f(x) = 3x-x^3$

10 $f(x) = 2x^3-9x^2+12x$ ***11*** $f(x) = \frac{1}{3}x^3-x^2-3x+3$

12 $f(x) = x(x-2)^2$ ***13*** $f(x) = 1+x-x^2-x^3$

14 Show that for all real x the function $f: x \to x^3-3x^2+3x-10$ is never decreasing. For what replacement for x is the function stationary?

15 A hyperbola has equation $y = c^2/x$, $x \neq 0$ and c constant. Show that the gradient of each tangent to the hyperbola is negative. Find the equation of the tangent at the point $A(c, c)$, and show that if this tangent cuts the axes at M and N, then A is the midpoint of MN.

8 *Stationary values*

The graph of the function f defined by $f(x) = 5x^3 - 3x^5$ is shown in Figure 7.

The derived function f' has formula $f'(x) = 15x^2 - 15x^4 = 15x^2(1-x)(1+x)$.

At the points A, O and B, where $x = 1$, 0 and -1 respectively, $f'(x) = 0$, and the tangents to the graph are parallel to the x-axis. At these points, f is neither increasing nor decreasing, and the function has *stationary values* $f(1)$, $f(0)$ and $f(-1)$.

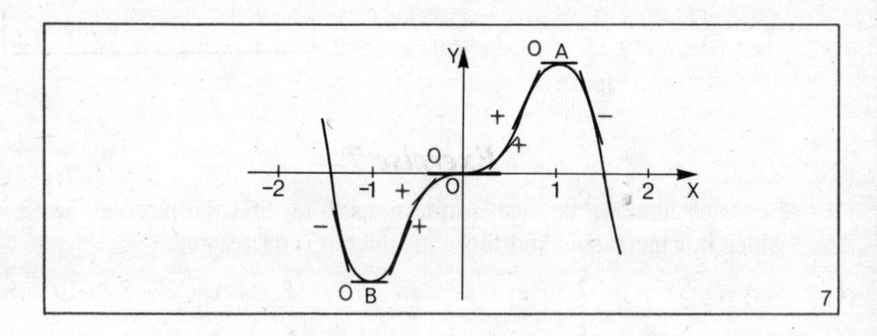

We shall examine the *nature* of these stationary values by considering the sign of $f'(x)$ in the neighbourhood of each.

(i) *The stationary value at* A

If $x < 1$, $f'(x)$ is positive.
At $x = 1$, $f'(1) = 0$.
If $x > 1$, $f'(x)$ is negative.

} Hence $f'(x)$ changes sign from positive through zero to negative, as shown by the gradients of the tangents.

We say that f has a *maximum turning value*, $f(1) = 2$, at $x = 1$.

(ii) *The stationary value at* B

If $x < -1$, $f'(x)$ is negative.
At $x = -1$, $f'(-1) = 0$.
If $x > -1$, $f'(x)$ is positive.

} Hence $f'(x)$ changes sign from negative through zero to positive, as shown by the gradients of the tangents.

We say that f has a *minimum turning value*, $f(-1) = -2$, at $x = -1$.

(iii) *The stationary value at* O

If $x < 0, f'(x)$ is positive.
At $x = 0, f'(0) = 0$.
If $x > 0, f'(x)$ is positive.

} Note how the graph bends across the tangent at O.

We say that the graph has a horizontal *point of inflexion* at O.

If $f'(a) = 0$, then $f(a)$ is a stationary value of f at $x = a$. The stationary value may be a *maximum turning value*, a *minimum turning value* or a horizontal *point of inflexion* on the graph of f.

Generalising, suppose that $x = a$ gives a stationary value $f(a)$ of a function f. If $f'(x)$ exists at every point in a neighbourhood of $x = a$ (i.e. some small interval on the x-axis containing a), then near $x = a$ there are four possibilities for the graph of f (for the types of function in this book). See Figure 8 and the tables of sign for $f'(x)$ below.

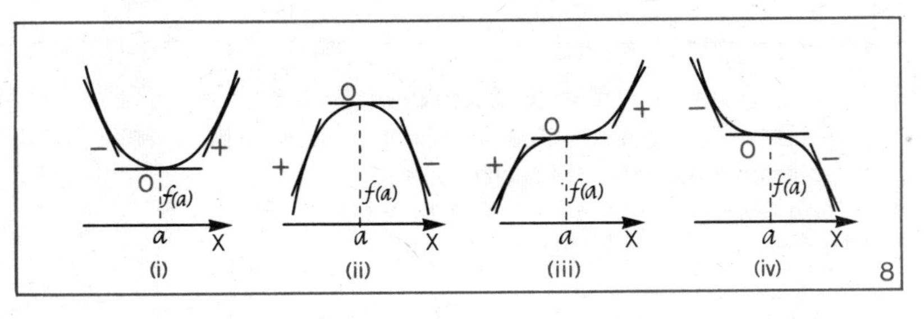

x	$a-$	a	$a+$
$f'(x)$	$-$	0	$+$

$f(a)$ is a minimum turning value of f.

x	$a-$	a	$a+$
$f'(x)$	$+$	0	$-$

$f(a)$ is a maximum turning value of f.

x	$a-$	a	$a+$
$f'(x)$	$+$	0	$+$

x	$a-$	a	$a+$
$f'(x)$	$-$	0	$-$

$(a, f(a))$ is a point of inflexion on graph of f.

In the above tables, $a-$ and $a+$ should be read 'a little less than a' and 'a little more than a' respectively.

Example. Find the stationary values of the function f defined by $f(x) = x^3(x-4)$, and determine the nature of each.

$$f(x) = x^3(x-4) = x^4 - 4x^3$$
$$\Rightarrow \quad f'(x) = 4x^3 - 12x^2 = 4x^2(x-3)$$

For stationary values of $f, f'(x) = 0$.

Here $\quad f'(x) = 0 \quad \Leftrightarrow \quad x = 0$ or $x = 3$.

So f has stationary values $f(0) = 0$ at $x = 0$, and $f(3) = -27$ at $x = 3$.

Nature of stationary values

	(i) $x = 0$			(ii) $x = 3$		
x	$0-$	0	$0+$	$3-$	3	$3+$
$4x^2$	$+$	0	$+$	$+$	$+$	$+$
$x-3$	$-$	$-$	$-$	$-$	0	$+$
$f'(x)$	$-$	0	$-$	$-$	0	$+$
Shape of graph	\	—	\	\	—	/
	Point of inflexion (0, 0)			*Min. turning value* -27		

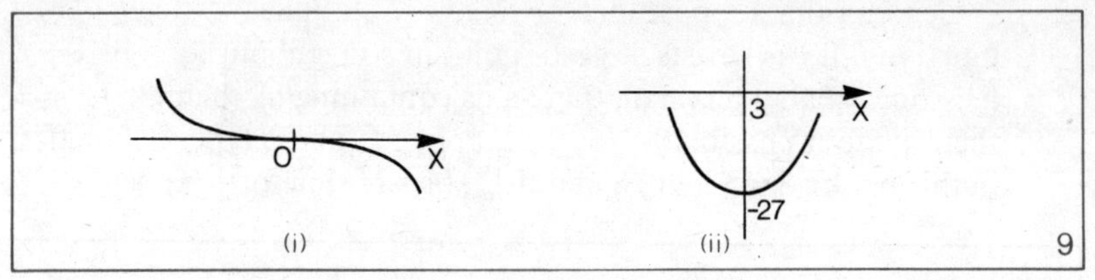

Exercise 8

Find the stationary values of the functions defined below, and determine the nature of each.

1 $f(x) = x^2$

2 $f(x) = x^2 - 2x$

3 $f(x) = 4 - x^2$

4 $f(x) = (x+1)(3-x)$

5 $f(x) = (4-x)^2$

6 $f(x) = x^3$

7 $f(x) = x^3 - 3x$

8 $f(x) = 2x^3 - 9x^2 + 12x$

9 $f(x) = x(x-2)^2$

10 $f(x) = x + \dfrac{1}{x}$

11 $f(x) = \frac{1}{4}x^4 - \frac{9}{2}x^2$

12 $f(x) = x^3(4-x)$

13 $f(x) = x^3 - 12x + 3$

14 $f(x) = 2x^4 - 2x^2$

15 $f(x) = 4x^3 - 15x^2 + 12x + 9$

9 *Curve sketching*

In the last section our attention was on values of a function, and, in particular, on stationary values. Now the emphasis is on the graph of a function. The language therefore becomes geometrical, and we refer to stationary points, maximum and minimum turning points, etc.

Function f	*Curve $y = f(x)$*
Stationary value $f(a)$ at $x = a$ given by $f'(x) = 0$.	*Stationary point (x, y) on the curve given by $\frac{dy}{dx} = 0$.*

The ability to sketch a curve is of great importance in understanding and applying calculus. In sketching the graph of a differentiable function some or all of the following may be helpful:

(i) The points where the curve cuts the x- and y-axes (if easily found).
(ii) Stationary points and their nature.
(iii) Values of y for large positive and negative x.

Example. Sketch the curve $y = x(x-3)^2$.

(i) *Points of intersection with axes*
If $x = 0$, then $y = 0$, giving $(0, 0)$.
If $y = 0$, then $x = 0$ or $x = 3$, giving $(0, 0)$ and $(3, 0)$.

(ii) *Stationary points and their nature*

$$y = x(x-3)^2 = x(x^2-6x+9) = x^3-6x^2+9x.$$

$$\Rightarrow \quad \frac{dy}{dx} = 3x^2-12x+9 = 3(x^2-4x+3) = 3(x-1)(x-3).$$

For stationary points on the curve, $\frac{dy}{dx} = 0$.

Here $\frac{dy}{dx} = 0 \quad \Leftrightarrow \quad x = 1$ or $x = 3$, giving stationary points $(1, 4)$ and $(3, 0)$.

x	$1-$	$x = 1$ 1	$1+$	$3-$	$x = 3$ 3	$3+$
$3(x-1)$ $(x-3)$	$-$ $-$	0 $-$	$+$ $-$	$+$ $-$	$+$ 0	$+$ $+$
$\frac{dy}{dx}$	$+$	0	$-$	$-$	0	$+$
Shape of graph	/	— Max.	\	\	Min. —	/
		Max. turning point $(1, 4)$			*Min. turning point* $(3, 0)$	

(iii) $y = x(x-3)^2$. For large x, $y \doteqdot x(x)^2 = x^3$.

So for x large and positive, y is large and positive;

for x large and negative, y is large and negative.

All of the above information enables us to sketch the curve as shown in Figure 10.

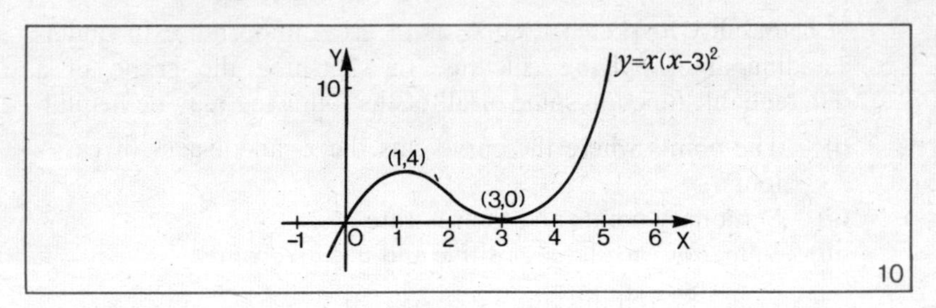

10

Exercise 9

Sketch the following curves:

1	$y = x^2-4$	*2*	$y = 8x-x^2$	*3*	$y = x^2+x-2$
4	$y = (5-x)^2$	*5*	$y = x^3$	*6*	$y = x^4$
7	$y = x(x-1)^2$	*8*	$y = \frac{1}{3}x^3-\frac{1}{2}x^2$	*9*	$y = 8-x^3$
10	$y = 3x-x^3$	*11*	$y = 3x^2-x^3$	*12*	$y = x^3(4-x)$
13	$y = x^4-2x^2$	*14*	$y = 3x^5-5x^3$	*15*	$y = 8+2x^2-x^4$

10 *Maximum and minimum values of a function in a closed interval*

The graph of the function f defined by $f(x) = 5x^3-3x^5$, which we studied in Section 8, is shown in Figure 11.

The maximum value of f in the closed interval $\{x: 0 \leqslant x \leqslant 2\}$ is obviously $f(1) = 2$.

But for the interval $\{x: 0 \leqslant x \leqslant \frac{1}{2}\}$ the maximum value of f is $f(\frac{1}{2}) = \frac{17}{32}$, and for the interval $\{x: -2 \leqslant x \leqslant 2\}$ the maximum value of f is $f(-2) = 56$ (see Figures 11(ii) and (iii)).

For the interval $\{x: -2 \leqslant x \leqslant 2\}$ the minimum value of f is $f(2) = -56$, and we can write $-56 \leqslant f(x) \leqslant 56$ for this interval.

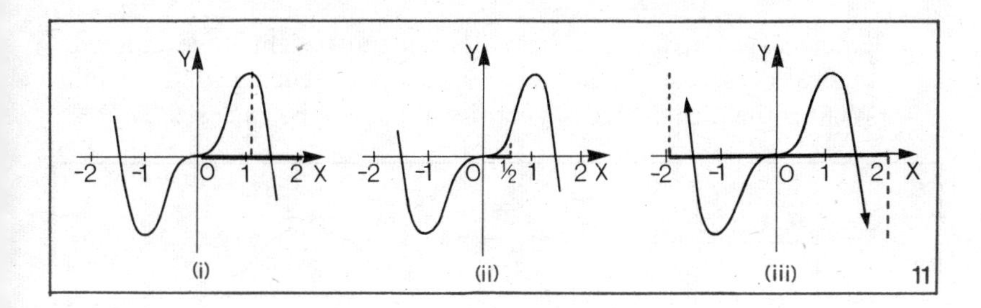

It should be noted, therefore, that the maximum or minimum value of a function f *in a closed interval* may not be the maximum or minimum turning value of f.

The maximum or minimum value of a function f in a closed interval is given by a stationary value of f in the interval, *or* by the value of f at one of the endpoints of the interval.

Exercise 10

Find the maximum and minimum values of the following functions in the given closed intervals. Express the answers in the form $a \leqslant f(x) \leqslant b$, and illustrate by sketches.

1 $f: x \to x^2, \{x: -4 \leqslant x \leqslant 4\}$

2 $f: x \to x^2 - 9, \{x: -6 \leqslant x \leqslant 6\}$

3 $f: x \to 2x - x^2, \{x: -1 \leqslant x \leqslant 1\frac{1}{2}\}$

4 $f: x \to x^2 - 6x + 9, \{x: 0 \leqslant x \leqslant 5\}$

5 $f: x \to 2x^3, \{x: -3 \leqslant x \leqslant 3\}$

6 $f: x \to x(4-x), \{x: -2 \leqslant x \leqslant 6\}$

7 $f: x \to x^3 - 6x^2, \{x: -1 \leqslant x \leqslant 3\}$

8 $f: x \to 2x^2 - x^4, \{x: -\frac{1}{2} \leqslant x \leqslant \frac{1}{2}\}$

11 *Problems involving maxima and minima*

Example. An area of hill farm alongside a straight stone wall is to be fenced against sheep, and 400 metres of fencing are available. What is the greatest rectangular area that can be enclosed?

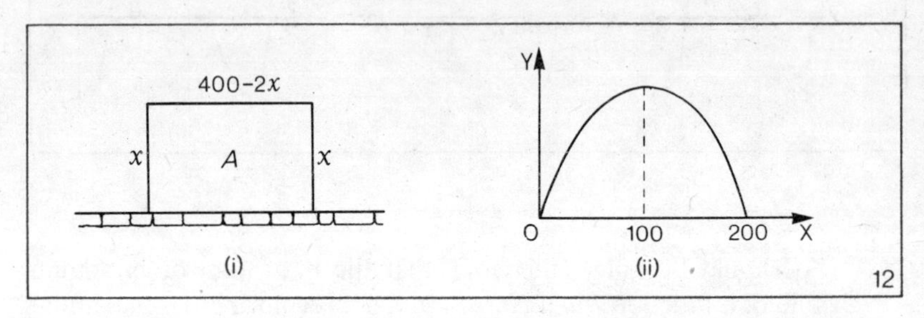

We first construct a mathematical model of the problem, and then analyse it.

If the breadth of the enclosure is x metres, the length is $(400-2x)$ metres, as shown in Figure 12(i).

Obviously, $x \geqslant 0$ and $400-2x \geqslant 0$, i.e. $0 \leqslant x \leqslant 200$.

The area of the enclosure in square metres is given by

$$A(x) = x(400-2x) = 400x - 2x^2,$$

and we have to find the maximum value of A.

$$A'(x) = 400 - 4x = 4(100-x).$$

For stationary values of A, $A'(x) = 0$.

Here $A'(x) = 0 \Leftrightarrow x = 100$

x	100–	100	100+
$4(100-x)$	+	0	–
$A'(x)$	+	0	–
Shape of graph	/	— Max.	\

(See graph of A in Figure 12(ii).)

Thus $x = 100$ gives a maximum turning value $A(100) = 20\,000$.

Also, at the endpoints of the interval $0 \leqslant x \leqslant 200$, $A(0) = 0$ and $A(200) = 0$.

Thus the required maximum area is 20 000 m^2, and this occurs when the breadth is 100 m and the length is 200 m.

Exercise 11

1 Repeat the worked example for a length of 600 metres of fencing.

2 The height h metres of a projectile above its point of projection after t seconds is $h(t) = 600t - 5t^2$. Find the greatest height reached.

3 The sum of two numbers x and y is 40, and their product is P. Write down the equation connecting x and y, and hence express P in terms of x. Find the greatest value of the product.

4 Repeat question ***3*** for two numbers whose sum is 28.

5 The perimeter of a rectangular enclosure is to be 100 metres.

a If the length is x metres and the breadth is y metres, write down the equation connecting x and y in its simplest form.

b Write down a formula for the area A m^2 of the enclosure, and then express A in terms of x. Hence find the dimensions of the enclosure for maximum area.

6 In Figure 13, PQRS is a rectangle 10 cm long and 6 cm wide. PW = QX = RY = SZ = x cm as shown.

a By subtracting the areas of the four triangles from the area of the rectangle, show that the area A cm^2 of quadrilateral WXYZ is given by $A(x) = 60 - 16x + 2x^2$.

b Find the minimum area of the quadrilateral.

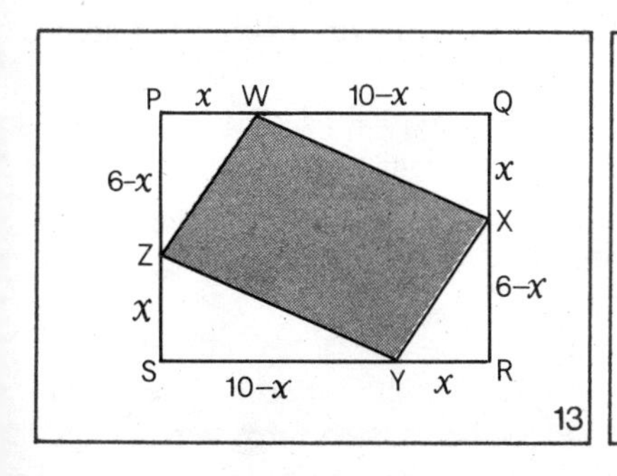

13

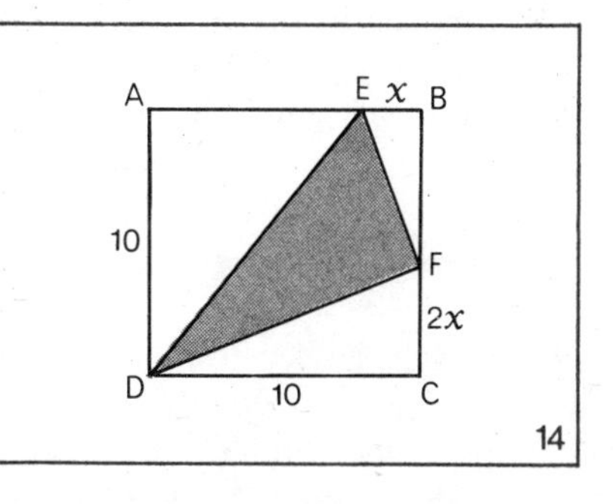

14

7 Figure 14 shows a square ABCD of side 10 cm. BE = x cm, and CF = $2x$ cm. Write down the lengths of AE and BF in terms of x.

Show that the area $A\,\text{cm}^2$ of triangle DEF is given by $A(x) = 50 - 10x + x^2$, and hence find x such that A is a minimum.

8 Squares of side x cm are cut from the corners of a cardboard square of side 12 cm, as shown in Figure 15. The flaps are then bent up and taped, to form a small tray.

a Show that the volume $V\,\text{cm}^3$ of the tray is $V(x) = 144x - 48x^2 + 4x^3$.
b Find x for the tray to have maximum volume, and calculate this volume.

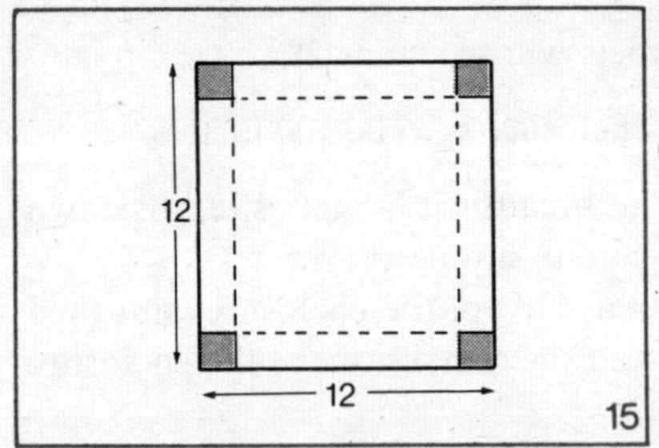

15

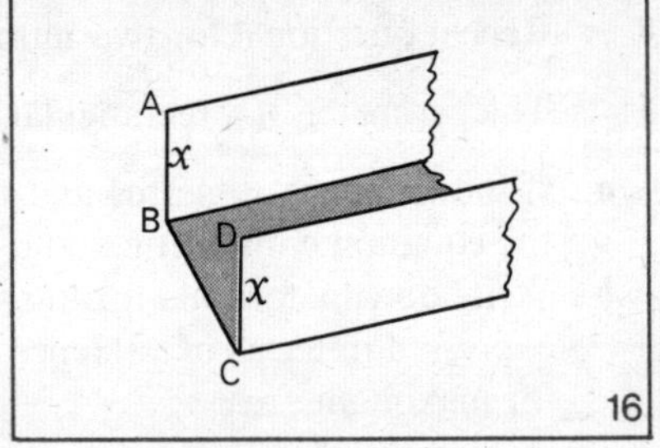

16

9 Figure 16 shows part of a long rectangular sheet of roofing metal, 40 cm wide, which is bent to form an open rectangular rain gutter.

a If AB = CD = x cm, show that the cross-section area $A\,\text{cm}^2$ of the gutter is such that $A(x) = 40x - 2x^2$.
b Find the dimensions of the cross-section for a maximum flow of water.

10 From a rectangular sheet of cardboard 8 cm long and 5 cm wide four equal squares of side x cm are cut from the corners, and the flaps are bent up to form an open box of height x cm. Find the dimensions of the box for maximum volume.

11 A right-angled triangle is formed by the x-axis, the y-axis and the line $y = 8 - 2x$. A point $P(x, y)$ is taken on the line, and perpendiculars are drawn from P to the axes to form a rectangle with diagonal OP. Find an expression for the area of this rectangle in terms of x, and hence find the coordinates of P so that the area of the rectangle is a maximum.

12a A sector of a circle with radius r cm has area $16\,\text{cm}^2$. Show that the perimeter P cm of the sector is $P(r) = 2\left(r + \frac{16}{r}\right)$.

b Find the minimum value of P.

13 A rectangular tin box with a square base of side x cm and height h cm is open at the top. Its volume is to be $32\,\text{cm}^3$.

a Write down the equation connecting x and h, and a formula for the surface area $A\,\text{cm}^2$ in terms of x and h.

b Deduce that $A(x) = x^2 + \frac{128}{x}$, and hence find the dimensions of the box so that it has a minimum surface area.

14 P is the point $(1, t^2+1)$ and Q is $(t+1, 2t^2+5)$, where $t > 0$. Find the gradient of PQ in terms of t, and then find t for PQ to have minimum gradient.

15 A solid circular cylinder is to be machined from a solid metal sphere of radius R mm, as indicated in Figure 17.

a Taking the radius of the cylinder to be x mm and the height $2y$ mm, show that the volume $V\,\text{mm}^3$ of the cylinder is $V(y) = 2\pi y(R^2 - y^2)$.
b Show that the height of the largest possible cylinder is $2R/\sqrt{3}$ mm.
c Show also that the ratio of the volume of this cylinder to the volume of the sphere is $1:\sqrt{3}$.

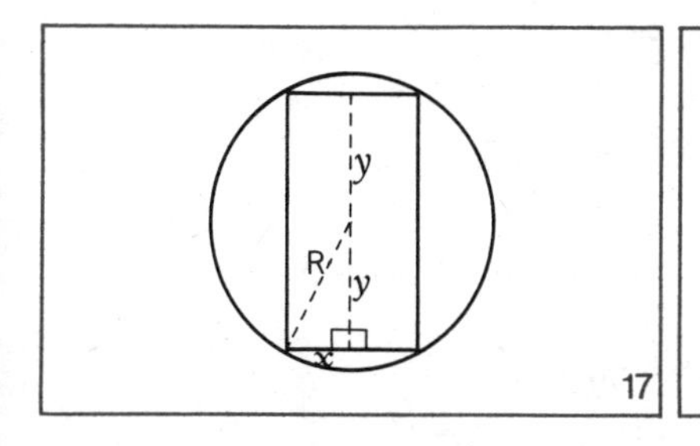

17

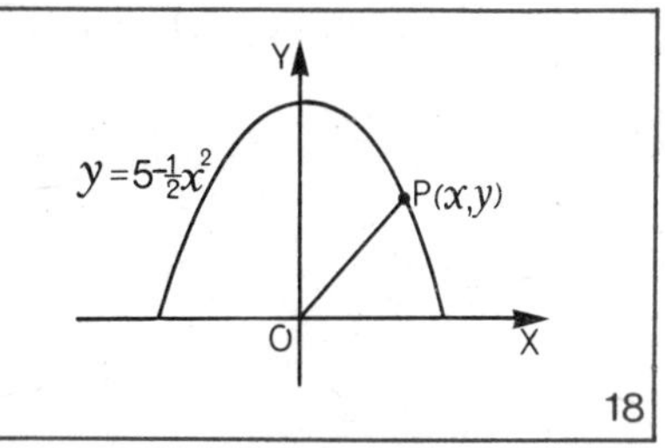

18

16 Figure 18 shows the part of the parabola $y = 5 - \frac{1}{2}x^2$ for which $y \geqslant 0$. If $P(x, y)$ is any point on the curve, show that $OP = \sqrt{(\frac{1}{4}x^4 - 4x^2 + 25)}$.

Taking $f(x) = \frac{1}{4}x^4 - 4x^2 + 25$, show that the stationary values of f are given by $x = 0$, $2\sqrt{2}$ and $-2\sqrt{2}$. Hence calculate the least distance from O to the parabola.

17 In the manufacture of closed cylindrical cans to hold a fixed volume V, the labour cost C varies directly as the total length of the seams in the cans, i.e. the sum of the height of the can and twice its circumference. If the radius of the base is r and the height is h, show that $C = k\left(\dfrac{V}{\pi r^2} + 4\pi r\right)$, where k is the constant of variation.

Prove that the labour cost is least when the height is equal to the circumference.

12 Further applications of the Leibniz notation $\dfrac{df}{dx}$

The notation $\dfrac{df}{dx}$ for the derivative of f is particularly convenient in applied mathematics. For example, let s metres be the *displacement* OP of an object P, moving in a straight line, from a fixed point O on the line at time t seconds (i.e. t seconds after an initial time $t = 0$).

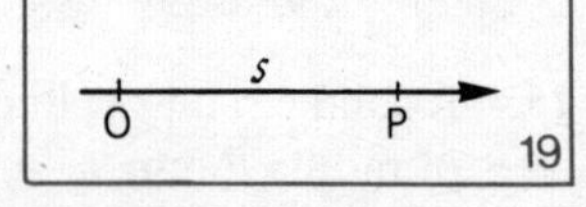

Then corresponding to an increment Δt seconds in the time we have an increment Δs metres in the displacement; i.e. the object has moved Δs metres in Δt seconds, so

$\dfrac{\Delta s}{\Delta t}$ = the average velocity in Δt seconds, and

$\dfrac{ds}{dt} = \lim\limits_{\Delta t \to 0} \dfrac{\Delta s}{\Delta t}$ = the velocity at time t seconds.

Denoting the *velocity* at time t seconds by v metres per second,

$$v = \frac{ds}{dt}.$$

v may be positive, negative or zero.

If $v > 0$, then $\dfrac{ds}{dt} > 0$ and s is increasing with t;

if $v < 0$, then $\dfrac{ds}{dt} < 0$ and s is decreasing with t.

The numerical value of v, denoted by $|v|$ or $\left|\frac{ds}{dt}\right|$, is called the *speed* at time t seconds.

If the velocity of the object is changing, the rate of change of the velocity is called the *acceleration* of the object. Denoting the acceleration at time t seconds by a metres per second per second,

$$a = \frac{dv}{dt}.$$

Example. The displacement s metres at time t seconds ($t \geqslant 0$) of an object moving in a straight line is given by the formula $s = 3 - 6t + 2t^3$.

a Calculate its displacement at $t = 1$ and $t = 2$.
b Find formulae for its velocity v m/s and its acceleration a m/s^2 at time t seconds.
c Show that its velocity is zero at $t = 1$.
d Calculate its velocity when its acceleration is zero.

a At $t = 1$, $s = 3 - 6 + 2 = -1$.
At $t = 2$, $s = 3 - 12 + 16 = 7$.

b $v = \frac{ds}{dt} = -6 + 6t^2$

$a = \frac{dv}{dt} = 12t$

c $v = 0 \quad \Leftrightarrow \quad -6 + 6t^2 = 0 \quad \Leftrightarrow \quad t = \pm 1$.
So the velocity is zero after 1 second.

d $a = 0 \quad \Leftrightarrow \quad 12t = 0 \quad \Leftrightarrow \quad t = 0 \quad \Leftrightarrow \quad v = -6$.

So the velocity is -6 m/s at the instant when the acceleration is zero. The *speed* is 6 m/s.

Exercise 12B

1 The displacement s metres at time t seconds of a body moving in a straight line is given by the formula $s = 5 + 12t - t^3$.

a Calculate its displacement at $t = 1$ and $t = 2$.
b Find formulae for its velocity v m/s and its acceleration a m/s^2 at time t seconds.
c Find t when the velocity is zero.
d Calculate the velocity when the acceleration is zero.

2 Repeat question *1* for the formula $s = 2t(t^2-3)$.

3 At time t seconds the velocity of an object is v m/s, where $v = 12t-3t^2$. Find the acceleration at each of the times when the velocity is zero.

4 The height h metres reached by a projectile after t seconds is given approximately by the equation $h = 150t-5t^2$.

Find its velocity v m/s after t seconds by calculating $\frac{dh}{dt}$, and deduce that the greatest height reached is approximately 1125 m.

5 A rectangular storage tank has a square base of side 5 m. If V m^3 is the volume of liquid when the depth is x metres, find a formula relating V and x, and calculate the rate at which V is increasing with respect to x as liquid flows in.

6 The area A mm^2 of a circle of radius r mm is given approximately by the formula $A = 3{\cdot}14r^2$. Find the rate at which the area is increasing with respect to r, when $r = 5$.

7 If a wheel rotates through an angle of θ radians in t seconds, its angular velocity is given by $\frac{d\theta}{dt}$ radians per second. For a certain wheel, $\theta = 15t+7t^2-\frac{1}{3}t^3$.

a Find a formula for its angular velocity, and calculate its angular velocity when $t = 3$.
b Find also the time that elapses before the angular velocity of the wheel is zero.

8 The gravitational attraction between two given particles is a force whose magnitude F varies inversely as the square of the distance x between them.

a Find a formula relating F and x in terms of k, the constant of variation.
b Find the rate of change of F with respect to x.

9 Given $pv = k$, where k is a constant, express p in terms of k and v, and hence show that $-v\frac{dp}{dv} = p$.

The current i in a circuit at time t is given by the formula $i = 4+3t-5t^2$, and the voltage V is given by $V = 2i+0{\cdot}02\dfrac{di}{dt}$.

Calculate V when $t = 0{\cdot}1$.

Summary

1 *Rate of change.* The rate of change of the function f at $x = a$ is given by $f'(a) = \lim_{h \to 0} \dfrac{f(a+h)-f(a)}{h}$.

2 *Derived function, or derivative.* The derived function, or derivative, of f is given by $f'(x) = \lim_{h \to 0} \dfrac{f(x+h)-f(x)}{h}$.

3 *Particular derivatives.* If $f(x) = x^n$, then $f'(x) = nx^{n-1}$, $n \in Q$.

4 *Gradient of tangent.* The gradient of the tangent to the curve $y = f(x)$ at the point (x, y) is denoted by $\dfrac{dy}{dx}$, where $\dfrac{dy}{dx} = f'(x)$.

5 *Increasing and decreasing functions.* For a differentiable function f to be increasing, $f'(x) > 0$; to be decreasing, $f'(x) < 0$.

6 *Stationary value of a function.* If $f'(a) = 0$, then $f(a)$ is a stationary value of f at $x = a$.

7 *Nature of a stationary value (maximum, minimum, point of inflexion).* The nature is found from the sign of $f'(x)$ in a neighbourhood of $x = a$.

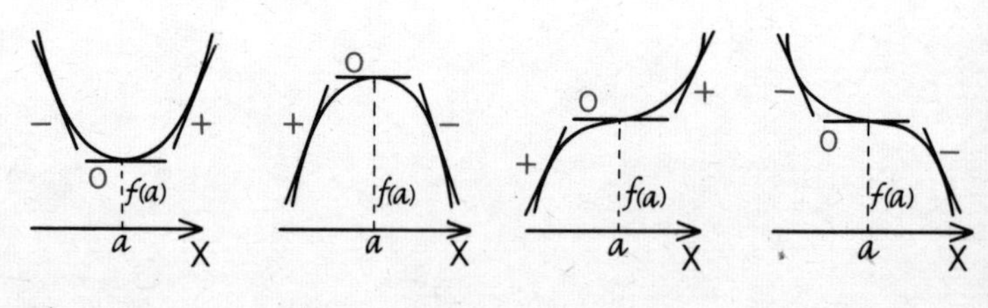

x	$a-$	a	$a+$
$f'(x)$	$-$	0	$+$

$f(a)$ is a minimum turning value of f.

x	$a-$	a	$a+$
$f'(x)$	$+$	0	$-$

$f(a)$ is a maximum turning value of f.

x	$a-$	a	$a+$
$f'(x)$	$+$	0	$+$

x	$a-$	a	$a+$
$f'(x)$	$-$	0	$-$

$(a, f(a))$ is a point of inflexion on graph of f.

Maximum and minimum values of a function in a closed interval occur at stationary values or at the endpoints of the interval.

8 *Curve sketching.* To sketch a curve, investigate:

(i) the points where the curve cuts the x- and y-axes

(ii) stationary points (given by $\frac{dy}{dx} = 0$), and their nature

(iii) values of y for large positive and negative x.

9 *Applications of maxima and minima.* Problems involving maxima and minima may be solved by differentiation provided that a mathematical model can be constructed and analysed.

10 *Velocity and acceleration.* If s metres denotes the displacement, from a fixed point O, of an object P moving in a straight line, at time t seconds, then

$v = \frac{ds}{dt}$ metres per second is the velocity of P, and

$a = \frac{dv}{dt}$ metres per second per second is the acceleration of P,

at time t.

2 The Integral Calculus

1 Introduction

The mathematical method which we develop in this chapter was known to the famous Greek mathematician Archimedes, who lived in Syracuse in the second century B.C., though it was not until the seventeenth century that it was thoroughly explored by Newton and Leibniz.

Archimedes was interested in calculating the area of a circle, the area cut off by a chord of a parabola, and so on. Since his time, his method has been put on a firm mathematical basis, and in this chapter we take a first step towards the development of that method.

At the beginning of Chapter 1 on Calculus, it was explained that the ideas of Calculus can be divided into two basic parts. In that chapter it was seen that the part which deals with *rate of change* requires the notion of a *limit*, and this was developed with the notation and ideas of *differentiation*.

In the present chapter we introduce the other part of Calculus, called *integration*, and we begin by linking it with Chapter 1.

2 Anti-differentiation; integration

In our mathematics course we have studied a number of operations which have inverses. For example:

Operation	*Inverse operation*
Addition	Subtraction
Multiplication	Division
Squaring a positive number	Taking the positive square root
Taking the logarithm of a number	Taking the corresponding anti-logarithm

Note to the Teacher on Chapter 2

There are two main processes in integration, (i) anti-differentiation and (ii) the limit of a sum process for definite integrals, these being tied together in the Fundamental Theorem of Calculus. It is not an easy matter to decide from the pedagogic point of view whether it is better to start with (i) or (ii). So far as the motivation of the integral notation $\int_a^b f(x)\,dx$ is concerned $\left(\text{i.e. as } \lim_{\text{each } \Delta x \to 0} \sum_{x=a}^{x=b} f(x).\Delta x\right)$, of the logical development of the subject $\Bigl($from $\int_a^b f(x)\,dx$ to the area function $S(x)$, to $S'(x) = f(x)$, and so to the fact that $S(x)$ is an anti-derivative of $f(x)\Bigr)$, and of uses of integration in applications, there is no doubt that it is better to start with (ii). However this approach involves meeting some rather demanding ideas right at the beginning of the course, and at the school stage experience has shown that it is better to start with (i), by tackling some introductory work on anti-differentiation.

We have, therefore, taken the first approach mentioned above, defining anti-derivatives in *Section* 2 and consolidating there the use of $\int x^n\,dx = \frac{x^{n+1}}{n+1} + C$ $(n \neq -1)$, and in *Section* 3 giving a few applications of anti-derivatives. *It should be noted, however, that the arrangement in the book makes it possible for teachers who wish to approach the subject by (ii) above, to do so; the order of Sections would then be 4, 5, 2, in such a way that the anti-differentiation in Section 2 could be applied in Exercises 4 and 5 of Section 5.* As with so many parts of calculus at school level, certain difficulties have to be ignored, such as the proof of the fact that any two anti-derivatives of $f(x)$ differ by a constant. The meaning of the constant of integration C often causes trouble and it is hoped that the examples in *Section* 3, in which particular anti-derivatives given by particular values of C are obtained, will help with the idea of C as a parameter and with the understanding of the following statement: If $F(x)$ is any one anti-derivative of $f(x)$, then the set of all anti-derivatives of $f(x)$ is $\{F(x)+C : C \in R\}$.

Sections 2 and 3 should provide a fairly solid, but satisfying, introduction to integration and help to develop confidence to face the introduction of the limit of a sum approach to area in *Section* 4 and, in *Section* 5, of the tie-up between anti-differentiation and the limit of a sum idea. It could be argued that all that matters, so far as many pupils at this stage are concerned, is that a good practical working knowledge of the Fundamental Theorem of Calculus is developed rather than a full understanding of any proof given. The work on area is taken further in *Section* 6 with examples on areas of regions bounded by two curves. Finally in *Section* 7 the traditional work on volumes of solids of revolution about the x- and y-axes is developed, giving further experience on curves and definite integrals.

Facility in many examples in calculus depends on a related skill in algebraic manipulation. In Exercise 4, for example, the use of the distributive law in $(x-1)(2x-1)(3x-1)$, the square $(x+1)^2$ and indices in $\frac{1+\sqrt{x}}{x^2}$ and $(u^{\frac{1}{2}}+1)^2$, etc., will be necessary. If difficulty is experienced, the need for such skill offers timely motivation for working a refresher set of related algebraic examples. In this chapter the only general anti-derivative required is $\int x^n\,dx = \frac{x^{n+1}}{n+1}+C$ $(n \neq -1)$ and pupils should have no difficulty in getting to know this result through experience of examples.

In calculus we find it useful to have a process that is, in a sense, the inverse of differentiation.

If $F(x) = \frac{1}{2}x^2$, and $f(x) = x$, then
$F'(x) = x = f(x)$ for all real x.

That is, by *differentiation* the function F has *derivative* f.

Can we now find an inverse operation, *anti-differentiation*, which produces a function F whose derivative is f?

Given $f(x) = x$, we see that possible formulae for F include:

$$\begin{aligned} F(x) &= \tfrac{1}{2}x^2, && \text{since } F'(x) = x = f(x). \\ F(x) &= \tfrac{1}{2}x^2+1, && \text{since } F'(x) = x = f(x). \\ F(x) &= \tfrac{1}{2}x^2-3, && \text{since } F'(x) = x = f(x). \\ F(x) &= \tfrac{1}{2}x^2+C, && \text{since } F'(x) = x = f(x). \end{aligned}$$

In fact, the set of all *anti-derivatives* F of f, where $f(x) = x$, is given by the formula

$$F(x) = \tfrac{1}{2}x^2+C$$

where each real number C gives a different F.

Again, if $f(x) = x^2$, then

$$F(x) = \tfrac{1}{3}x^3+C, \text{ since } F'(x) = x^2 = f(x).$$

And if $f(x) = x^3$, then

$$F(x) = \tfrac{1}{4}x^4+C, \text{ since } F'(x) = x^3 = f(x).$$

From the pattern emerging it appears that if $f(x) = x^n$, the set of anti-derivatives of f is defined by

$$F(x) = \frac{x^{n+1}}{n+1} + C, \; n \neq -1.$$

This is true, since $F'(x) = \dfrac{(n+1)x^n}{n+1} = x^n = f(x)$.

The integral notation

The set of all anti-derivatives of $f(x)$ is denoted by $\int f(x)\,dx$, which is read 'the integral of $f(x)$ with respect to x'. $\int f(x)\,dx$ is called the *indefinite integral* of $f(x)$, and $f(x)$ is the *integrand*.

For the function f defined by $f(x) = x$, we write

$$\int x\,dx = \tfrac{1}{2}x^2 + C$$

For the function f defined by $f(x) = x^n$,

$$\int x^n\,dx = \frac{x^{n+1}}{n+1} + C,\ n \neq -1$$

In the general case, if $F(x)$ is any anti-derivative of $f(x)$,

$$\int f(x)\,dx = F(x) + C$$

since the set of all $F(x)+C$, for all real C, is the set of anti-derivatives of $f(x)$.

The process of finding anti-derivatives is called *integration*, and in using the process we are said to *integrate* the function. C is called the *constant of integration*. The significance of the symbol $\int$ will be explained in Sections 4 and 5.

Example 1. Given $F'(x) = 4x-1$, and $F(3) = 20$, find the function F.

$$F'(x) = 4x-1$$

$$\Rightarrow \quad F(x) = \int(4x-1)\,dx = \frac{4x^2}{2} - x + C = 2x^2 - x + C$$

$$\Rightarrow \quad F(3) = 18 - 3 + C = 15 + C$$

But $\quad F(3) = 20$, so $C = 5$.

Hence $\quad F(x) = 2x^2 - x + 5$.

Example 2. Integrate $(3x-2)^2$.

$$\int(3x-2)^2\,dx$$

$$= \int(9x^2 - 12x + 4)\,dx$$

$$= \frac{9x^3}{3} - \frac{12x^2}{2} + 4x + C$$

$$= 3x^3 - 6x^2 + 4x + C$$

Example 3. Find $\int\left(\sqrt{x} + \frac{1}{\sqrt{x}}\right)dx$.

$$\int\left(\sqrt{x} + \frac{1}{\sqrt{x}}\right)dx$$

$$= \int(x^{1/2} + x^{-1/2})\,dx$$

$$= \frac{x^{3/2}}{\frac{3}{2}} + \frac{x^{1/2}}{\frac{1}{2}} + C$$

$$= \tfrac{2}{3}x^{3/2} + 2x^{1/2} + C$$

Exercise 1

Integrate:

1 *a* x *b* x^2 *c* x^3 *d* x^5 *e* x^8

2 *a* $2x$ *b* $3x^2$ *c* $8x^3$ *d* $-6x^2$ *e* 5

3 *a* x^{-2} *b* x^{-3} *c* x^{-4} *d* $\dfrac{1}{x^5}$ *e* $-\dfrac{6}{x^4}$

4 *a* $x^{1/2}$ *b* $x^{3/2}$ *c* $x^{-1/2}$ *d* $x^{2/3}$ *e* $\dfrac{2}{\sqrt{x}}$

Find:

5 $\int x\,dx$ **6** $\int(2x-3)dx$ **7** $\int(1-x)dx$

8 $\int(x^2-1)dx$ **9** $\int(4-4x^3)dx$ **10** $\int(3x^2+4x+5)dx$

11 $\int(6x^2-1)dx$ **12** $\int(10x^4+3x^2)dx$ **13** $\int(x-2)(x+2)dx$

14 $\int(x-3)^2\,dx$ **15** $\int(1-3x)^2\,dx$ **16** $\int x(x+1)(x-2)dx$

17 Given the following, find the function F in each case.

a $F'(x) = 2x$, and $F(4) = 10$ *b* $F'(x) = 1-2x$, and $F(3) = 4$

c $F'(x) = 6x^2$, and $F(0) = 0$ *d* $F'(x) = x-\dfrac{2}{x^2}$, and $F(2) = 9$

e $F'(x) = 1-\dfrac{1}{\sqrt{x}}$, and $F(4) = 1$ *f* $F'(x) = 3(x^2-3)$, and $F(1) = 12$

Find:

18 $\displaystyle\int 2x\left(3x-\frac{1}{x}\right)dx$ **19** $\int x^{-1}(2x+x^2)dx$ **20** $\displaystyle\int\left(x-\frac{1}{x}\right)^2 dx$

21 $\int x(1-x)^2\,dx$ **22** $\int(x^{1/2}-2x^{-1/2})dx$ **23** $\int x^{1/2}(5x+3)dx$

24 $\displaystyle\int\frac{x^3+x}{x}dx$ **25** $\displaystyle\int\frac{x^4+1}{x^2}dx$ **26** $\displaystyle\int\frac{(x^2+1)^2}{x^2}dx$

27 $\displaystyle\int\frac{x^2+2}{\sqrt{x}}dx$ **28** $\int x^{1/2}(1+x^{1/2})dx$ **29** $\displaystyle\int\frac{1}{\sqrt{x}}(1+\sqrt{x})^2\,dx$

30 $\displaystyle\int\left(2x-\frac{1}{x^3}\right)^2 dx$ **31** $\displaystyle\int\frac{x(5x-3)}{\sqrt{x}}dx$ **32** $\displaystyle\int\left(\sqrt{x}+\frac{1}{\sqrt{x}}\right)^3 dx$

3 *Some applications of integration*

Suppose we are told that the gradient of the tangent at each point (x, y) on a curve $y = f(x)$ is x. Can we find $f(x)$?

Using the notation of differential calculus for gradient,

$$\frac{dy}{dx} = x \quad \Rightarrow \quad y = \int x\,dx$$

$$\Rightarrow \quad y = \tfrac{1}{2}x^2 + C, \qquad C \in R \qquad \ldots (1)$$

This is the equation of a *family* of parabolas, some of which are shown in Figure 1 for $C = 1, 0, -1, -2$ in the interval $[-3, 3]$. The given information about the curve is therefore insufficient to define it exactly. If, in addition, we are told that the curve passes through the point (2, 3), then from equation (1),

$$3 = 2 + C \quad \Rightarrow \quad C = 1$$

So $y = \frac{1}{2}x^2 + 1$ is the equation of this curve.

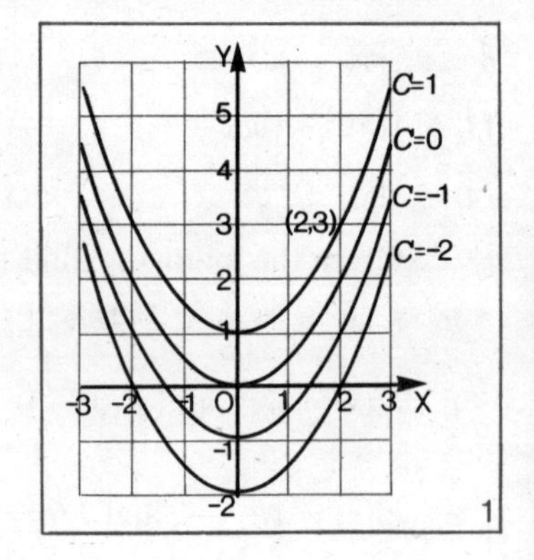

1

Exercise 2

1 The gradient of a family of straight lines is 2. Write down the equation of the family and sketch four members of the family.

2 Find the equations of the curves which satisfy the following conditions. At each point (x, y),

a $\dfrac{dy}{dx} = 4x$; and the curve passes through the point (1, 3)

b $\dfrac{dy}{dx} = 2x - 1$; and the curve passes through the point (2, 8)

c $\dfrac{dy}{dx} = 3x^2 - 10x$; and the curve passes through the point $(-1, 0)$

d $\dfrac{dy}{dx} = 6x^2 - 6x + 3$; and the curve passes through the point (0, 0).

3 The gradient of the tangent to a curve at each point (x, y) is given by $\dfrac{dy}{dx} = 3x(2-x)$. If the curve passes through the point $(-1, 10)$, find its equation.

4 At each point (x, y) of a curve, $\dfrac{dy}{dx} = 1 - \dfrac{4}{x^2}$. The curve passes through the point (2, 5). Find the equation of the curve.

5 The velocity-time graph of a body moving with velocity v at time t is a straight line whose equation is $v = 5 - 2t$. Given that $v = \dfrac{ds}{dt}$, and $s = 0$ when $t = 0$, find by integration a formula for the displacement s at time t.

6 *a* The velocity v m/s of a body after t seconds is given by $v = -\int 10\,dt$ and $v = 15$ when $t = 0$. Find a formula for v in terms of t.

b Given that $v = \dfrac{ds}{dt}$, where s metres is the displacement at time t seconds, and that $s = 0$ when $t = 0$, find an equation for s in terms of t.

7 The velocity v m/s of a body starting from rest is given by $v = \int (t^2 + 2t)dt$, where t seconds is the time from rest. Find a formula for v in terms of t, and use it to find the velocity of the body after 3 seconds.

8 Using the information in question 7, find the displacement s metres at the end of 3 seconds, given that $v = \dfrac{ds}{dt}$, and $s = 0$ at $t = 0$.

9 If $\dfrac{dM}{dx} = \frac{1}{2}wl - wx$ and w, l are positive constants, find M in terms of w, l, x, given that $M = 0$ at $x = 0$. What is the maximum value of M?

10 The gradient of a curve at each point (x, y) is given by $\dfrac{dy}{dx} = 3x^2 - 8x + 5$. If the curve passes through the point (2, 0), show that it also passes through the point (3, 4).

11 The gradient of a curve at each point (x, y) is given by $\frac{dy}{dx} = 1-2x$. If the maximum value of y is $6\frac{1}{4}$, find the equation of the curve.

12 A light metal beam AB is l mm in length. It is fixed at end A and carries a load at end B. The sag of the beam, y mm, at distance x mm from A (measured along the beam) is given by $\frac{dy}{dx} = kx(l-\frac{1}{2}x)$, where k is a constant. Find the sag at B (where $x = l$).

4 *Area as the limit of a sum*

The closed curve C in Figure 2(i) surrounds a region of the plane. How can we find the area S of this region?

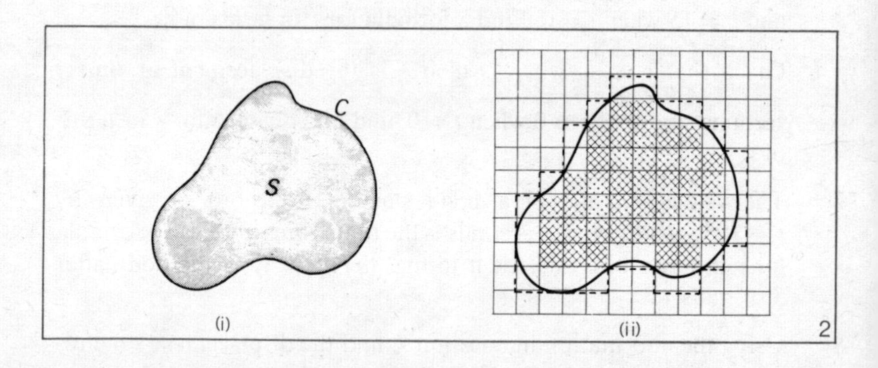

As in Book 1, Arithmetic, we can find approximations to the area by placing a grid of squares over it. From Figure 2(ii), by counting squares, $S > 41$ and $S < 66$, i.e. $41 < S < 66$. By taking a finer 'mesh' of squares we can obtain a better approximation for S.

We now study another way of approximating to S, using rectangles, in which S can be found by a limit process.

Figure 3(i) shows part of the curve $y = f(x)$ from $x = a$ to $x = b$. We will find an expression for the area S bounded by the curve, the x-axis, and the lines $x = a$ and $x = b$.

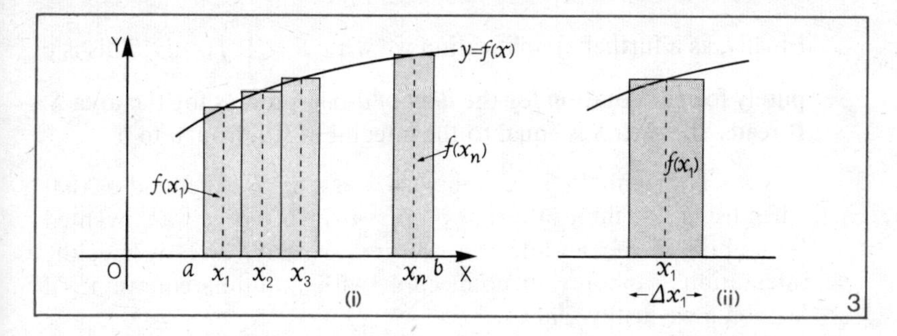

The interval $[a, b]$ is divided into n intervals of lengths Δx_1, $\Delta x_2, \ldots, \Delta x_n$. $x_1, x_2, \ldots, x_n$ are the x-coordinates of n points on the x-axis, one from each of the n small sub-intervals, and n rectangles are drawn as indicated in Figure 3(i).

The first rectangle is drawn to a larger scale in Figure 3(ii). The height of this rectangle is $f(x_1)$, the value of f at $x = x_1$.

The area of the first rectangle $= f(x_1) . \Delta x_1$
The area of the second rectangle $= f(x_2) . \Delta x_2$
The area of the third rectangle $= f(x_3) . \Delta x_3$
....
The area of the last rectangle $= f(x_n) . \Delta x_n$

Using the Greek capital letter Σ (sigma) to denote 'the sum of', we have

$$S \doteqdot \sum_{i=1}^{i=n} f(x_i) . \Delta x_i$$

In order to emphasise that the sum extends over the interval $[a, b]$, we often write this sum as

$$S \doteqdot \sum_{x=a}^{x=b} f(x) . \Delta x$$

For differentiable functions, it can be shown that $\sum_{x=a}^{x=b} f(x) . \Delta x$ can be made as close as we wish to S by taking n sufficiently large, which is equivalent to making Δx sufficiently small.

We define $$S = \lim_{\Delta x \to 0} \sum_{x=a}^{x=b} f(x) . \Delta x .$$

Finally, as a further simplification we write $S = \int_a^b f(x)\,dx$. This is a purely formal notation for the *limit of a sum* process for the area S. It reads 'the area S is equal to the integral of $f(x)$ from a to b'.

Note. The symbol $\int$ is an elongated S and helps to remind us that when using it, a limit of a sum process is involved. It was invented by Leibniz in the seventeenth century. We have already met the integration symbol $\int$ in connection with anti-differentiation; in Section 5 we justify that use.

Example. Show by shading in sketches the areas associated with:

(i) $\int_1^3 x\,dx$ (ii) $\int_{-1}^1 x^3\,dx$ (iii) $\int_0^{\pi/2} \cos x\,dx$

The answers are given in Figure 4.

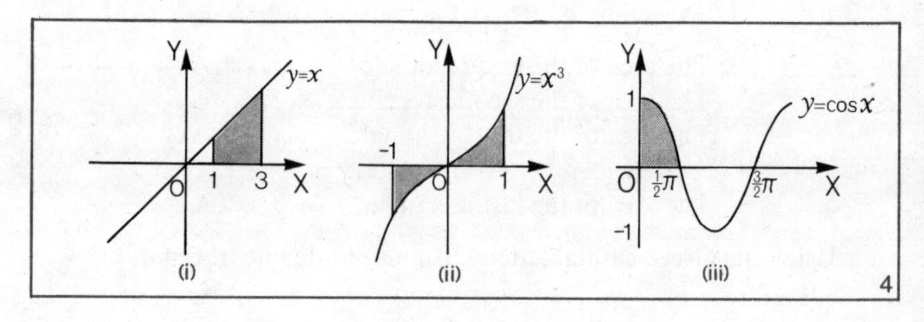

Exercise 3

Show by shading in sketches the areas associated with:

1 $\int_0^4 x\,dx$ **2** $\int_1^4 (x+2)dx$ **3** $\int_{-2}^2 (2x+1)dx$ **4** $\int_0^4 x^2\,dx$

5 $\int_{-3}^3 x^2\,dx$ **6** $\int_1^4 x^3\,dx$ **7** $\int_0^{\pi/2} \sin x\,dx$ **8** $\int_{\pi/2}^{3\pi/2} \cos x\,dx$

9 Write down integrals to represent the shaded regions in Figure 5.

10 Show, by shading, the area given by $\int_0^4 \sqrt{(16-x^2)}dx$. Calculate this area. *Hint:* $y = \sqrt{(16-x^2)} \Rightarrow y^2 = 16-x^2 \Rightarrow x^2+y^2 = 16$.

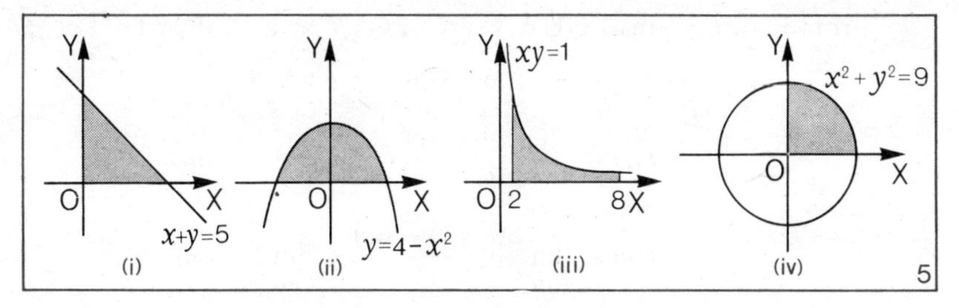

11 Use Figure 6 to calculate $\int_1^3 \sqrt{(x-1)(3-x)}dx$.

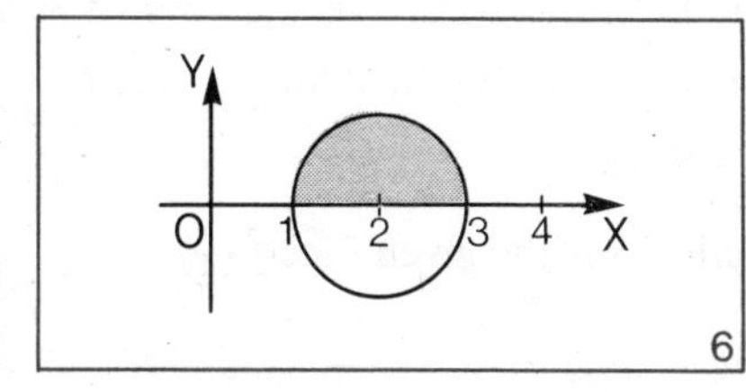

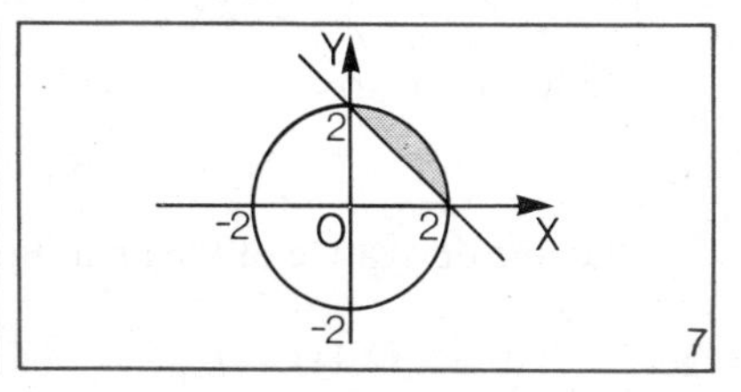

12a Use formulae for the areas of a circle and a triangle to calculate the area of the shaded segment in Figure 7.

b Express this area as the difference of two integrals.

5 *Calculating the area under a curve*

In this section we establish an important method of calculating areas, using anti-derivatives.

Figure 8 shows part of the graph of f. We will find the area bounded by the curve, the x-axis, and the lines $x = a$ and $x = b$. Denoting this area by $S(b)$, we have

$$S(b) = \int_a^b f(x)dx \qquad \dots (1)$$

$$S(a) = \int_a^a f(x)dx = 0 \quad \dots (2)$$

$$S(c) = \int_a^c f(x)dx$$

$$S(c+h) = \int_a^{c+h} f(x)dx.$$

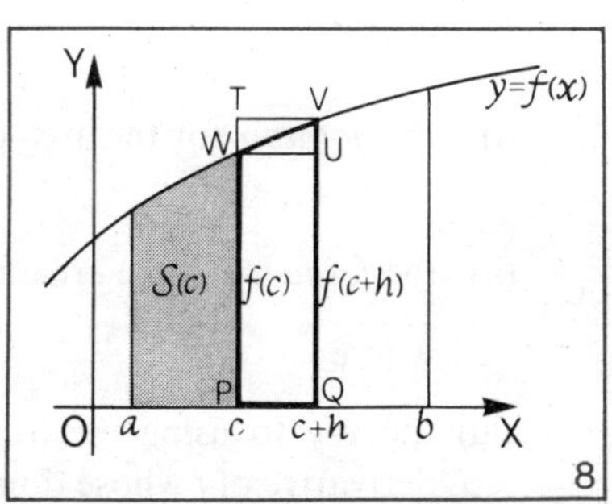

From Figure 8, area PQUW < area PQVW < area PQVT

$$\Rightarrow \quad f(c).h < S(c+h)-S(c) < f(c+h).h$$

$$\Rightarrow \quad f(c) < \frac{S(c+h)-S(c)}{h} < f(c+h), \qquad h \neq 0$$

$$\Rightarrow \quad f(c) \leqslant \lim_{h\to 0} \frac{S(c+h)-S(c)}{h} \leqslant \lim_{h\to 0} f(c+h)$$

$$\Rightarrow \quad f(c) \leqslant S'(c) \leqslant f(c)$$

$$\Rightarrow \quad S'(c) = f(c)$$

Since this is true for each c in the interval $[a, b]$, then for $a \leqslant x \leqslant b$,

$$S'(x) = f(x),$$

i.e. the derivative of the area function S is the given function f.

$$\begin{aligned}
\text{Now} \quad & S'(x) = f(x) \\
\Rightarrow \quad & S(x) = \int f(x)dx \\
& = F(x)+C, \quad \text{where } F \text{ is any anti-derivative of } f \\
\Rightarrow \quad & S(a) = F(a)+C = 0, \quad \text{by (2)} \\
\Rightarrow \quad & C = -F(a)
\end{aligned}$$

$$\begin{aligned}
\text{So} \quad & S(x) = F(x)-F(a) \\
\Rightarrow \quad & S(b) = F(b)-F(a) \\
\Rightarrow \quad & \int_a^b f(x)dx = F(b)-F(a), \quad \text{by (1).}
\end{aligned}$$

This result is known as the Fundamental Theorem of Calculus; it links differentiation and integration. We have now united the three different aspects of integration developed in Sections 2, 4 and 5, namely:

(i) the *notation* for the area under a curve, $\int_a^b f(x)dx$

(ii) the *formula* for the area under a curve, $\int_a^b f(x)dx = F(b)-F(a)$, where

(iii) the *key* to using the formula is the fact that F is any anti-derivative of f whose domain includes interval $[a, b]$.

Note. A convenient abbreviation for $F(b)-F(a)$ is $\Big[F(x)\Big]_a^b$, the 'square brackets' notation. We then have the area between the curve $y=f(x)$, the x-axis, and the lines $x=a$ and $x=b$ given by the *definite integral,*

$$\int_a^b f(x)dx = \Big[F(x)\Big]_a^b = F(b)-F(a).$$

a and b are called the *lower and upper limits of integration.* The interval $[a,b]$ is called the *interval of integration.*

Example 1. Evaluate the definite integral $\int_{-1}^{3}(3x+1)^2\,dx$.

$$\int_{-1}^{3}(3x+1)^2\,dx \doteq \int_{-1}^{3}(9x^2+6x+1)dx = \Big[3x^3+3x^2+x\Big]_{-1}^{3}$$
$$=(81+27+3)-(-3+3-1)=112$$

Example 2. Calculate the area between the curve $y=x^3$, the x-axis and the lines $x=2$ and $x=4$.

$$\text{The area} = \int_2^4 x^3\,dx = \left[\frac{x^4}{4}\right]_2^4 = 64-4=60$$

Example 3. Show that $\int_a^b f(x)dx = -\int_b^a f(x)dx$.

$$\int_a^b f(x)dx = F(b)-F(a) = -[F(a)-F(b)] = -\int_b^a f(x)dx$$

Exercise 4 (Definite integrals)

Evaluate the following definite integrals:

1 $\int_0^1 4x\,dx$ 2 $\int_1^2 6x^2\,dx$ 3 $\int_0^4 x^{1/2}\,dx$ 4 $\int_1^3 \frac{dx}{x^2}$

5 $\int_1^2 x^3\,dx$ 6 $\int_0^1 (2x+3)dx$ 7 $\int_0^4 (7-x)dx$

8 $\int_0^2 (3x^2+4x+1)dx$ *9* $\int_1^2 (6x^2+4)dx$ *10* $\int_{-1}^1 (4x^3+3x^2)dx$

11 $\int_{-1}^1 (x+1)^2\,dx$ *12* $\int_{-1}^2 (3x-1)^2\,dx$ *13* $\int_{-1}^2 (1+3v)(1-v)dv$

14 $\int_1^3 \left(x^2-\frac{1}{x^2}\right)dx$ *15* $\int_{-1}^1 \left(t^2+\frac{2}{t^2}\right)dt$ *16* $\int_1^2 \left(6u^2-\frac{2}{u^2}\right)du$

17 $\int_0^a t(t^2-2)dt$ *18* $\int_0^p (u^{1/2}+1)^2\,du$ *19* $\int_1^4 \frac{1+\sqrt{x}}{x^2}dx$

20 $\int_{-1}^1 12x(x+1)(x-1)dx$ *21* $\int_0^1 (x-1)(2x-1)(3x-1)dx$

22 Find a in each case, given that:

a $\int_0^a x^{1/2}\,dx = 18$ *b* $\int_0^a x(1-x)dx = 0$

23 The pressure p and volume v of a gas are related by the equation $pv^{1\cdot 5} = 128$. Evaluate $\int_1^4 p\,dv$.

24 Verify each of the following:

a $\int_1^2 4x^3\,dx+\int_2^3 4x^3\,dx = \int_1^3 4x^3\,dx$ *b* $\int_0^1 \pi x^2\,dx = \pi\int_0^1 x^2\,dx$

c $\int_2^4 (2x+3)dx = \int_2^4 2x\,dx+\int_2^4 3\,dx$

d $\int_0^3 (3x^2+1)dx = -\int_3^0 (3x^2+1)dx$ *e* $\int_{-1}^1 x^4\,dx = \int_{-1}^1 t^4\,dt$

Note: In question **24** we verified in special cases the following general results:

(i) $\int_a^b f(x)dx+\int_b^c f(x)dx = \int_a^c f(x)dx$

(ii) $\int_a^b kf(x)dx = k\int_a^b f(x)dx$

(iii) $\int_a^b [f(x)+g(x)]dx = \int_a^b f(x)dx + \int_a^b g(x)dx$

(iv) $\int_a^b f(x)dx = -\int_b^a f(x)dx$ (Worked Example 3 on page 225.)

(v) $\int_a^b f(x)dx$ depends on a and b, not on x; x is a *dummy variable.*

Area between curve and x-axis

Figure 9 shows the graph of the function $f: R \to R$ defined by $f(x) = x(x-1)(x-2)$.

A_1 and A_2 represent the areas cut off above and below the x-axis as shown.

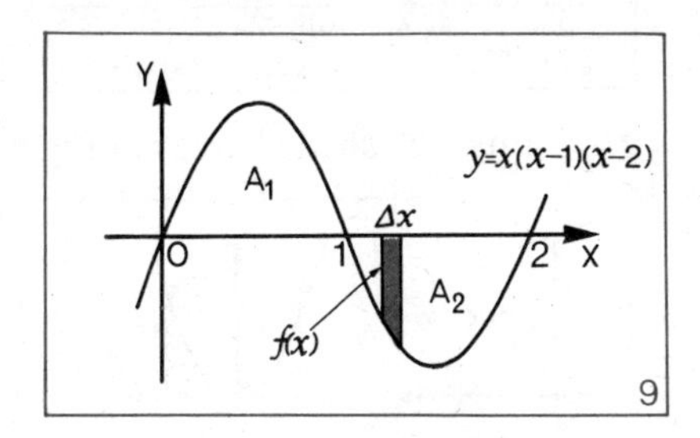

$$A_1 = \int_0^1 x(x-1)(x-2)dx$$
$$= \int_0^1 (x^3 - 3x^2 + 2x)dx$$
$$= [\tfrac{1}{4}x^4 - x^3 + x^2]_0^1$$
$$= (\tfrac{1}{4} - 1 + 1) - 0$$
$$= \tfrac{1}{4}$$

$$A_2 = \int_1^2 x(x-1)(x-2)dx$$
$$= \int_1^2 (x^3 - 3x^2 + 2x)dx$$
$$= [\tfrac{1}{4}x^4 - x^3 + x^2]_1^2$$
$$= (4 - 8 + 4) - (\tfrac{1}{4} - 1 + 1)$$
$$= -\tfrac{1}{4}$$

To see why A_2 is negative, we have to remember that

$$\int_a^b f(x)dx = \lim_{\Delta x \to 0} \sum_{x=a}^{x=b} f(x).\Delta x.$$

In the region A_2, each $f(x)$ is negative, so that each product $f(x).\Delta x$ is negative. Consequently $\int_1^2 x(x-1)(x-2)dx$ represents the magnitude of A_2, but with the negative sign. For this reason, it is essential to draw a sketch graph of such a function f for the appropriate interval, and to calculate the areas of the separate regions A_1, A_2, etc. In the above example, the total area cut off between the curve and the x-axis is $\frac{1}{4}+\frac{1}{4}$, i.e. $\frac{1}{2}$.

Exercise 5 (Area under a curve)

1 Calculate the shaded areas in Figure 10 by integration, and check your results by using the appropriate goemetrical formulae.

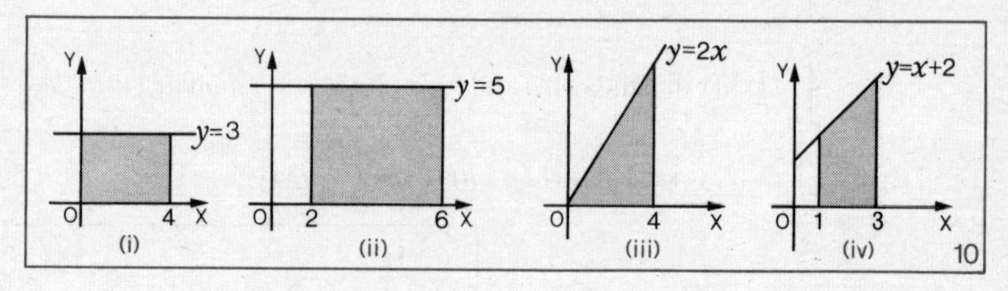

10

2 Find the shaded areas in Figure 11 by means of integration.

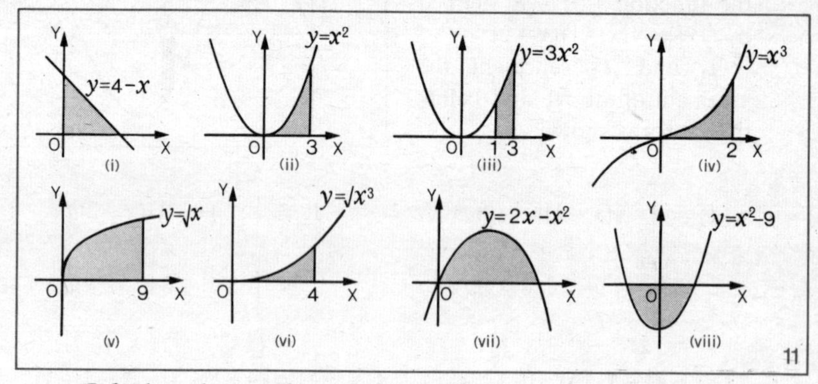

11

Calculate the area bounded by each of these curves and the x-axis:

3 $y = 1 - x^2$ *4* $y = x^2 - 3x$ 5 $y = x^2(3 - x)$

6 Sketch the curve $y = (x-2)^2$, and calculate the area enclosed by the curve, the x-axis and the y-axis.

7 The rectangle formed by the coordinate axes and the lines $x = 3$ and $y = 9$ is divided into two parts by the parabola $y = x^2$. Make a sketch, and find the area of each part.

8 Calculate $\int_0^2 (x^3 - 3x^2 + 2x)dx$. Explain your answer with reference to a sketch.

9 Find the points of intersection of the curve $y = x^3 - 4x$ with the x-axis. Calculate the area enclosed between the curve and the x-axis.

10 Repeat question **9** for the curve $y = x(1-x^2)$.

11 Repeat question **9** for the curve $y = x^3 - x^2 - 6x$.

12 In Figure 12(i), the curve $y = x^2 - 4x + 3$ cuts the y-axis at A and the x-axis at B and C. Find the coordinates of A, B and C, and calculate the total area of the shaded parts.

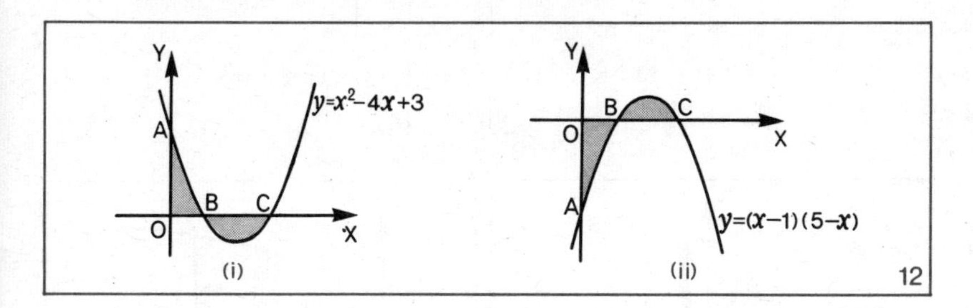

13 Repeat question **12** for the curve $y = (x-1)(5-x)$ shown in Figure 12(ii).

14 Show that the area bounded by the parabola $y = (x+2)(x-4)$ and the x-axis is divided in the ratio 7:20 by the y-axis.

15a Sketch the curve $y = \dfrac{1}{x^2}$ in the interval from $x = 1$ to $x = 4$ and find the area enclosed by the curve, the x-axis and the ordinates at $x = 1$ and $x = 4$.

b Find the real number k such that the line $x = k$ divides this area into two equal parts.

16a Find the coordinates of the maximum and minimum turning points M and N of the curve $y = x(x^2 - 9x + 24)$.

b Calculate the area enclosed by arc MN, the ordinates through M and N, and the x-axis.

6 *The area between two curves*

To calculate the shaded area between the curve $y = 4 - x^2$ and the line $y = 3x$ in Figure 13, we first find the interval of integration.

At the points of intersection of the line and curve,

$$3x = 4-x^2 \Leftrightarrow x^2+3x-4 = 0 \Leftrightarrow (x+4)(x-1) = 0 \Leftrightarrow x = -4 \text{ or } 1.$$

The area of the darker strip of width Δx is approximately $[(4-x^2)-3x].\Delta x$. Hence the shaded area

$$= \lim_{\Delta x \to 0} \sum_{x=-4}^{x=1} (4-x^2-3x).\Delta x = \int_{-4}^{1} (4-x^2-3x)dx$$

$$= [4x-\tfrac{1}{3}x^3-\tfrac{3}{2}x^2]_{-4}^{1} = (4-\tfrac{1}{3}-\tfrac{3}{2})-(-16+\tfrac{64}{3}-24)$$

$$= 2\tfrac{1}{2}-\tfrac{1}{3}+40-\tfrac{64}{3} = 20\tfrac{5}{6}.$$

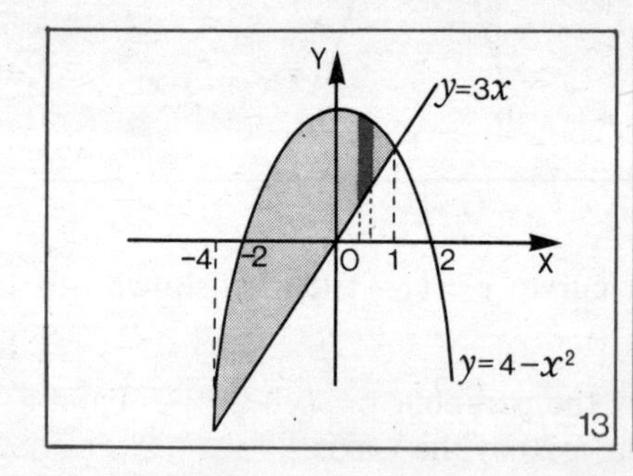

13

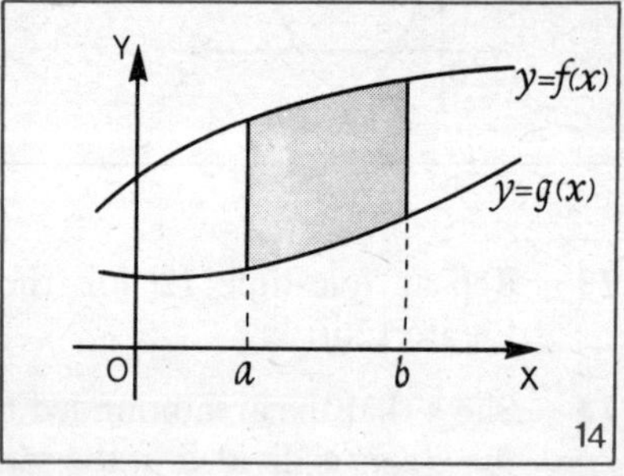

14

In general, the area enclosed between the curves $y=f(x)$ and $y=g(x)$ in the interval $[a,b]$ is given by $\int_a^b [f(x)-g(x)]dx$, when $f(x) \geqslant g(x)$ and $a \leqslant x \leqslant b$ (see Figure 14).

Exercise 6

In this Exercise, illustrate every answer with a sketch.

Calculate the areas enclosed by the lines and curves whose equations are given in the following questions.

1 $y = x$, $y = 3x$ and $x = 5$

2 $y = x^2$ and $y = x$

3 $y = x^3$ and $y = x$

4 $y = x^2$ and $y = x^3$

5 $y = x^2$ and $y^2 = x$

6 $y^2 = x$ and $y = \frac{1}{2}x$

7 $y = x^2-2x$ and $y = 2x$

8 $y = x^2-2x$ and $y = 6x-x^2$

9 $y = (x+2)^2$ and $y = 10-x^2$

10 $y = x^2$ and $y = x+2$

11 $y = x^2 - 7x + 10$ and $y = 2 - x$ *12* $y = 4 - x^2$ and $y = x^2(x^2 - 4)$

13a Calculate the area S bounded by the curve $y = x(6-x)$ and the x-axis in the first quadrant.

b Find the points of intersection of this curve and the curve $y = x^2$, and show that the curve $y = x^2$ divides S in the ratio 1:3.

14a Find the equation of the tangent to the parabola $y = x(4-3x)$ at the point A(1, 1) on the curve.

b Calculate the area bounded by the tangent, the y-axis and the arc OA of the curve, where O is the origin.

15a Find the point of intersection of the tangents to the parabola $y = x^2 + 2$ at the points P(-2, 6) and Q(1, 3).

b Hence calculate the area bounded by the tangents and the arc PQ.

7 *Volumes of solids of revolution*

If a plane figure is rotated through one complete revolution about a line, called the *axis of rotation*, it is said to sweep out, or generate, a *solid of revolution*. For example, in Figure 15 a cylinder is generated by rotating a rectangle about one of its sides; a cone by rotating a right-angled triangle about one of the sides containing the right angle, and a sphere by rotating a semicircle about its diameter.

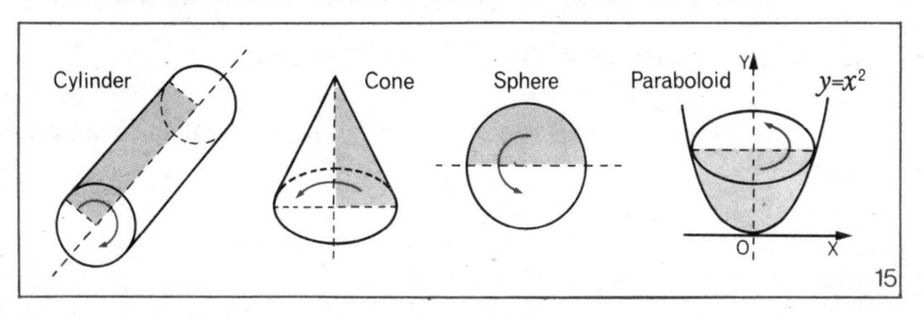

(i) Volume of solid of revolution about the x-axis

The formula $V = \frac{1}{3}\pi r^2 h$ for the volume of a right circular cone, height h and radius of base r, was obtained in an intuitive manner in Book 5. We can now use a calculus method to obtain this formula.

The special case in which $r = h = 1$ is considered, but the method is quite general. The cone can be generated by rotating the right-angled triangle shown in Figure 16(ii) through one complete revolution about the x-axis.

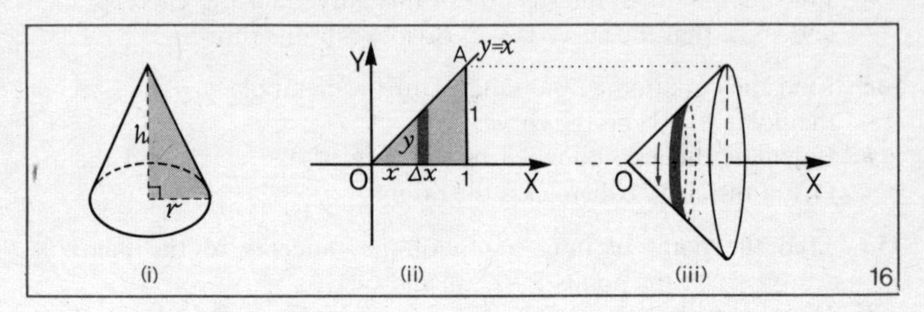

This triangular area lies between the line $y = x$, the x-axis and the line $x = 1$. The volume obtained by rotating the strip of width Δx is approximately the volume of a circular cylinder of radius y and height Δx, i.e. $\pi y^2 . \Delta x$, or $\pi x^2 . \Delta x$ since $y = x$ (see Figure 16(ii)).

Hence, an approximating sum to the volume of the cone is $\sum_{x=0}^{x=1} \pi x^2 . \Delta x$. Taking the limit of this sum as in Section 4,

$$\text{volume } V = \lim_{\Delta x \to 0} \sum_{x=0}^{x=1} \pi x^2 . \Delta x = \pi \int_0^1 x^2\,dx = \pi[\tfrac{1}{3}x^3]_0^1 = \tfrac{1}{3}\pi.$$

In general, if the area under the curve $y = f(x)$ for which $a \leqslant x \leqslant b$ is rotated through a complete revolution about the x-axis, the

$$\text{volume } V = \lim_{\Delta x \to 0} \sum_{x=a}^{x=b} \pi[f(x)]^2 \Delta x \text{ (see Figure 17).}$$

Hence the volume of the solid of revolution about the x-axis is given by

$$V = \pi \int_a^b [f(x)]^2 dx = \pi \int_a^b y^2\,dx$$

Example. The area bounded by the curve $y = x^2$, the x-axis, and the line $x = 2$ is rotated through 360° about the x-axis. Find the volume of the solid formed (see Figure 18).

$$V = \pi \int_0^2 y^2\,dx = \pi \int_0^2 x^4\,dx = \pi[\tfrac{1}{5}x^5]_0^2 = \tfrac{32}{5}\pi$$

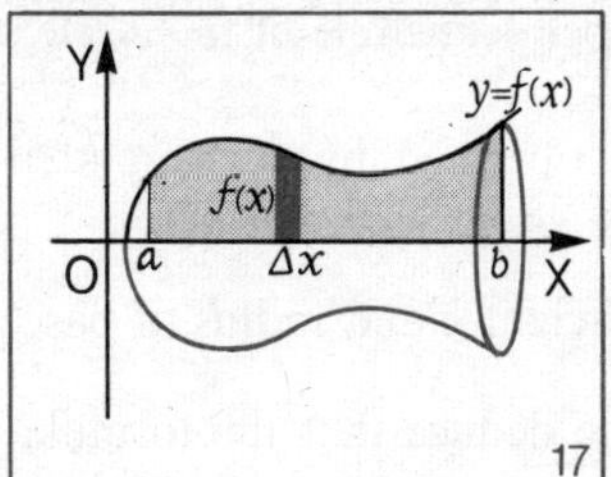

17

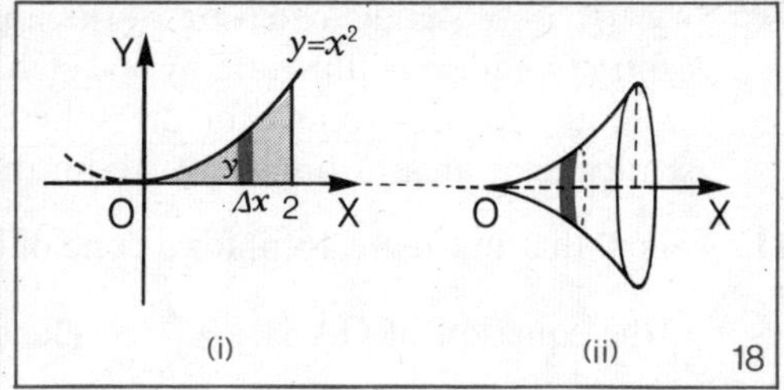

(i) (ii) 18

Exercise 7

1 If each of the shaded areas in Figure 19 is rotated through 2π radians about the x-axis, calculate the volume of the solid of revolution generated.

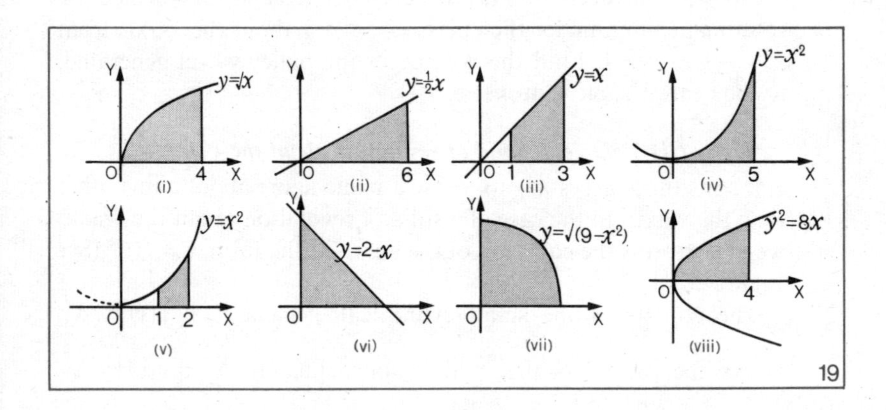

19

Calculate the volumes of the solids of revolution formed when the regions bounded by the following curves and the x-axis are rotated through one revolution about the x-axis.

2 $y = \frac{4}{x}$, $x = 1$ and $x = 4$

3 $y = 2\sqrt{x}$, $x = 0$ and $x = 1$

4 $y = 3x - 1$, $x = 1$ and $x = 3$

5 $x + 2y = 2$, $x = 0$ and $x = 2$

6 $y = \frac{1}{x^2}$, $x = \frac{1}{3}$ and $x = \frac{1}{2}$

7 $y = x^2 + 1$, $x = 0$ and $x = 1$

8 $y = x(x-1)$

9 $y = x(2-x)$

10 $y = \sqrt{(r^2 - x^2)}$ is the equation of the upper semicircle of the circle, centre O, radius r. Illustrate by a sketch.

Show that the volume of the sphere obtained by rotating the semicircular area through 360° about the x-axis is $\frac{4}{3}\pi r^3$.

11 Show that in Figure 16(ii) for a cone of height h and radius of base r, the equation of OA is $y = \frac{r}{h}.x$. Hence deduce that the formula for the volume of the cone is $V = \frac{1}{3}\pi r^2 h$.

12 Origin O, A(1, 3), B(4, 0) are the vertices of triangle OAB. This triangular area is rotated through one revolution about the x-axis. Calculate the volume of the solid of revolution generated.

13 A pulley-wheel of external diameter 12 cm and thickness 2 cm has a groove 1 cm deep. It may be considered as being formed by rotating the area under the curve $y = x^2 + 5$ about the x-axis from $x = -1$ to $x = 1$. Find the volume of the pulley-wheel generated, to the nearest cubic centimetre.

(ii) Volume of solid of revolution about the y-axis

It is sometimes necessary to rotate a plane figure about a line other than the x-axis. In the case of a solid of revolution about the y-axis, we first express the equation of the curve in the form $x = f(y)$. (See Figure 20.)

The volume of the small cylindrical element is $\pi[f(y)]^2.\Delta y$.

Hence the volume of the solid is approximately $\sum_{y=c}^{y=d} \pi[f(y)]^2.\Delta y$.

Taking the limit of this sum when $\Delta y \to 0$, we obtain

$$V = \pi \int_c^d [f(y)]^2\,dy = \pi \int_c^d x^2\,dy$$

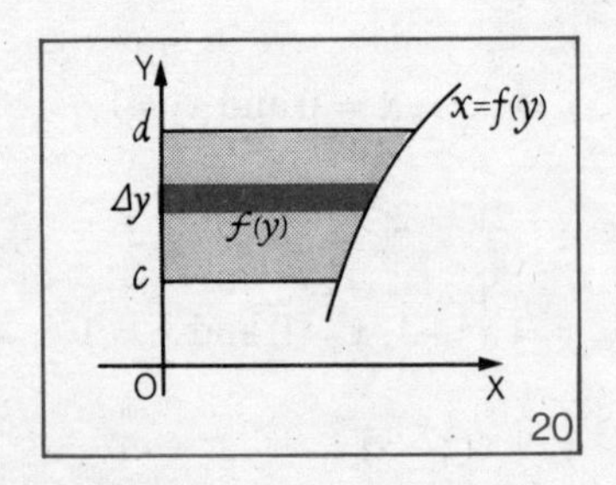

20

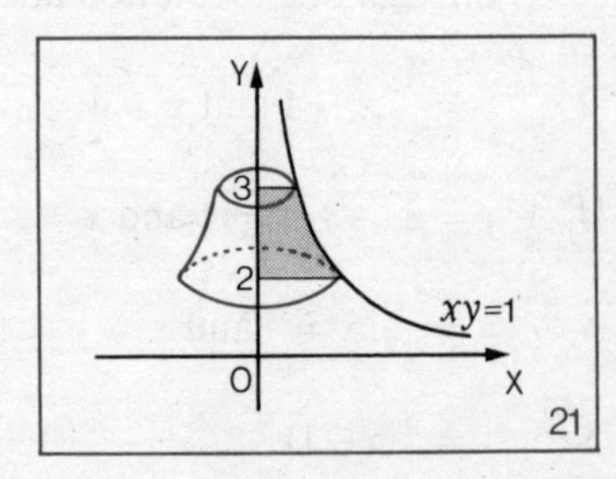

21

Example. Find the volume of the solid formed by rotating the area bounded by the y-axis, the hyperbola $xy = 1$ and the lines $y = 2$ and $y = 3$ through 2π radians about the y-axis as shown in Figure 21.

$$V = \pi \int_2^3 x^2\,dy = \pi \int_2^3 \frac{dy}{y^2} = \pi \int_2^3 y^{-2}\,dy = \pi\left[-\frac{1}{y}\right]_2^3$$
$$= \pi[(-\tfrac{1}{3})-(-\tfrac{1}{2})] = \pi(-\tfrac{1}{3}+\tfrac{1}{2}) = \tfrac{1}{6}\pi$$

Exercise 8

Calculate the volumes of the solids formed when the regions in the first quadrant bounded by the y-axis and the following curves are rotated through 2π radians about the y-axis.

1. $x = \sqrt{y}$, and $y = 4$
2. $x = y^2$, and $y = 1$
3. $xy = 1$, $y = 3$ and $y = 6$
4. $x = y^{1/3}$, and $y = 8$
5. $y = x^2 - 1$, $y = 0$ and $y = 1$
6. $y = 4 - x^2$, $y = -4$ and $y = 4$
7. $xy^2 = 2$, $y = 2$ and $y = 4$
8. $y = x + 1$, $y = 1$ and $y = 4$
9. $x = y^2 + 1$, $y = -1$ and $y = 1$
10. $x^2 = y(1-y)^2$, $y = 0$, $y = 1$

(iii) Volume of solid of revolution for the area between two curves

Example. Calculate the volume of the solid of revolution generated by the rotation of the area between the curve $y = x^2$ and the line $y = 2x + 3$ about the x-axis.

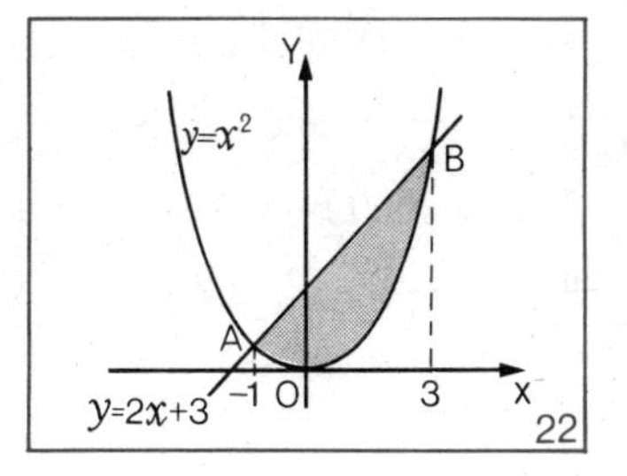

The curve and the line intersect where

$$x^2 = 2x + 3$$
$$\Leftrightarrow \quad x^2 - 2x - 3 = 0$$
$$\Leftrightarrow \quad (x+1)(x-3) = 0$$
$$\Leftrightarrow \quad x = -1 \text{ or } 3$$

Volume of revolution of area under AB $= \pi \displaystyle\int_{-1}^{3} (2x+3)^2\,dx$

Volume of revolution of area under arc AOB $= \pi \displaystyle\int_{-1}^{3} (x^2)^2\,dx$

Volume of revolution of shaded area

$$= \pi \int_{1}^{3} (2x+3)^2\, dx - \pi \int_{-1}^{3} (x^2)^2\, dx$$

$$= \pi \int_{-1}^{3} (4x^2+12x+9-x^4) dx = \pi[\tfrac{4}{3}x^3+6x^2+9x-\tfrac{1}{5}x^5]_{-1}^{3}$$

$$= \pi[(36+54+27-48\tfrac{3}{5})-(-\tfrac{4}{3}+6-9+\tfrac{1}{5})] = 72\tfrac{8}{15}\pi.$$

In general, the volume of the solid of revolution formed by rotating the area between two curves about the x-axis is

$$V = \pi \int_a^b [(f(x))^2-(g(x))^2]\, dx = \pi \int_a^b (y_1{}^2 - y_2{}^2)\, dx.$$

where $y_1 = f(x)$ and $y_2 = g(x)$.

Exercise 9B

Calculate the volumes of the solids formed when the areas enclosed by the following curves are rotated through 2π radians about the x-axis.

1 $y = x$, and $y = x^2$

2 $y = x$, and $y = x^3$

3 $y = 2x$, and $y = 2x^2$

4 $y = x^2$, and $y^2 = x$

5 $y = \frac{1}{2}x$, and $y = \sqrt{x}$

6 $y = x^2$, and $y = x^4$

7 $y = \sqrt{(2-x^2)}$, and $y = 1$

8 $y = x^2+3$, and $y = 4$

9 $y = 1+x^2$, and $y = 9-x^2$

10 $y = x^2$, and $y = 6x-x^2$

Summary

1 *Anti-derivative; integration*

A function F such that $F'(x) = f(x)$, for all x in the domain of f, is called an *anti-derivative* of f.

$\int f(x)dx = F(x)+C$ is the *indefinite integral* of $f(x)$; C is the *constant of integration*.

If $f(x) = x^n$, $\int x^n\,dx = \dfrac{x^{n+1}}{n+1}+C, \quad n \neq -1.$

2 *Area as the limit of a sum*

In Figure 1, area $S = \lim\limits_{\Delta x \to 0} \sum\limits_{x=a}^{x=b} f(x).\Delta x = \int_a^b f(x)dx$

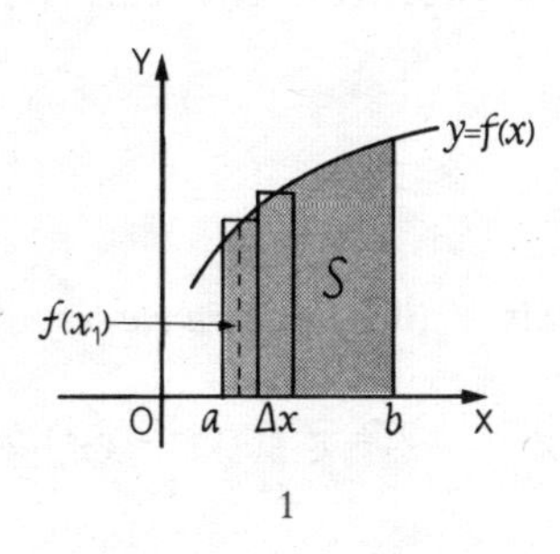

1

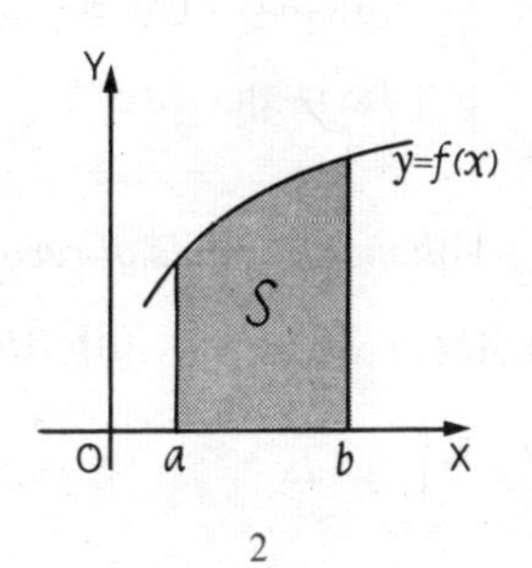

2

3 *Area formula*

In Figure 2, $S = \int_a^b f(x)dx = \Big[F(x) \Big]_a^b = F(b) - F(a)$

$\int_a^b f(x)dx$ is called a *definite integral.*

4 *Area under a curve*

In Figure 3, the magnitude of the shaded area is $\int_a^b f(x)dx$.

In (i) the sign is positive; in (ii) the sign is negative.

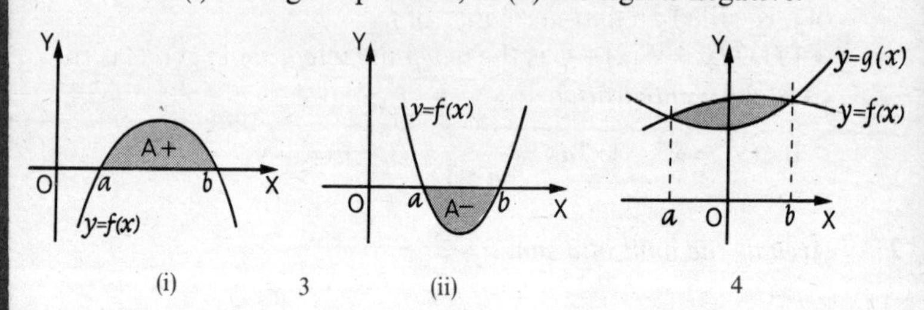

(i) 3 (ii) 4

The shaded area in Figure 4 between the two curves is $\int_a^b [f(x)-g(x)]dx$.

5 *Volumes of solids of revolution*

(i) *About x-axis*

$$V = \pi\int_a^b y^2\,dx$$

(ii) *About y-axis*

$$V = \pi\int_c^d x^2\,dy$$

(iii) *Between curves*

$$V = \pi\int_a^b ({y_1}^2 - {y_2}^2)dx$$

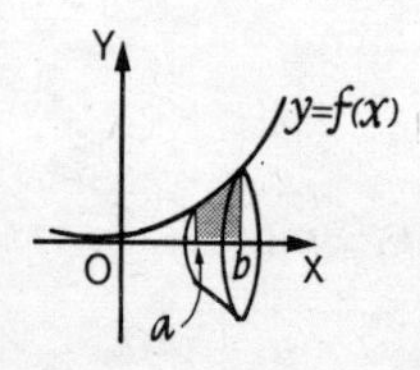

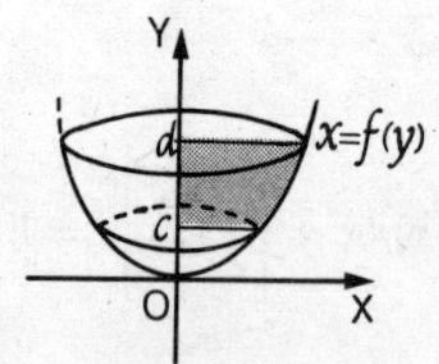

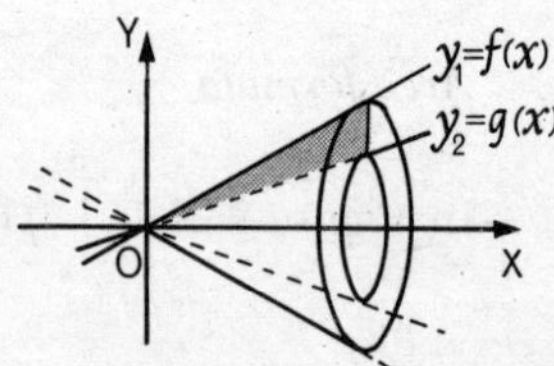

Revision Exercises

Revision Exercise on Chapter 1

An Introduction — The Differential Calculus

Revision Exercise 1

1 The distance s metres travelled by a car in t seconds is given by the formula $s = 3t^2$. Calculate the speed after 2 seconds by evaluating $\lim_{h \to 0} \dfrac{f(2+h)-f(2)}{h}$.

2 Differentiate the following functions from first principles, i.e. by using the definition $f'(x) = \lim_{h \to 0} \dfrac{f(x+h)-f(x)}{h}$.

a $f(x) = 5x$ *b* $f(x) = 3x^2$ *c* $f(x) = x^2+2x$

d $f(x) = \dfrac{2}{x}$ *e* $f(x) = \dfrac{1}{3x^2}$ *f* $f(x) = 4x+\dfrac{1}{x}$

3 Differentiate the following with respect to the appropriate variables:

a $2x^6+4x^3+2$ *b* $\frac{2}{5}x^5+\frac{7}{2}x^2$ *c* $2t+\dfrac{1}{2t}$

d $2-u-\dfrac{2}{u^2}$ *e* $\dfrac{2}{r^2}+\dfrac{3}{r^3}$ *f* $x-\dfrac{1}{x^{1/3}}$

g $\sqrt{x}-\dfrac{4}{\sqrt{x}}$ *h* $p^2+p+\dfrac{2}{5p}$ *i* $x^{3/2}(x^{1/2}+1)$

4 Differentiate the following:

a $(x+1)(3x+1)$ *b* $(x^2+1)^2$ *c* $(1-x)^3$

d $\dfrac{u+1}{u}$ *e* $\dfrac{r^2+1}{2r}$ *f* $\dfrac{(2x+1)^2}{x}$

g $\dfrac{(t+1)(t-2)}{t^2}$ *h* $\dfrac{x^2-3x+4}{x^{1/3}}$ *i* $\dfrac{x^{7/2}+x^{3/2}+1}{x^{1/2}}$

5 Given $f(x) = 4x^2+8x+2$, find $f'(0)$, $f'(-2)$ and $f'(\frac{1}{2})$. Find x for which $f'(x) = 0$, and the corresponding value of f.

6 The equation of a curve is $y = 4x^2-6x+2$. Find the value of $\dfrac{dy}{dx}$ at: *a* $x = 1$ *b* $x = -1$ *c* $x = 0$.

Find also the equations of the tangents to the curve at the points with these x-coordinates.

7 A curve has equation $y = 2x^{3/2}+1$. Calculate the gradient of the curve at: *a* $x = 4$ *b* $x = 16$.

Find the point on the curve at which the gradient is 6, and then find the equation of the tangent to the curve at this point.

8 Find the equations of the tangents to the curve $y = x^2-2x+2$ at the points given by $x = 0$ and $x = 2$. Prove that these tangents intersect on the x-axis.

9 Show that the point A$(\frac{1}{2}, 2\frac{1}{2})$ lies on the curve $y = x+\dfrac{1}{x}$. The tangent at A meets the x- and y-axes at C and D. Show that the length of CD is $\frac{4}{3}\sqrt{10}$ units.

10 Find the intervals in which each of the following functions is increasing or decreasing.

a $f(x) = 2x^2-10x$ *b* $f(x) = x^3-3x^2-6$.

11 Prove that for all real values of x, the function $f: x \to \frac{1}{5}x^5+\frac{2}{3}x^3+x$ is increasing. Find the value of x at which the rate of increase of f with respect to x is least.

12 Find the stationary values of the following functions, and determine the nature of each.

a $f(x) = 4-x-3x^2$ *b* $g(x) = x^3+3x^2-9x-1$

c $h(x) = x(x-3)^2$ *d* $q(x) = 2x+\dfrac{1}{x^2}$

13 Find the stationary points and the nature of each for the following curves, and then sketch the curves:

a $y = x^2 - 6x + 8$ *b* $y = x^3 + 9$ *c* $y = (x^2 - 1)^2$
d $y = x^3(8 - x)$ *e* $y = (x - 1)^3$ *f* $y = x - 2x^{1/2}, x > 0$

14 ABCD is a square of side 6 cm. E is a point on AB such that BE = $2x$ cm, and F is a point on BC such that BF = $2x$ cm. Show that if the area of triangle DEF is A cm^2, then $A = 12x - 2x^2$. Hence find the maximum area of the triangle.

15 The area of a rectangle is 16 cm^2. If one of its sides is x cm long, show that the perimeter P cm is given by $P = 2x + \frac{32}{x}$. Find the dimensions of the rectangle which has minimum perimeter.

16 A ball thrown upwards with a velocity of 20 m/s travels according to the equation $h = 20t - 5t^2$, where h is the height in metres above the starting point after t seconds.

a Find h when t is 0, 2 and 4.
b Find the velocity after 1, 2 and 3 seconds.
c Explain your results in ***a*** and ***b***.

17 The distance s metres of a train from its starting point after t seconds is given by the formula $s = t^3 - 2t^2 + 3t$. Calculate:

a the average speed of the train over the interval $t = 1$ to $t = 2$
b the speed of the train at the end of 2 seconds.

18 A spherical balloon of radius r mm is inflated. Find the rate of change of its volume with respect to the radius at $r = 20$.

Revision Exercise on Chapter 2

The Integral Calculus

Revision Exercise 2

1 Integrate:

a x^2+x+1 *b* x^5-x^3+7 *c* $3-2x-6x^2$

d $(x-1)(3x-5)$ *e* $(3x+4)^2$ *f* $x(x-2)(x+2)$

2 Integrate:

a $x^{1/2}-x^{-1/2}$ *b* $x^{1/3}(x^{2/3}-x^{-1/3})$ *c* $(x^{1/2}+1)^2$

d $\dfrac{1}{x^2}-\dfrac{1}{x^3}$ *e* $\dfrac{x^2-3}{x^2}$ *f* $\dfrac{x^{1/2}+x^{-1/2}}{x}$

3 Given $f'(x)$, find $f(x)$ for each of the following:

a $f'(x)=8x-3$, and $f(-1)=10$ *b* $f'(x)=x^2-\dfrac{1}{x^2}$, and $f(1)=\frac{1}{3}$

4 Find:

a $\displaystyle\int\left(3x^2+\frac{2}{x^3}\right)dx$ *b* $\displaystyle\int\left(3x^{1/2}+\frac{1}{3x^{1/3}}\right)dx$ *c* $\displaystyle\int\left(\frac{1}{4x^2}+\frac{4}{x^3}\right)dx$

d $\displaystyle\int\left(3x+\frac{1}{x^3}\right)^2 dx$ *e* $\displaystyle\int(2\sqrt{x}-1)^2\,dx$ *f* $\displaystyle\int\frac{3x^2+4x-5}{\sqrt{x}}dx$

5 The gradient of the tangent to a curve at each point (x, y) is given by $\dfrac{dy}{dx}=4x-\dfrac{4}{x^2}$. If the curve passes through the point $(2, 11)$, find its equation.

6 Evaluate the following definite integrals:

a $\displaystyle\int_0^1(9x^2+1)dx$ *b* $\displaystyle\int_1^2(3x^2+4x-5)dx$ *c* $\displaystyle\int_{-1}^0(2-2x)dx$

d $\displaystyle\int_0^1(x+1)^3\,dx$ *e* $\displaystyle\int_1^4\left(\sqrt{x}-\frac{1}{\sqrt{x}}\right)dx$ *f* $\displaystyle\int_1^4\left(3\sqrt{x}+\frac{1}{3\sqrt{x}}\right)dx$

7 Find a, given:

a $\int_1^a (2x+1)dx = 4$ *b* $\int_{-1}^{2a} \frac{dx}{x^2} = \frac{1}{2}$ *c* $\int_a^4 \sqrt{x}\, dx = 0$

8 Show by shading in sketches the areas associated with the following and then calculate the areas:

a $\int_{-1}^{2} (2x+4)dx$ *b* $\int_{-6}^{6} x^2\, dx$ *c* $\int_{-2}^{2} (x^3-4x)dx$

9 Find the area enclosed between each of these curves and the x-axis:

a $y = (x-1)(x+2)$ *b* $y = (2-x)(4+x)$ *c* $y = x(x-2)(2x-1)$

10a Find the area enclosed by the x-axis, the parabola $y = x^2$ and the line $x = 4$.

b Find the ratio in which the line $x = 2$ divides this area.

11 Sketch the following pairs of curves, find their points of intersection and calculate the area enclosed by the curves in each case.

a $y = x^2$ and $y = 9$ *b* $y = x^2$ and $y = 2-x^2$

c $y = 5+2x-x^2$ and $y = 1-x$ *d* $y = x(x-3)$ and $y = 2x(3-x)$

e $y = 10-x^2$ and $y = (x-2)^2$ *f* $y = 9-x^2$ and $y = x^2(x^2-9)$

12 The tangents at the points A(0, 1) and B(2, 5) on the parabola $y = x^2+1$ meet at C. Calculate the area between the tangents and the arc AB of the parabola.

13 Find the volumes of the solids of revolution formed by rotating through 2π radians about the x-axis the areas enclosed by the x-axis and the following curves:

a $y = x^2$, $x = 1$ *b* $y^2 = 4x$, $x = 3$ *c* $y = x^2+4$, $x = 1$, $x = -1$

14 Find the volumes of the solids of revolution formed by rotating through 2π radians about the y-axis the areas enclosed by the y-axis and the following curves:

a $y = x^2$, $y = 6$ *b* $xy = 4$, $y = 1$, $y = 2$

15 Find the volume of the solid formed when the area enclosed by the axes and the part of the ellipse $\frac{1}{4}x^2+\frac{1}{9}y^2 = 1$ lying in the first quadrant is rotated through 2π radians about:

a the x-axis *b* the y-axis.

16 Find the volumes of the solids of revolution formed by rotating the area between the curves:

a $y^2 = x$ and $y = x$ through 2π radians about the x-axis
b $y^2 = x$ and $y = x^2$ through 2π radians about the y-axis.

17 Find the volumes of the solids of revolution formed by rotating the area enclosed by the lines $x = 0$, $y = 0$, $x = 2$ and $2x + y = 5$ through 2π radians about: ***a*** the x-axis ***b*** the y-axis.

Answers

Answers

Algebra — Answers to Chapter 1

Page 5 Exercise 1

1	$\{(1, 0)\}$	*2*	$\{(-1, 2)\}$	*3*	$\{(\frac{1}{2}, \frac{1}{3})\}$	*4*	$\{(0, 5)\}$
5	$\{(4, 0)\}$	*6*	$\{(2, 5)\}$	*7*	$\{(1\frac{1}{3}, -2\frac{1}{3})\}$	*8*	$\{(0, 2)\}$
9	$\{(3, -2)\}$	*10*	$\{(1\frac{1}{2}, \frac{1}{2})\}$	*11*	$\{(3, 1)\}$	*12*	$\{(-2, 5)\}$

Page 6 Exercise 2

1	$\{(2, 1, 1)\}$	*2*	$\{(3, 2, 1)\}$	*3*	$\{(1, 1, 1)\}$	*4*	$\{(1, 3, 2)\}$
5	$\{(3, 1, -2)\}$	*6*	$\{(-2, -1, 1)\}$	*7*	$\{(5, 0, 1)\}$	*8*	$\{(4, -1, 0)\}$
9	$\{(4, -7, 2)\}$	*10*	$\{(1, 2, 3)\}$	*11*	$\{(2, 3, -1)\}$	*12*	$\{(0, -1, 0)\}$
13	$\{(1, 3, -2)\}$	*14*	$\{(-1, -2, -3)\}$	*15*	$\{(5, 2, 1)\}$	*16*	$\{(2, 3, -1)\}$
17	$\{(1, -2, -3)\}$	*18*	$\{(-2, 4, 3)\}$	*19*	$a = 1, b = -1, c = 2; y = x^2 - x + 2$		

20 $a = 2, b = 3, c = -4; y = 2x^2 + 3x - 4$

21 $a = -2, b = -2, c = 1; x^2 + y^2 - 2x - 2y + 1 = 0$

22 $a = -4, b = -2, c = -20; x^2 + y^2 - 4x - 2y - 20 = 0$

23 $a = 4, b = -7, c = 2; u_n = 4n^2 - 7n + 2$

24 $p = 3, q = 2, r = 1; u_n = 3n^2 + 2n + 1$ *25* $\{(1, 2, 1)\}$ *26* $\{(60, 40, 20)\}$

Page 9 Exercise 3

1	$\{(-2, 4), (3, 9)\}$	*2*	$\{(-3, 9), (1, 1)\}$	*3*	$\{(0, 0), (6, 12)\}$

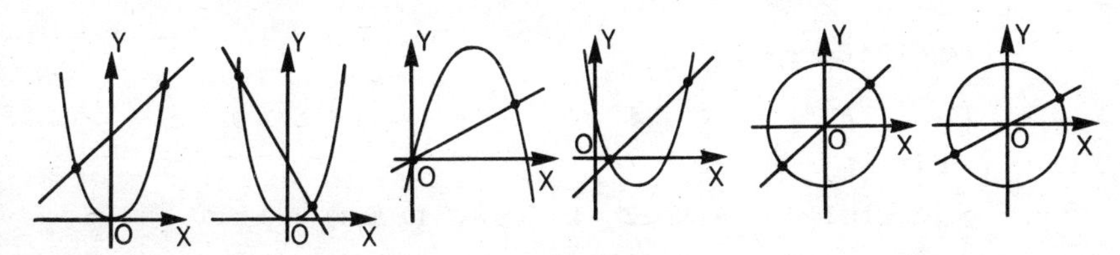

4	$\{(1, 0), (6, 5)\}$	*5*	$\{(-3, -3), (3, 3)\}$	*6*	$\{(-4, -2), (4, 2)\}$
7	$\{(6, 9), (-6, 9)\}$	*8*	$\{(5, 4), (5, -4)\}$	*9*	$\{(-4, 16), (2, 4)\}$
10	$\{(1, 4), (5, 20)\}$	*11*	$\{(-8, -1), (1, 8)\}$	*12*	$\{(-2, -2), (1, 4)\}$
13	$\{(-4, -3), (3, 4)\}$	*14*	$\{(0, 1), (2, 0)\}$	*15*	$\{(1, 2), (4, -4)\}$

16 $\{(-1, -1), (0, -\frac{1}{2})\}$ *17* $\{(-1, 3), (2, -3)\}$ *18* $\{(-2, 10)\}$

19 $\{(-2, 5), (5, -9)\}$ *20* $\{(0, -10), (5, -5)\}$ *21* $\{(-1, -2), (3, 4)\}$

22 $\{(\frac{1}{2}, 2), (2, 1)\}$ 23 $\{(3\frac{1}{2}, 1), (5, 2)\}$

24 $\{(-2\frac{1}{2}, -2\frac{3}{4}), (2, 4)\}$ 25 $\{(-2, 1), (3, -1)\}$

26 $\{(-\frac{1}{3}, 0), (\frac{2}{3}, 1\frac{1}{3})\}$ *27a* $\{(0, 0), (6, 12)\}$ *b*

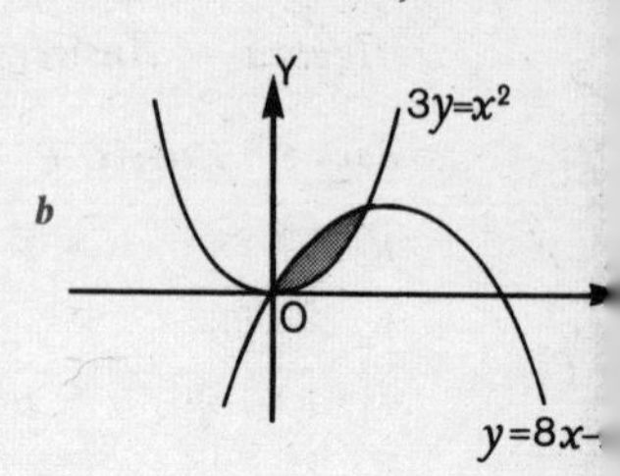

Algebra — Answers to Chapter 2

Page 13 Exercise 1

1 a 2, 4, 6, 8 *b* 1, 4, 9, 16 *c* 0, 1, 2, 3 *d* 3, 5, 7, 9 *e* 0, 3, 8, 15
f 2, 6, 12, 20 *g* 3, 9, 27, 81 *h* −1, 1, −1, 1 *i* $\frac{1}{2}, \frac{2}{3}, \frac{3}{4}, \frac{4}{5}$

2 *a* 5, 6; $u_n = n$ *b* 10, 12; $u_n = 2n$ *c* 15, 18; $u_n = 3n$
d 6, 7; $u_n = n+1$ *e* 25, 36; $u_n = n^2$ *f* 125, 216; $u_n = n^3$
g 32, 64; $u_n = 2^n$ *h* 16, 32; $u_n = 2^{n-1}$ *i* 9, 11; $u_n = 2n-1$

3 *a* $u_n = \frac{1}{2^n}$ *b* $u_n = \frac{1}{2^{n-1}}$ *c* $u_n = \frac{1}{n}$ *d* $u_n = \frac{1}{n(n+1)}$ *e* $u_n = \frac{1}{3n}$ *f* $u_n = \frac{1}{3^n}$

4 *a* $u_n = (-1)^n$ *b* $u_n = (-1)^{n-1}$ or $(-1)^{n+1}$ *c* $u_n = (-2)^{n-1}$ *d* $u_n = (-\frac{1}{3})^{n-1}$

5 *a* 4, 7, 10, 13 *b* 65th 6 *a* 1, 7, 17, 31 *b* 9th

7 *a* 2, 6, 12, 20 *b* 11th 8 *a* 1, 2, 4, 8 *b* 1, 2, 4, 7. First three terms are the same. 9 *a* $u_n = 2n-1$ *b* (*1*) 1, 3, 5 (*2*) 1, 3, 5 *c* Different formulae can give part at least of the same sequence.

Page 15 Exercise 2

1 *a* 2 *b* 4 *c* 5 *d* −1 *e* −17 *f* 0·5 *g* $-\frac{1}{2}$
h 1·4 *i* −20

2 *a* 5, 8, 11, 14 *b* 0, −2, −4, −6 *c* 10, 20, 30, 40
d 11, 13, 15, 17 *e* 24, 23, 22, 21 *f* 18, 23, 28, 33

3 *a* 200 *b* 46 *c* 41 *d* −42 *e* 39 *f* −90

4 *a* $u_n = 2n-1$ *b* $u_n = 3n+1$ *c* $u_n = 4n-3$ *d* $u_n = 2n+3$ *e* $u_n = 3n-1$
f $u_n = n-6$

5 $a = 5, d = 4$ 6 $a = 24, d = -6$ 7 $a = -3, d = -2$

Page 17 *Exercise 3*

1 400 *2* 860 *3* 300 *4* −225 *5* 80 *6* 5050 *7* 820

8 2550 *9* 121 *10* 1010 *11* 1665 *12a* 10 *b* 15

13 12 *14a* 4 or 9; some terms are negative *b* 13 *15* 472

16 5, 9, 12, 14; 5, 4, 3, 2; $u_n = 6-n$

17 −7, −11, −12, −10; −7, −4, −1, 2; $u_n = 3n-10$

18b 21, 28, 36; no *c* 210 *19* £20 500; £20 600

20 second is better after 8 years

Page 19 *Exercise 4*

1 a 3 *b* $\frac{1}{2}$ *c* −2 *d* 3 *e* $\frac{2}{3}$ *f* $\frac{1}{10}$ *g* −1 *h* $\frac{1}{2}$ *i* $\sqrt{3}$

2 a 1, 3, 9, 27 *b* 3, −6, 12, −24 *c* 6, −3, $1\frac{1}{2}$, $-\frac{3}{4}$

3 a 16 *b* 486 *c* $1{\cdot}2^7$ *d* $100 \times 1{\cdot}1^{20}$

4 a $u_n = 2^{n-1}$ *b* $u_n = \dfrac{1}{2^n}$ *c* $u_n = 2(-3)^{n-1}$

d $u_n = 3 \times 2^{n-1}$ *e* $u_n = 4 \times \dfrac{1}{2^{n-1}} = 2^{3-n}$ *f* $u_n = 9 \times \dfrac{1}{3^{n-1}} = 3^{3-n}$

5 ±2; 96 *6* 2; 800 *7* $-\frac{1}{3}; \frac{4}{9}$ *8* 2^{63} (more than a million, million, million) *9 a* 1, 2, 4, 8, 16 *b* 1·28 cm *c* $5{\cdot}6 \times 10^7$ km

Page 20 *Exercise 5*

1 a £5, £105 *c* $100(1{\cdot}05)^3$, $100(1{\cdot}05)^4$, $100(1{\cdot}05)^n$

2 a £r, £(100+r) *c* $100\left(1+\dfrac{r}{100}\right)^3, 100\left(1+\dfrac{r}{100}\right)^4, 100\left(1+\dfrac{r}{100}\right)^n.$

3 1, 1·05, 1·10, 1·16, 1·21, 1·27; 1·62 metres

4 1 000 000, 950 000, 904 000, 859 000, 817 000

5 0·8 *a* £600, £480, £384, £307, £245

6 £520

Page 23 *Exercise 6*

1 255 *2* 728 *3* $1\frac{31}{32}$ *4* $1\frac{40}{81}$ *5* 22 *6* 122

7 $\dfrac{1-x^n}{1-x}$ *8* $\dfrac{1-(-y)^n}{1+y}$ *10a* n *b* 1 if n is odd, 0 if n is even

11a 4 *b* 8 *12* $11\frac{13}{16}$ m *13* $89\frac{71}{81}$ cm *14* $(1+\sqrt{5})'' - 1$ *15* $2'' - 1$; 6

16 18·20 *17* £572 *18* £10 800 *19* £1490 *20* £12 400

(Assuming 3-figure logarithm table accuracy)

Page 27 Exercise 7

1 $\frac{1}{3}$; yes; $1\frac{1}{2}$ *2* 2; no *3* $\frac{1}{4}$; yes; $5\frac{1}{3}$ *4* $\frac{1}{2}$; yes; 16

5 $-\frac{1}{2}$; yes; $\frac{2}{3}$ *6* $\frac{3}{4}$; yes; 38·4 *7* 0·5; yes; 0·2 *8* -5; no

9 $\frac{2}{3}$; yes; 6 *10* 1; no *11* -1; no *12* $-\frac{9}{10}$; yes; $5\frac{5}{19}$

13 $\frac{1}{3}$ *14* $\frac{1}{11}$ *15a* 5, 7·5, 8·75, 9·375, 9·6875 *b* 10

16 $\frac{1}{2}$ *17* -20 *18* no; $r < -1$ *19* $90\frac{10}{11}$, $111\frac{1}{9}$ *20* $\frac{1}{4}, \frac{1}{16}, \frac{1}{4}, \frac{1}{3}$ *21* $\frac{1}{2}, \frac{1}{8}, \frac{1}{4}, \frac{2}{3}$

22a $r = 2x^{-5/3}$ *b* $r = \frac{1}{16}$; $85\frac{1}{3}$ *23* 432 cm *24* 10 m *25* 5 *26* 6

Algebra — Answers to Chapter 3

Page 31 Exercise 1

1 a (*1*) 3 (*2*) 4 *b* 5 6 7 8 *c* 3 7 11 *d* (*1*) 1 (*2*) 11 (*3*) 8

e (*1*) first row, fourth column (*2*) third row, first column (*3*) second row, second column (*4*) third row, third column (*5*) first row, second column (*6*) second row, first column

2 a 2, 2; -4 *b* 2, 3; b *c* 1, 3; 0 *d* 3, 2; 3 *e* 3, 3; -1 *f* 2, 4; 2

3 a $\begin{pmatrix} 3 & 2 \\ 4 & 5 \end{pmatrix}$ *b* $\begin{pmatrix} 2 & -1 \\ 1 & 2 \end{pmatrix}$ *c* $\begin{pmatrix} 2 & 1 \\ 0 & 3 \end{pmatrix}$

5 a $\begin{pmatrix} x' \\ y' \end{pmatrix} = \begin{pmatrix} 3 & 5 \\ 2 & 4 \end{pmatrix}\begin{pmatrix} x \\ y \end{pmatrix}$ *b* $\begin{pmatrix} x' \\ y' \end{pmatrix} = \begin{pmatrix} 2 & 3 \\ 3 & -2 \end{pmatrix}\begin{pmatrix} x \\ y \end{pmatrix}$ *c* $\begin{pmatrix} x' \\ y' \end{pmatrix} = \begin{pmatrix} 3 & -4 \\ -2 & 1 \end{pmatrix}\begin{pmatrix} x \\ y \end{pmatrix}$

d $\begin{pmatrix} x' \\ y' \end{pmatrix} = \begin{pmatrix} \cos\alpha & \sin\alpha \\ -\sin\alpha & \cos\alpha \end{pmatrix}\begin{pmatrix} x \\ y \end{pmatrix}$ *e* $\begin{pmatrix} x' \\ y' \end{pmatrix} = \begin{pmatrix} 1 & 0 \\ 2 & 1 \end{pmatrix}\begin{pmatrix} x \\ y \end{pmatrix}$ *f* $\begin{pmatrix} x' \\ y' \end{pmatrix} = \begin{pmatrix} 0 & 1 \\ 1 & 0 \end{pmatrix}\begin{pmatrix} x \\ y \end{pmatrix}$

6 $\begin{pmatrix} 0 & 1 & 0 & 0 & 0 \\ 1 & 0 & 1 & 1 & 1 \\ 0 & 1 & 0 & 0 & 1 \\ 0 & 1 & 0 & 0 & 1 \\ 0 & 1 & 1 & 1 & 0 \end{pmatrix}$ *7* $\begin{pmatrix} 0 & 2 & 1 & 0 \\ 2 & 0 & 1 & 1 \\ 1 & 1 & 0 & 3 \\ 0 & 1 & 3 & 0 \end{pmatrix}$ *8* $\begin{pmatrix} 0 & 1 \\ 1 & 10 \end{pmatrix}, \begin{pmatrix} 0 & 0 \\ 0 & 1 \end{pmatrix}$

Page 33 Exercise 2

1 a 2×4 *b* 3×2 *c* 3×3 *d* 4×2 *e* 1×3 *f* 3×1

2 a 8 *b* 6 *c* 9 *d* 8 *e* 3 *f* 3

3 a 9 *b* 12 *c* 1 *d* n *e* mn *f* n^2

5 $A = C$, $F = G$, $H = L$

6 $1\times 3, 1\times 3, 1\times 3;\ 2\times 1, 2\times 1, 2\times 1;\ 2\times 1, 2\times 2, 2\times 2;\ 2\times 2, 2\times 2, 1\times 2$

7 ***a*** 4, −3 ***b*** 4, −1 ***c*** 5, 2 ***d*** 2, −3

8 ***a*** 7 ***b*** 3 ***c*** 2 ***d*** 1 ***e*** 8 **9** ***a*** 3, 3 ***b*** 2, 1 or 3, 4

10*a* $\begin{pmatrix} c_{11} & c_{12} & c_{13} \\ c_{21} & c_{22} & c_{23} \\ c_{31} & c_{32} & c_{33} \end{pmatrix}$ ***b*** $\begin{pmatrix} d_{11} & d_{12} & d_{13} & d_{14} \\ d_{21} & d_{22} & d_{23} & d_{24} \\ d_{31} & d_{32} & d_{33} & d_{34} \end{pmatrix}$

11*a* $\begin{pmatrix} 3 & 5 \\ 1 & 4 \\ 4 & 0 \\ 2 & 7 \end{pmatrix}$ 4×2 ***b*** $\begin{pmatrix} 2 & 4 & 1 \\ -1 & 8 & -2 \end{pmatrix}$ 2×3 ***c*** $\begin{pmatrix} a & h & g \\ h & b & f \\ g & f & c \end{pmatrix}$ 3×3 ***d*** $\begin{pmatrix} 1 & 3 & 5 & 7 \\ 2 & 4 & 6 & 8 \end{pmatrix}$ 2×4

e $\begin{pmatrix} -1 \\ 0 \\ 1 \end{pmatrix}$ 3×1 ***f*** $(u \quad v \quad w)$ 1×3 **12** 5, −4

Page 35 Exercise 3

1 $\begin{pmatrix} 7 \\ 6 \end{pmatrix}$ **2** $\begin{pmatrix} 1 \\ 7 \end{pmatrix}$ **3** $\begin{pmatrix} 9a \\ -2b \end{pmatrix}$ **4** $\begin{pmatrix} m+1 \\ n+2 \end{pmatrix}$ **5** $\begin{pmatrix} p+r \\ q+s \end{pmatrix}$ **6** $\begin{pmatrix} 0 \\ 0 \end{pmatrix}$

7 $(3 \quad 9)$ **8** $(-3 \quad 5)$ **9** $(7i \quad 2j \quad 4k)$ **10** $\begin{pmatrix} 3 & 3 \\ 4 & 6 \end{pmatrix}$ **11** $\begin{pmatrix} 12 & 7 \\ 5 & 12 \end{pmatrix}$

12 $\begin{pmatrix} -2 & 3 \\ 4 & -1 \end{pmatrix}$ **13** $\begin{pmatrix} 6 & 0 \\ 0 & 2 \end{pmatrix}$ **14** $\begin{pmatrix} 3 & 1 & 4 \\ -2 & 2 & -4 \end{pmatrix}$ **15** $\begin{pmatrix} 3a & 3b \\ -a & 0 \end{pmatrix}$ **16** $\begin{pmatrix} 3x & 2y \\ -2x & 4y \end{pmatrix}$

17 $\begin{pmatrix} a+f & b+g \\ c+h & d+k \end{pmatrix}, \begin{pmatrix} f+a & g+b \\ h+c & k+d \end{pmatrix}$. Commutative law of addition.

18*a* (1) $\begin{pmatrix} 4 & 6 \\ 6 & 1 \end{pmatrix}$ (2) $\begin{pmatrix} -1 & 7 \\ 6 & -2 \end{pmatrix}$ (3) $\begin{pmatrix} 2 & 9 \\ 7 & -3 \end{pmatrix}$ (4) $\begin{pmatrix} 2 & 9 \\ 7 & -3 \end{pmatrix}$

b Yes. Associative law of addition.

20 $A+B = B+A = \boldsymbol{O}$, so A and B are additive inverses.

21*a* $\begin{pmatrix} -3 \\ -2 \end{pmatrix}$ ***b*** $\begin{pmatrix} -5 \\ 3 \\ -4 \end{pmatrix}$ ***c*** $\begin{pmatrix} -4 & 7 \\ 5 & -8 \end{pmatrix}$ ***d*** $\begin{pmatrix} 3 & -1 & 0 \\ -4 & 2 & 1 \end{pmatrix}$ **22** $\begin{pmatrix} -2 & 1 & -4 \\ -7 & -6 & 3 \end{pmatrix}$

Page 38 Exercise 4

1 ***a*** $\begin{pmatrix} 3 \\ 1 \end{pmatrix}$ ***b*** $\begin{pmatrix} 5 \\ 2 \end{pmatrix}$ ***c*** $\begin{pmatrix} 0 \\ 4 \end{pmatrix}$ ***d*** $\begin{pmatrix} -3a \\ 4b \end{pmatrix}$ ***e*** $\begin{pmatrix} x-1 \\ 1-2y \end{pmatrix}$ ***f*** $\begin{pmatrix} a-c \\ b-d \end{pmatrix}$

2 ***a*** $\begin{pmatrix} 2 & 4 \\ 1 & 2 \end{pmatrix}$ ***b*** $\begin{pmatrix} 2 & 0 \\ 5 & 7 \end{pmatrix}$ ***c*** $\begin{pmatrix} -3 & 3 \\ -1 & -2 \end{pmatrix}$ ***d*** $\begin{pmatrix} 2x & 3 \\ 2 & 6y \end{pmatrix}$

3 *a* $\begin{pmatrix}-1 & 5\\ 3 & 5\end{pmatrix}$ *b* $\begin{pmatrix}6 & 4\\ 2 & 4\end{pmatrix}$ *c* $\begin{pmatrix}4 & 7\\ 2 & 5\end{pmatrix}$ *d* $\begin{pmatrix}3 & -1\\ 3 & 3\end{pmatrix}$ *e* $\begin{pmatrix}7 & -1\\ -1 & -1\end{pmatrix}$ *f* $\begin{pmatrix}4 & 0\\ -4 & -4\end{pmatrix}$

g $\begin{pmatrix}5 & 9\\ 5 & 9\end{pmatrix}$ *h* $\begin{pmatrix}7 & -1\\ -1 & -1\end{pmatrix}$. $C-B=(A+C)-(A+B)$ **4** $\begin{pmatrix}-6\\ 7\end{pmatrix}$

5 $-2, 3, -3$ **6** *a* $\begin{pmatrix}-1 & 2\\ 2 & 0\end{pmatrix}$ *b* $\begin{pmatrix}2 & -4\\ -1 & -3\end{pmatrix}$ *c* $\begin{pmatrix}7 & -2\\ 3 & 4\end{pmatrix}$ *d* $\begin{pmatrix}4 & 1\\ -1 & 13\end{pmatrix}$

7 *a* $p=5, q=9, r=0, s=2$ *b* $p=5, q=-7, r=-1, s=1$

8 *a* $\begin{pmatrix}4 & 0\\ 4 & 9\\ 9 & 12\end{pmatrix}$ *b* $\begin{pmatrix}-1 & -4\\ 0 & 5\\ 7 & 1\end{pmatrix}$ *c* $\begin{pmatrix}0 & -2\\ 3 & 9\\ 12 & 7\end{pmatrix}$ *d* $\begin{pmatrix}0 & -2\\ 3 & 9\\ 12 & 7\end{pmatrix}$

Page 40 Exercise 5

1 *a* $\begin{pmatrix}6\\ 9\end{pmatrix}$ *b* $\begin{pmatrix}6\\ 2\\ 0\end{pmatrix}$ *c* $(15 \quad -5 \quad 10)$ *d* $\begin{pmatrix}6 & 2\\ 4 & 8\end{pmatrix}$ *e* $\begin{pmatrix}6 & -3\\ 0 & 3\end{pmatrix}$

f $\begin{pmatrix}-3 & -2\\ 0 & 1\end{pmatrix}$ *g* $\begin{pmatrix}6 & 2 & -4\\ 2 & 8 & 0\end{pmatrix}$ *h* $\begin{pmatrix}-6 & 3 & 0\\ -15 & 0 & 9\end{pmatrix}$ *i* $\begin{pmatrix}5a & 10b & 5c\\ 10a & -5b & -5c\end{pmatrix}$

2 *a* $\begin{pmatrix}6 & 8\\ -2 & -4\end{pmatrix}$ *b* $\begin{pmatrix}9 & 12\\ -3 & -6\end{pmatrix}$ *c* $\begin{pmatrix}-15 & -20\\ 5 & 10\end{pmatrix}$ *d* $\begin{pmatrix}-3 & -4\\ 1 & 2\end{pmatrix}$ *e* $\begin{pmatrix}-3 & -4\\ 1 & 2\end{pmatrix}$

d and ***e***.

4 *a* $\begin{pmatrix}5 & 3 & 4\\ 6 & 5 & 3\end{pmatrix}$ *b* $\begin{pmatrix}10 & 6 & 8\\ 12 & 10 & 6\end{pmatrix}$ *c* $\begin{pmatrix}6 & 8 & 2\\ 4 & 0 & 6\end{pmatrix}$ *d* $\begin{pmatrix}4 & -2 & 6\\ 8 & 10 & 0\end{pmatrix}$

e $\begin{pmatrix}10 & 6 & 8\\ 12 & 10 & 6\end{pmatrix}$ *f* $\begin{pmatrix}18 & 24 & 6\\ 12 & 0 & 18\end{pmatrix}$ *g* $\begin{pmatrix}18 & 24 & 6\\ 12 & 0 & 18\end{pmatrix}$ *h* $\begin{pmatrix}16 & -8 & 24\\ 32 & 40 & 0\end{pmatrix}$

i $\begin{pmatrix}16 & -8 & 24\\ 32 & 40 & 0\end{pmatrix}$

5 *a* $2\begin{pmatrix}2 & 3\\ 4 & 1\end{pmatrix}$ *b* $3\begin{pmatrix}2 & -1\\ -3 & 0\end{pmatrix}$ *c* $-5\begin{pmatrix}1 & -2 & 0\\ -3 & 0 & 2\end{pmatrix}$ *d* $\frac{1}{2}\begin{pmatrix}8 & 12 & 6\\ 2 & 4 & 0\end{pmatrix}$

6 *a* $\begin{pmatrix}4 & -6\\ 8 & 2\end{pmatrix}$ *b* $\begin{pmatrix}-8 & 2\\ 6 & -4\end{pmatrix}$ *c* $\begin{pmatrix}6 & -9\\ 12 & 3\end{pmatrix}$ *d* $\begin{pmatrix}10 & -15\\ 20 & 5\end{pmatrix}$

e $\begin{pmatrix}-2 & -2\\ 7 & -1\end{pmatrix}$ *f* $\begin{pmatrix}-4 & -4\\ 14 & -2\end{pmatrix}$ *g* $\begin{pmatrix}6 & -4\\ 1 & 3\end{pmatrix}$ *h* $\begin{pmatrix}12 & -8\\ 2 & 6\end{pmatrix}$

i $\begin{pmatrix}-4 & -4\\ 14 & -2\end{pmatrix}$ *j* $\begin{pmatrix}12 & -8\\ 2 & 6\end{pmatrix}$ *k* $\begin{pmatrix}10 & -15\\ 20 & 5\end{pmatrix}$ *l* $\begin{pmatrix}4 & -6\\ 8 & 2\end{pmatrix}$

7 *a* $\begin{pmatrix}8 & -1 & -9\\ 21 & 2 & 4\end{pmatrix}$ *b* $\begin{pmatrix}7 & 0 & -1\\ 7 & 3 & 4\end{pmatrix}$

8 *a* $\begin{pmatrix}2 & -1\\ 4 & 3\end{pmatrix}$ *b* $\begin{pmatrix}3 & 2\\ -1 & 3\end{pmatrix}$ *c* $\begin{pmatrix}2 & 1\\ 1 & 5\end{pmatrix}$ *d* $\begin{pmatrix}4 & -3\\ -4 & -2\end{pmatrix}$

9 a $\begin{pmatrix} 1 & 8 & -18 \\ 5 & -1 & -7 \end{pmatrix}$ *b* $\begin{pmatrix} 7 & -6 \\ 1 & -4 \end{pmatrix}$ *10* $p = -1, q = 4, r = 3, s = -2$

Page 44 Exercise 6

1 a (11) *b* (26) *c* (9) *d* (13) *e* (15) *f* $(8x-5y-z)$

2 a 4 *b* 3 *c* -6 *d* ± 3

3 a $\begin{pmatrix} 9 \\ 8 \end{pmatrix}$ *b* $\begin{pmatrix} 10 \\ 4 \end{pmatrix}$ *c* $\begin{pmatrix} 1 \\ 8 \end{pmatrix}$ *d* $\begin{pmatrix} 7 \\ 5 \end{pmatrix}$ *e* $\begin{pmatrix} -7 \\ 7 \end{pmatrix}$ *f* $\begin{pmatrix} 4 \\ 6 \end{pmatrix}$

g $\begin{pmatrix} 3x-2y \\ 4x+5y \end{pmatrix}$ *h* $\begin{pmatrix} 7x \\ 0 \end{pmatrix}$ *i* $\begin{pmatrix} 0 \\ -3a \end{pmatrix}$

4 $AB = (1 \quad -3)$, $BC = \begin{pmatrix} 11 \\ -4 \end{pmatrix}$, $AC = (-3)$, $CA = \begin{pmatrix} 12 & 20 \\ -9 & -15 \end{pmatrix}$

5 a $\begin{pmatrix} 1 \\ 22 \end{pmatrix}$ *c* (26) *d* $\begin{pmatrix} 11 \\ 13 \\ 7 \end{pmatrix}$ *f* $\begin{pmatrix} 9 \\ 8 \\ 7 \end{pmatrix}$ *h* (17 24) *i* (1 0)

6 a 4, −4 *b* 3, −1 *c* 2, 3 *d* 5, −2 *e* 8, 5 *f* 2, −1

7 a $\begin{pmatrix} 4 & 7 \\ 9 & 7 \end{pmatrix}$ *b* $\begin{pmatrix} 5 & 5 \\ 16 & 7 \end{pmatrix}$ *c* $\begin{pmatrix} 13 & 18 \\ 21 & 16 \end{pmatrix}$ *d* $\begin{pmatrix} 11 & 12 \\ 22 & 24 \end{pmatrix}$ *e* $\begin{pmatrix} 3 & -4 & 2 \\ -6 & -2 & 1 \end{pmatrix}$

f $\begin{pmatrix} -3 & 2 & 1 & -2 \\ -7 & -3 & 4 & -3 \end{pmatrix}$

8 a $\begin{pmatrix} 1 & -2 \\ -3 & 6 \end{pmatrix}$ *b* $\begin{pmatrix} 1 & -2 \\ -3 & 6 \end{pmatrix}$ *c* $\begin{pmatrix} a & b \\ c & d \end{pmatrix}$ *d* $\begin{pmatrix} a & b \\ c & d \end{pmatrix}$ *e* $\begin{pmatrix} 1 & 0 \\ 0 & 1 \end{pmatrix}$ *f* $\begin{pmatrix} 0 & 1 \\ 1 & 0 \end{pmatrix}$

9 $AB = \begin{pmatrix} 0 & 0 \\ 0 & 0 \end{pmatrix}$, $BA = \begin{pmatrix} -4 & 8 \\ -2 & 4 \end{pmatrix}$ *10* $\begin{pmatrix} 4 & -5 \\ 25 & -1 \end{pmatrix}$, $\begin{pmatrix} -13 & -14 \\ 70 & -27 \end{pmatrix}$

11 $a = 1, b = -1, c = -2, d = 3$ *12* $p = 1\frac{1}{2}, q = -\frac{1}{2}, r = -2, s = 1$

13a 9 *b* (9); 9 routes from A to C *c* (12); 12 routes from A to C

14b 8 from x_1, 7 from x_2; matrix product $\begin{pmatrix} 4 & 4 & 0 \\ 0 & 2 & 5 \end{pmatrix}$

Page 47 Exercise 6B

1 a $\begin{pmatrix} 4 & 1 \\ -1 & -4 \end{pmatrix}$ *b* $\begin{pmatrix} 1 & 2 \\ 7 & -1 \end{pmatrix}$ *c* $\begin{pmatrix} 5 & -1 \\ 0 & 23 \end{pmatrix}$ *d* $\begin{pmatrix} 11 & 9 \\ 8 & 17 \end{pmatrix}$ *e* $\begin{pmatrix} 4 & 13 \\ 5 & -1 \end{pmatrix}$ *f* $\begin{pmatrix} 10 & -1 \\ 1 & -7 \end{pmatrix}$

2 a $\begin{pmatrix} 10 & 21 \\ 5 & -24 \end{pmatrix}$ *b* $\begin{pmatrix} 10 & 21 \\ 5 & -24 \end{pmatrix}$ *c* $\begin{pmatrix} 31 & 2 \\ 33 & -9 \end{pmatrix}$ *d* $\begin{pmatrix} 31 & 2 \\ 33 & -9 \end{pmatrix}$. Associative law.

3 a $\begin{pmatrix} 8 & 14 \\ 4 & -5 \end{pmatrix}$ *b* $\begin{pmatrix} 8 & 14 \\ 4 & -5 \end{pmatrix}$ *c* $\begin{pmatrix} 11 & 1 \\ 8 & -8 \end{pmatrix}$ *d* $\begin{pmatrix} 11 & 1 \\ 8 & -8 \end{pmatrix}$. Distributive law.

4 a $\begin{pmatrix} 4 & 11 \\ 10 & 25 \\ 16 & 39 \end{pmatrix}$ *b* $\begin{pmatrix} 7 & 13 \\ 4 & 11 \\ 16 & 34 \end{pmatrix}$ *c* $\begin{pmatrix} 4 & 3 \\ 4 & 6 \\ 11 & 10 \end{pmatrix}$ *5* Each side equals $\begin{pmatrix} 11 & 24 \\ 14 & 36 \\ 32 & 73 \end{pmatrix}$.

No. Not comformable for multiplication.

6 a $\begin{pmatrix}3 & 5\\1 & 2\end{pmatrix}$ *b* $\begin{pmatrix}1 & -3\\-3 & 4\end{pmatrix}$ *c* $\begin{pmatrix}-12 & 11\\-5 & 5\end{pmatrix}$ *d* $\begin{pmatrix}3 & 5\\-5 & 8\end{pmatrix}$ *e* $\begin{pmatrix}9 & 0\\0 & 9\end{pmatrix}$. No.

7 a $\begin{pmatrix}3 & -1\\1 & 0\end{pmatrix}$ *b* $\begin{pmatrix}8 & -3\\3 & -1\end{pmatrix}$ *c* $\begin{pmatrix}7 & 0\\0 & 7\end{pmatrix}$ *d* $\begin{pmatrix}-4 & -2\\16 & -20\end{pmatrix}$ *e* $\begin{pmatrix}10 & -9\\-6 & 7\end{pmatrix}$. No.

8 a $(A+B)(A+B) = (A+B)A+(A+B)B = A^2+BA+AB+B^2$; and in general, $AB \neq BA$.

b $(A+B)(A-B) = (A+B)A+(A+B)(-B) = A^2+BA-AB-B^2$; and in general, $AB \neq BA$.

9 $\begin{pmatrix}k^2 & 2k\\0 & k^2\end{pmatrix}$, $\begin{pmatrix}k^3 & 3k^2\\0 & k^3\end{pmatrix}$, $\begin{pmatrix}k^4 & 4k^3\\0 & k^4\end{pmatrix}$. $A^n = \begin{pmatrix}k^n & nk^{n-1}\\0 & k^n\end{pmatrix}$

12 Yes. *13* $\begin{pmatrix}1 & 1\\0 & -1\end{pmatrix}$, $\begin{pmatrix}-1 & 1\\0 & 1\end{pmatrix}$ *14* $p = -2$ and $q = 13$

15 $\begin{pmatrix}-\frac{1}{2} & \frac{1}{2}\sqrt{3}\\-\frac{1}{2}\sqrt{3} & -\frac{1}{2}\end{pmatrix}$, $\begin{pmatrix}-1 & 0\\0 & -1\end{pmatrix}$ *16* $\begin{pmatrix}5 & 0 & 5\\12 & 1 & 13\\8 & -1 & 7\end{pmatrix}$, $\begin{pmatrix}8 & 10\\2 & 5\end{pmatrix}$

17 $\begin{pmatrix}4 & 1 & 5\\3 & -1 & 5\\7 & 2 & 8\end{pmatrix}$, $\begin{pmatrix}7 & 4 & 12\\-1 & -2 & -2\\4 & 0 & 6\end{pmatrix}$

Page 50 Exercise 7

7 $\begin{pmatrix}1 & -1\\-1 & 2\end{pmatrix}$ *8* $\begin{pmatrix}5 & -3\\-3 & 2\end{pmatrix}$ *9* $\begin{pmatrix}7 & -3\\-9 & 4\end{pmatrix}$ *10* $\begin{pmatrix}2 & 3\\1 & 2\end{pmatrix}$

11 $\begin{pmatrix}4 & 5\\7 & 9\end{pmatrix}$ *12* $\begin{pmatrix}-4 & 7\\-3 & 5\end{pmatrix}$ *13* $\begin{pmatrix}1 & 1\\0 & 1\end{pmatrix}$ *14* $\begin{pmatrix}\cos\theta & -\sin\theta\\\sin\theta & \cos\theta\end{pmatrix}$

15 $\begin{pmatrix}11 & -12\\-10 & 11\end{pmatrix}$ *16b* $\frac{1}{2}\begin{pmatrix}1 & 0\\0 & 1\end{pmatrix}$ *c (1)* $\frac{1}{3}\begin{pmatrix}1 & 0\\0 & 1\end{pmatrix}$ *(2)* $\frac{1}{a}\begin{pmatrix}1 & 0\\0 & 1\end{pmatrix}$

17 $M^{-1} = \begin{pmatrix}2 & -5\\-1 & 3\end{pmatrix}$. $M^2 = \begin{pmatrix}14 & 25\\5 & 9\end{pmatrix}$, $(M^{-1})^2 = \begin{pmatrix}9 & -25\\-5 & 14\end{pmatrix}$ which is inverse of M^2.

Page 52 Exercise 8

1 $\frac{1}{2}\begin{pmatrix}2 & -3\\-2 & 4\end{pmatrix}$ *2* $\frac{1}{2}\begin{pmatrix}3 & -1\\-4 & 2\end{pmatrix}$ *3* $\frac{1}{4}\begin{pmatrix}2 & -2\\-1 & 3\end{pmatrix}$ *4* none

5 $\begin{pmatrix}-9 & 4\\16 & -7\end{pmatrix}$ *6* $\frac{1}{5}\begin{pmatrix}4 & -3\\-1 & 2\end{pmatrix}$ *7* none *8* $\frac{1}{3}\begin{pmatrix}9 & -7\\-6 & 5\end{pmatrix}$

9 $\begin{pmatrix}0 & 1\\1 & -1\end{pmatrix}$ *10* $\frac{1}{13}\begin{pmatrix}5 & 3\\-1 & 2\end{pmatrix}$ *11* $-\frac{1}{14}\begin{pmatrix}4 & -6\\-3 & 1\end{pmatrix}$ *12* $\frac{1}{7}\begin{pmatrix}2 & -3\\1 & 2\end{pmatrix}$

13 $\frac{1}{8}\begin{pmatrix}2 & 1\\1 & 3\end{pmatrix}$ *14* $\frac{1}{2}\begin{pmatrix}1 & 4\\1 & 2\end{pmatrix}$ *15* $-\frac{1}{3}\begin{pmatrix}3 & -2\\0 & -1\end{pmatrix}$

16a $\begin{pmatrix}7 & 19\\1 & 3\end{pmatrix}$ **b** $\begin{pmatrix}4 & 11\\2 & 6\end{pmatrix}$ **c** $\frac{1}{2}\begin{pmatrix}1 & -3\\0 & 2\end{pmatrix}$ **d** $\begin{pmatrix}3 & -5\\-1 & 2\end{pmatrix}$

e $\frac{1}{2}\begin{pmatrix}3 & -19\\-1 & 7\end{pmatrix}$ **f** $\frac{1}{2}\begin{pmatrix}6 & -11\\-2 & 4\end{pmatrix}$ **g** $\frac{1}{2}\begin{pmatrix}3 & -19\\-1 & 7\end{pmatrix}$ **h** $\frac{1}{2}\begin{pmatrix}6 & -11\\-2 & 4\end{pmatrix}$

17 $(AB)^{-1} = B^{-1}A^{-1}$, $A^{-1}B^{-1} = (BA)^{-1}$

18 $a+3b = 1$ and $2a+4b = 0 \Rightarrow a = -2$ and $b = 1$.

$c+3d = 0$ and $2c+4d = 1 \Rightarrow c = \frac{3}{2}$ and $d = -\frac{1}{2}$.

19a $\begin{pmatrix}\frac{1}{2} & -1\\-\frac{1}{2} & 2\end{pmatrix}$ **b** $\begin{pmatrix}1 & -2\\0 & \frac{1}{2}\end{pmatrix}$ **c** $\begin{pmatrix}1 & -\frac{1}{2}\\-2 & 1\frac{1}{2}\end{pmatrix}$ **d** $\begin{pmatrix}\frac{1}{2} & \frac{1}{4}\\\frac{1}{2} & \frac{3}{4}\end{pmatrix}$

20 $X = \begin{pmatrix}-3 & 4\\9 & -5\end{pmatrix}$

Page 55 Exercise 9

1 $\{(8, 3)\}$ **2** $\{(4, 4)\}$ **3** $\{(3, -2)\}$ **4** $\{(4, 7)\}$ **5** $\{(4, -3)\}$

6 $\{(2, -3)\}$ **7** $\{(7, -2)\}$ **8** $\{(3, 4)\}$ **9** $\{(4, -1)\}$ **10** $\{(3, 2)\}$

11 $\{(\frac{1}{5}, -1)\}$ **12** $\{(15, 20)\}$ **13a** $\{(1, 2)\}$ **b** $\{(4, -1)\}$ **c** $\{(1, -\frac{1}{2})\}$

14 There is no solution in either case. The matrices have no inverses. In ***a***, the straight lines given by the equations are coincident, and in ***b*** they are parallel.

15 $\begin{pmatrix}-1 & 1\\3 & -1\end{pmatrix}$ **16** $\begin{pmatrix}11 & 16\\7 & 10\end{pmatrix}$ **17** $\begin{pmatrix}7 & -5\\-10 & 7\end{pmatrix}$ **18** $\frac{1}{13}\begin{pmatrix}-4 & -27\\7 & -8\end{pmatrix}$

19 $\begin{pmatrix}0 & 1\\1 & 0\end{pmatrix}$ **20a** $\begin{pmatrix}\cos\theta & -\sin\theta\\\sin\theta & \cos\theta\end{pmatrix}$ **b** $x = x'\cos\theta - y'\sin\theta$, $y = x'\sin\theta + y'\cos\theta$

Algebra — Answers to Chapter 4

Page 60 Exercise 1

1 a (ii), (iv), (v) **b** (ii) **2 a** 5, 2, 17, 10 **b** ± 7 **3 a** $8, 4, 2, 1, \frac{1}{2}, \frac{1}{4}$ **b** 6

4 a $\{1, 2, 3, 4, 5\}$ **5 a** $\{0, 1, 4, 9\}$ **6 a** $7, 0, -5, -8, -9, -8, -5, 0, 7$

4 b

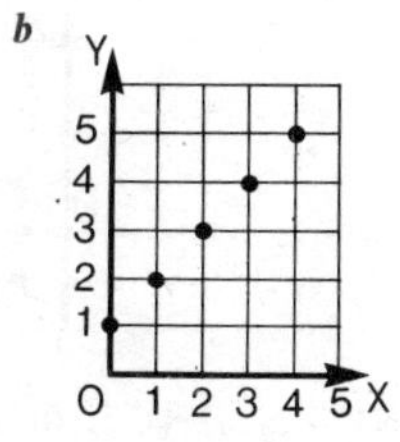

5 b $\{y : y \geqslant 0, y \in R\}$

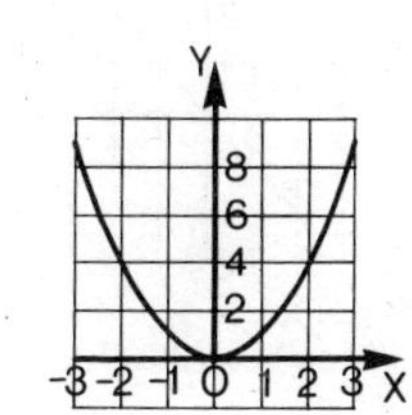

6 b $\{y : -9 \leqslant y \leqslant 7, y \in R\}$

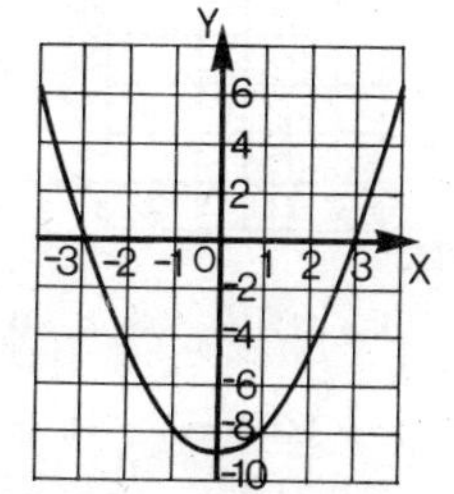

7 (i), (iii), (iv), (vii) **8a**(*1*) (0,0), (1,1), (2,2), (3,3), (4,4), (5,0), (6,1), (7,2), (8,3), (9,4), (10,0)}

(*2*)

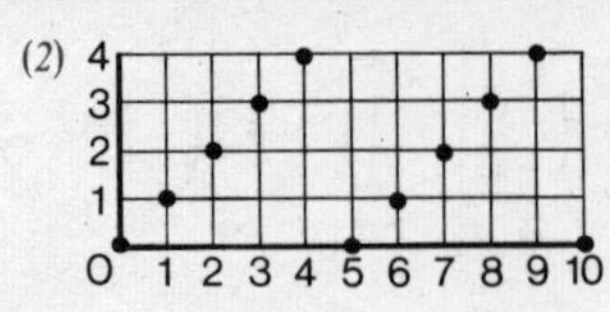

b Yes; each element of A is related to exactly one element of B.

9 ***a*** $(3,1), (-2,0), (0,-6), (4,-5)$

b $x = x' - 1, y = y' + 3$

10 ***a*** $-1, 3$ ***b*** $-3, -4, -3, 5$ **11** ***a*** $-2, 0, 2$ ***b*** $-3, 3, 15, -15$

c

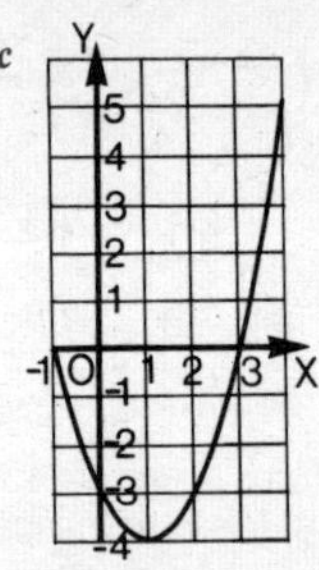

$\{y: -4 \leqslant y \leqslant 5, y \in R\}$

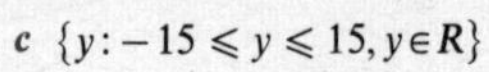

c $\{y: -15 \leqslant y \leqslant 15, y \in R\}$

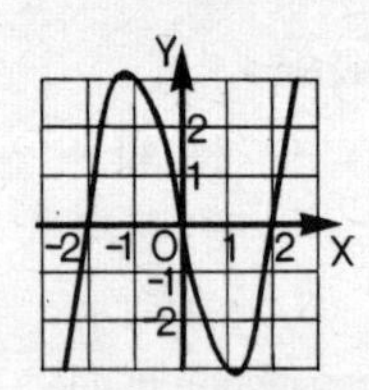

d $\{x: -2 < x < 0\} \cup \{x: 2 < x \leqslant 3\}$

12a 3, 2·8, 2·2, 0, 2·8, 2·2, 0

b

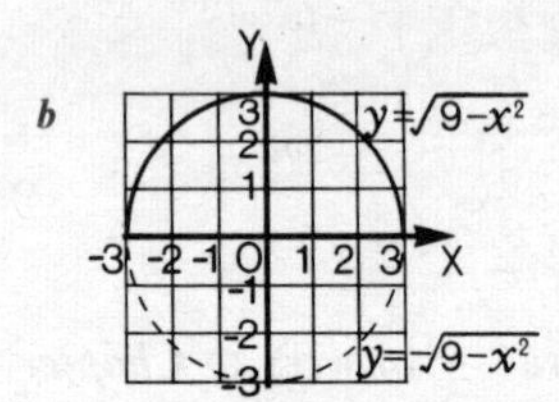

c a circle, centre O, radius 3

Page 63 Exercise 1B

1 ***a*** 4 in each case **2** $-3, -2, -1, 0, 1, 2, 3$

3 ***b***

b

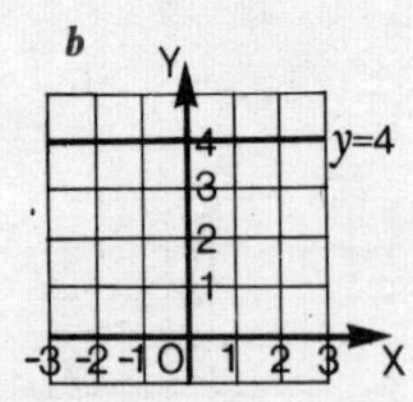

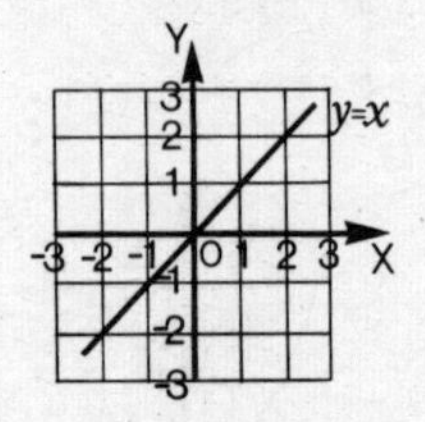

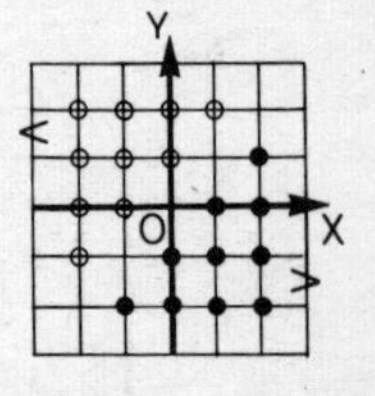

3 ***a*** (*1*) $\{(-1,-2), (0,-1), (0,-2), (1,0), (1,-1), (1,-2), (2,1), (2,0), (2,-1), (2,-2)\}$

(*2*) $\{(-2,-1), (-1,0), (-2,0), (0,1), (-1,1), (-2,1), (1,2), (0,2), (-1,2), (-2,2)\}$

4 a $\{(-5,-1), (-3,0), (-1,1), (1,2), (3,3), (5,4)\}$ *b*

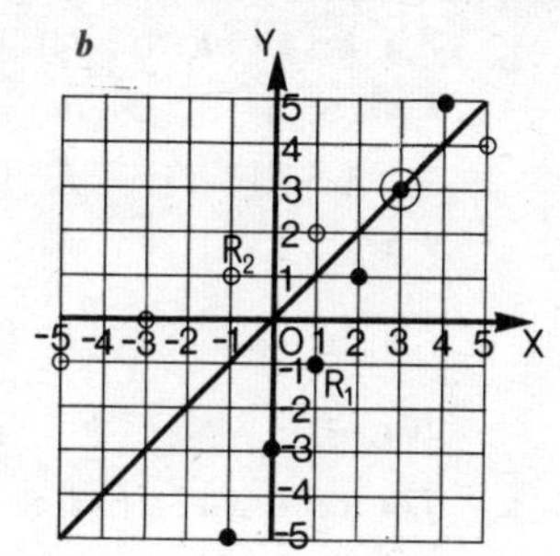

c $y = x$ *d* yes

5 2, 1, 0, 1, 2, 3

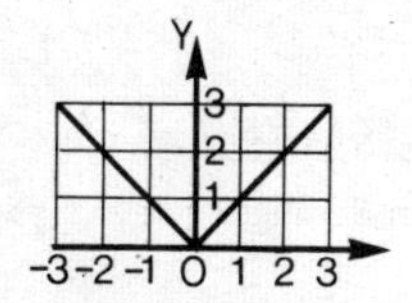

6 a

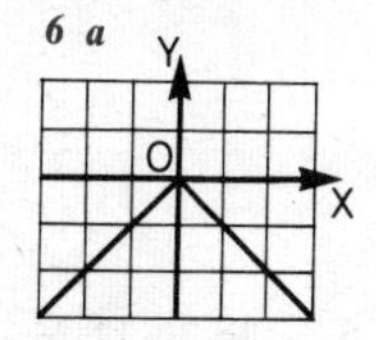

b

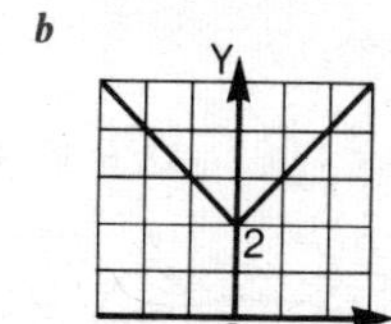

c

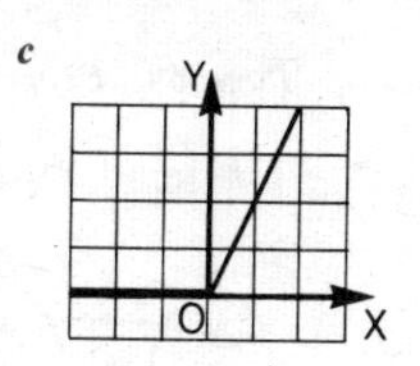

7

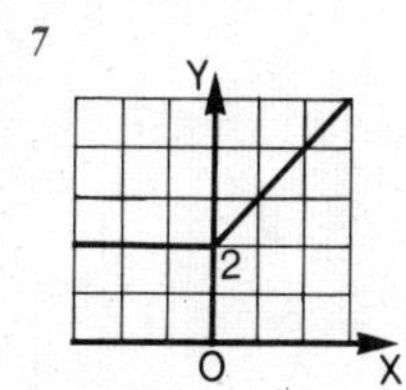

8

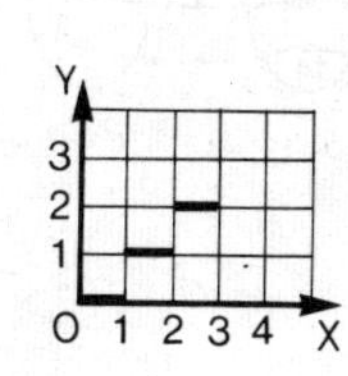

9 a gradient of straight line graph of f *10* $a = 3, b = -4; x = 1\frac{1}{3}$

11 $f(x) = 2ax + (a+b)$, a linear function

12a $f'(x) = 1$ *b* $f'(x) = 2x$ *c* $f'(x) = 3x^2$

Page 67 Exercise 2

1 a $q, x; r, z$ *b* x, x, z *c* domain $\{a, b, c\}$, range $\{x, z\}$

2 a $2 \to 3 \to 5$
$1 \to 2 \to 3$
$0 \to 1 \to 1$
$-1 \to 0 \to -1$

b

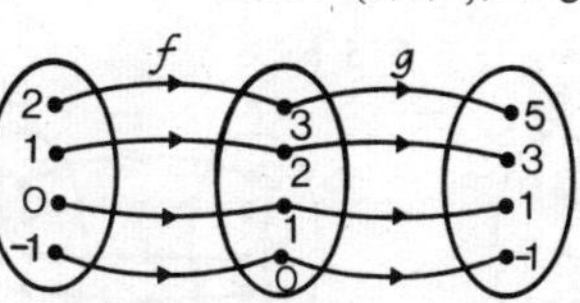

3 a 21, -9 *b* $3(2x+1)$ *4* 19, -11, $6x+1$

5 a 9, 1, $(x+2)^2$ *b* 3, 11, x^2+2 *6 a* 8, -8, $(x+1)^3$ *b* 2, -26, x^3+1

7 a $4x^2+4$ *b* $2(x^2+4)$ *c* $4x$ *d* $(x^2+4)^2+4 = x^4+8x^2+20$

8 a $(x+1)^2$ *b* $2x+7$ *c* x^2-x+1 *d* $2x^4+1$

e $4x^2-4x$ *f* $1/(x^4+2x^2+2)$ *g* $6x-27x^3$ *h* $\sin(x^2)°$

9 a x^2+1 *b* $2x+4$ *c* x^2+x *d* $(2x^2+1)^2$

e $2x^2-3$ *f* $\left(\frac{1}{x^2+1}\right)^2+1$ *g* $6x-3x^3$ *h* $\sin^2 x°$

10a 63 *b* $-\frac{1}{3}, 1$ *11* (3, 8), (5, 6), (3, 5). Yes

12a $4x^2+4x+3$ *b* (*1*) R (*2*) $\{x: x \geqslant 2, x \in R\}$ *c* domain R, range $\{x: x \geqslant 2, x \in R\}$

d -3 or 2

Page 69 Exercise 3

1 b 16, 17, 17; 4, 17, 17 *c* 36, 37, 37; 6, 37, 37

2

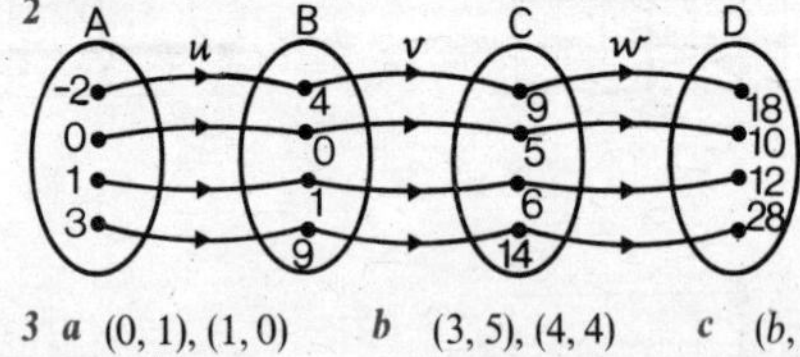

a 9, 5, 6, 14; 18, 10, 12, 28

b 4, 0, 1, 9; 18, 10, 12, 28

3 a (0, 1), (1, 0) *b* (3, 5), (4, 4) *c* $(b, a+1), (b+1, a)$

4 a (b, a) *b* $(b, -a)$ *c* $(-a, -b)$ *d* $(b, -a)$

5 b 25, 10 *6 b* -4, 62 *7 a* u, u, v, v

8 a $\begin{pmatrix} 1-4x \\ 2y+9 \end{pmatrix}$ *b* $\begin{pmatrix} 7-4x \\ 2y+6 \end{pmatrix}$

Page 72 Exercise 4

1 2 *3 a* *b*

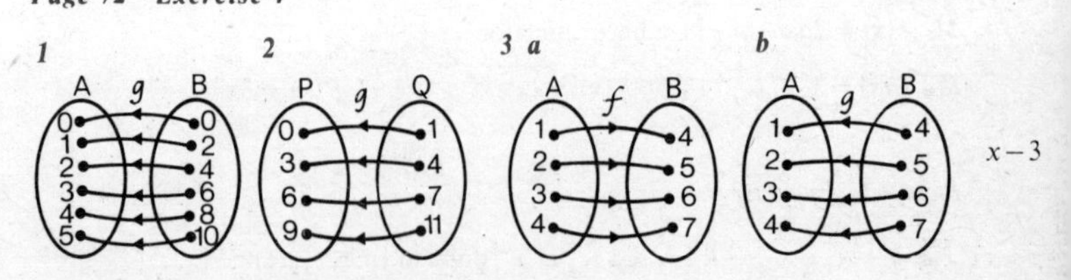

4 a

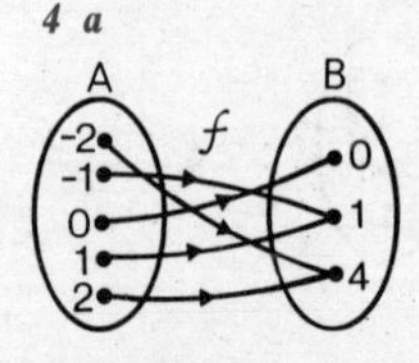

b 1 and 4 in *B* are related to more than one element of *A*.

5 a

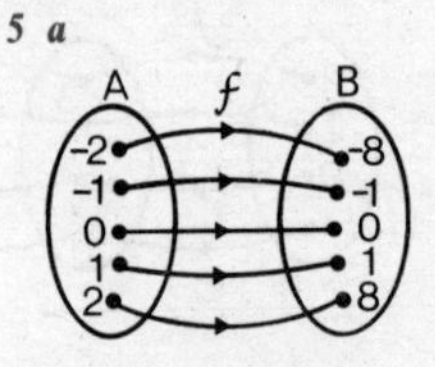

b Yes; $g(y) = \sqrt[3]{y}$

6 a a, b, c *b* p, q, r

c $A:B$

7 a (iii) only

Page 75 Exercise 5

1 $f^{-1}(x) = \frac{1}{2}(x-5)$ *2* $f^{-1}(x) = \frac{1}{3}(x+1)$ *3* $f^{-1}(x) = \frac{1}{3}(1-x)$

4 $f^{-1}(x) = 1-x$ *5* $f^{-1}(x) = \frac{1}{2}(2x+1)$ *6* $f^{-1}(x) = 2(1-x)$

7 $f^{-1}(x) = 2(x-3)$ *8* $f^{-1}(x) = 2x-9$ *9* $f^{-1}(x) = \frac{1}{2}(3x+5)$

10 $f^{-1}(x) = \frac{1}{3}(1-2x)$ *11* $f^{-1}(x) = \frac{1}{2}(10-3x)$ *12* $f^{-1}(x) = \sqrt[3]{(x+4)}$

13a $f^{-1}(x) = x-1,\ g^{-1}(x) = \frac{1}{2}x,\ h^{-1}(x) = \sqrt{x}$ *b* (*1*) 4 (*2*) 4 (*3*) 2

14a $f^{-1}(x) = \frac{1}{2}x,\ g^{-1}(x) = x-2$ *b* $2(x+1), \frac{1}{2}(x-2), \frac{1}{2}(x-2)$ *c* functions are equal

15a 2 *b* 4 *c* $\sqrt[3]{4}$ *d* 2

16 $f^{-1}(x) = 3-x^2; \{x : x \leqslant 3, x \in R\}$ *17* $f^{-1}(x) = (x+3)^2; \{x : x \geqslant 0, x \in R\}$

18 $f^{-1}(x) = \frac{1}{4}(x-5)^2; \{x : x \geqslant 0, x \in R\}$ *19* $f^{-1}(x) = \dfrac{1-x}{x}; \{x : x \in R, x \neq -1\}$

20 $f^{-1}(x) = \dfrac{3x-2}{x}; \{x : x \in R, x \neq 3\}$ *21* $f^{-1}(x) = \dfrac{4x}{x-1}; \{x : x \in R, x \neq 4\}$

22a

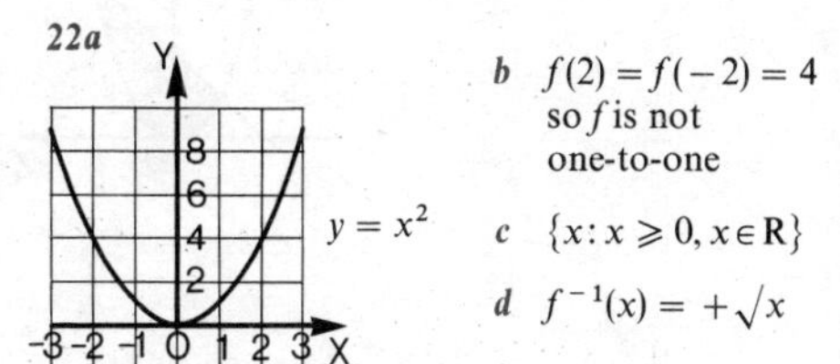

b $f(2) = f(-2) = 4$ so f is not one-to-one

c $\{x : x \geqslant 0, x \in R\}$

d $f^{-1}(x) = +\sqrt{x}$

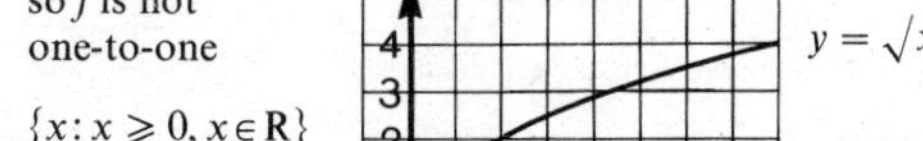

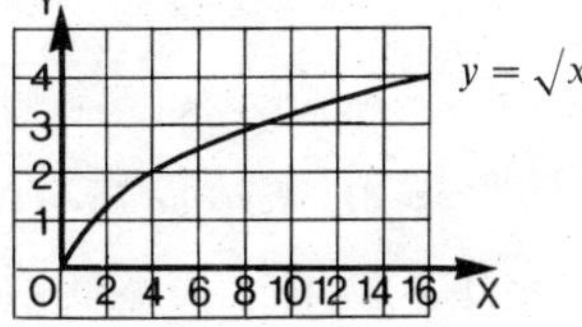

23 (i) *a* yes *b* yes (ii) *a* yes *b* yes (iii) *a* yes *b* no *c* no (iv) *a* yes *b* no *c* yes; take $x \geqslant 0$ or $x \leqslant 0$, for example (v) *a* yes *b* no *c* yes; take $x \geqslant 0$ or $x \leqslant 0$ (vi) *a* no *d* no

No two points on the graph may have the same x-coordinate or the same y-coordinate.

Algebra — Answers to Revision Exercises

Page 80 Revision Exercise 1

1 The solution set, $\{(7, 2)\}$, of the first system is the same as the second.

2 $(-2, 1)$ *3* $a = 10, b = -3; 1$ *4* $(2, 5), (3, 4), (-1, 2)$

5 $\{(2\frac{1}{2}, \frac{1}{2}, 2)\}$ *6* $\{(3, -4, 5)\}$ *7* $\{(-1, 2, 0)\}$ *8* $\{(1, -3, 5)\}$

9 $\{(2, 3, 0)\}$ *10* $\{(5, 4, -2)\}$ *11* $g = 3, f = -3, c = -7$

12 $a = -2, b = -5, c = 6$ *13* $a = 2, b = -1, c = -3$

14a $\{(-2, 0), (3, 5)\}$ *b* $\{(-4, 3), (3, -4)\}$ *c* $\{(1, 0), (0, 1)\}$

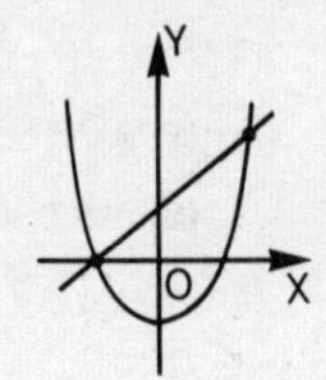

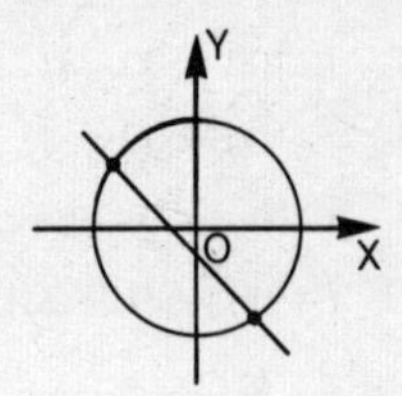

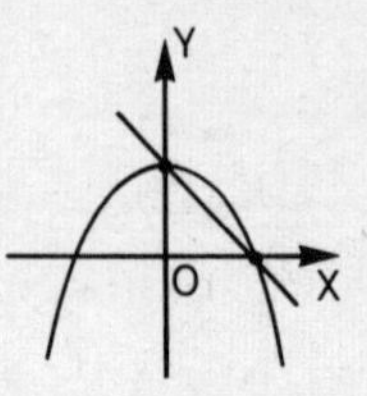

15 $\{(\frac{3}{5}, 2\frac{4}{5}), (1, 2)\}$ *16* $\{(1\frac{1}{2}, -2\frac{1}{2}), (3, -2)\}$ *17* $\{(-1\frac{1}{3}, 2\frac{2}{3}), (2, 1)\}$

18 $\{(-4, -3), (2, 1)\}$ *19* $\{(-7, 5), (-1, 1)\}$ *20* $\{(\frac{5}{11}, -\frac{7}{11}), (-2, 1)\}$

21 $\{(-4, -2, 6), (6, -2, -4)\}$ *22* $x = 2, y = 1, z = 3, u = -1$

23 $\{(\frac{1}{2}, \frac{1}{2}), (4, -3)\}$ *24a* $\{(-1, 1), (2, 4)\}$ *b*

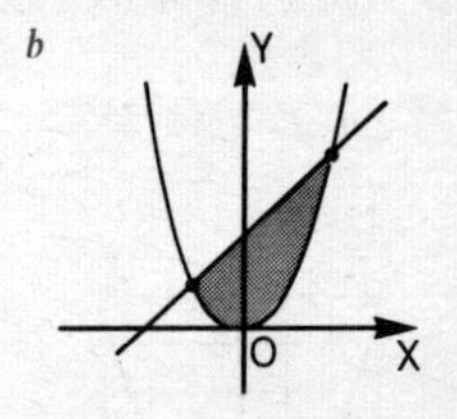

Page 81 Revision Exercise 2

1 a 3, 7, 11 *b* 0, 4, 8 *c* 0, 3, 8 *d* 0, 1, 4

2 a 36th *b* 17th *c* 12th *d* 9th *3 a* $u_n = 5n$ *b* $u_n = 10^{3-n}$ *c* $u_n = n(-1)^n$

4 a 1, 3, 5, 7 *b* 1, 3, 5, 19. Although the first 3 terms are the same, the sequences are different.

5 a 37 *b* 26th *6 a* -210 *b* 5 or 8; in the latter some positive and negative terms cancel each other. *7* 10 000 *8* $r = 2$ or -2 *9 a* 3.2^{n-1}; $3(2^n - 1)$ *b* 7

10 4, 2 *12* 189 cm *13* $-21, 15\frac{3}{4}$ *14* $4r^2 - 4r - 3 = 0$: r must lie between 1 and -1. *15a* arithmetic; 10 302 *c* geometric *d* geometric *e* geometric; $1\frac{1}{3}$ *17a* (*1*) £110, £121, £133·10 (*2*) 1·1; £176(7) *18* 8

Page 83 Revision Exercise 3

1 a $\begin{pmatrix} -1 & 7 & 1 \\ 3 & 8 & -8 \end{pmatrix}$ *b* $\begin{pmatrix} 1 & -7 & -1 \\ -3 & -8 & 8 \end{pmatrix}$ *2* $A = \begin{pmatrix} 2 & 3 & 4 & 5 \\ 4 & 5 & 6 & 7 \end{pmatrix}$

3 a $C = \begin{pmatrix} c_{11} & c_{12} & c_{13} \\ c_{21} & c_{22} & c_{23} \\ c_{31} & c_{32} & c_{33} \end{pmatrix}$ *b* $\begin{pmatrix} 3 & 5 & 7 \\ 4 & 6 & 8 \\ 5 & 7 & 9 \end{pmatrix}$

4 a $\begin{pmatrix} 2 & 0 \\ 2 & 5 \end{pmatrix}$ *b* $\begin{pmatrix} 2 & -2 \\ 0 & -5 \end{pmatrix}$ *c* $\begin{pmatrix} 0 & 1 \\ 5 & -8 \end{pmatrix}$ *d* $\begin{pmatrix} -8 & 9 \\ 9 & -1 \end{pmatrix}$

5 $a = 1\frac{1}{2}, b = -6, c = -2, d = -\frac{2}{3}$

6 ***a*** $\begin{pmatrix} 9 & 4 \\ 1 & -6 \end{pmatrix}$ ***b*** $\begin{pmatrix} 9 & 1 \\ 4 & -6 \end{pmatrix}$ ***c*** $\begin{pmatrix} 5 & -2 \\ 3 & 1 \end{pmatrix}$ ***d*** $\begin{pmatrix} 4 & 3 \\ 1 & -7 \end{pmatrix}$ ***e*** $\begin{pmatrix} 9 & 1 \\ 4 & -6 \end{pmatrix}$

$(A+B)' = A'+B'$

7 AB, BC, AC, CA; $AB = (-1 \quad 8)$, $BC = \begin{pmatrix} 13 \\ -5 \end{pmatrix}$, $AC = (16)$, $CA = \begin{pmatrix} 10 & 15 \\ 4 & 6 \end{pmatrix}$

8 $M^2 = \begin{pmatrix} 0 & 1 \\ -1 & 0 \end{pmatrix}$, $N^2 = \begin{pmatrix} \frac{1}{2} & -\frac{1}{2}\sqrt{3} \\ \frac{1}{2}\sqrt{3} & \frac{1}{2} \end{pmatrix}$, $N^3 = \begin{pmatrix} 0 & -1 \\ 1 & 0 \end{pmatrix}$ ***9*** ***a*** ***10*** all true

11a $M^{-1} = \begin{pmatrix} \frac{1}{2} & -\frac{1}{2}\sqrt{3} \\ \frac{1}{2}\sqrt{3} & \frac{1}{2} \end{pmatrix}$ ***b*** $x = \frac{1}{2}x' - \frac{1}{2}\sqrt{3}y'$, $y = \frac{1}{2}\sqrt{3}x' + \frac{1}{2}y'$

12 $p = 1, q = 3, r = \frac{1}{2}, s = 2$ ***13a*** $a = 1, b = -4$ ***b*** $\begin{pmatrix} 0 & -2 \\ 2 & 4 \end{pmatrix}, \begin{pmatrix} -2 & -1 \\ 5 & 2 \end{pmatrix}$

14 $x = -5, y = -3\frac{1}{2}$ ***15*** $V = \begin{pmatrix} -1 & 1 \\ 3 & -2 \end{pmatrix}$; 1, 1

16c $\cos(\alpha+\beta) = \cos\alpha\cos\beta - \sin\alpha\sin\beta$; $\sin(\alpha+\beta) = \sin\alpha\cos\beta + \cos\alpha\sin\beta$

17

	b_1	b_2
a_1	1	0
a_2	1	1
a_3	0	1

	c_1	c_2	c_3
b_1	1	1	0
b_2	0	1	1

; product = $\begin{pmatrix} 1 & 1 & 0 \\ 1 & 2 & 1 \\ 0 & 1 & 1 \end{pmatrix}$, 8

18 (3, 11); $x = 4, y = -3$

19a $\frac{1}{2}\begin{pmatrix} 3 & -1 \\ -4 & 2 \end{pmatrix}, \frac{1}{5}\begin{pmatrix} 4 & -3 \\ -1 & 2 \end{pmatrix}$ ***b*** (*1*) $\begin{pmatrix} -1 & 3 \\ 4 & -1 \end{pmatrix}$ (*2*) $\begin{pmatrix} 1 & 0 \\ -2 & 2 \end{pmatrix}$

20a 1, 2 ***b*** no real numbers ***21*** $P^{-1} = \frac{1}{ad-bc}\begin{pmatrix} d & -b \\ -c & a \end{pmatrix}$

Page 86 Revision Exercise 4

1 ***a*** (i), $\{q, r\}$; (iii), $\{p\}$; (iv), $\{p, q, r\}$ ***b*** (iv)

2

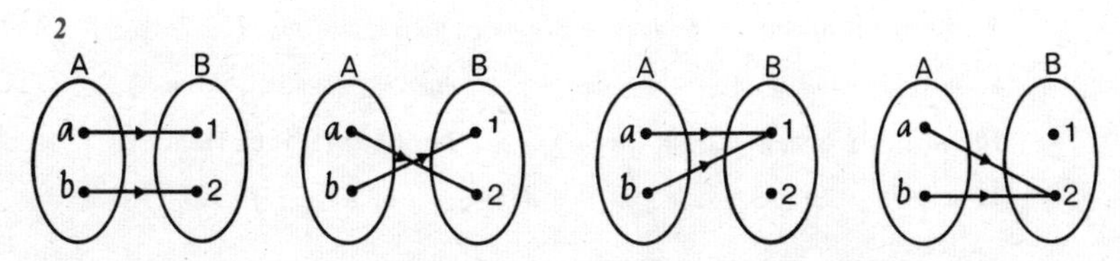

3 ***a*** $f(5) = 9$, $f(-5)$ does not exist as $-5 \notin A$, $f(0) = 24$, $f(1) = 25$ ***b*** -1 or 3

4 $a = -1\frac{1}{2}, b = 1\frac{1}{2}; -3$ **5** ***a*** $\{x : x \in R, x \neq \pm 2\}$ ***b*** $-1, \frac{1}{7}, -\frac{4}{5}$ c 4 or -4

6 a $(h \circ k)(x) = 1-2x^2$, $(k \circ h)(x) = (1-2x)^2$ **b** 2, −2 **c** $-\frac{1}{2}$, $1\frac{1}{2}$

7 $g \circ f = \{(1, 1), (2, 4), (3, 3), (4, 2)\}$ and $f \circ g = \{(1, 2), (2, 1), (3, 3), (4, 4)\}$

8 a 21, 51 **b** $(g \circ f)(x) = 2x^2+3$, $(f \circ g)(x) = 4x^2-4x+3$, $2a(a-2)$

9 a $(g \circ f)(x) = (3x-2)^3$ **b** $(f \circ g)(x) = 3x^3-2$ **c** $(f \circ f)(x) = 9x-8$

d $(g \circ g)(x) = x^9$

10a 1 **b** $\frac{5}{6}\pi$, $\frac{3}{2}\pi$ **11a** −13, 2 **b** $(h \circ g \circ f)(x) = 3-4x^2$ **c** 8, $2(3-x)^2$

12a 6, −80 **b** $3(5-4x+x^2)$, $1-9x^2$

13a $R, R, f^{-1}(x) = \frac{1}{3}(2-x)$ **b** $R, R, f^{-1}(x) = 2x+1$ **c** $R, R, f^{-1}(x) = 1-4x$

d $\{x : x \in R, x \neq 0\}, \{x : x \in R, x \neq 0\}, f^{-1}(x) = \dfrac{1}{2x}$

e $\{x : x \in R, x \neq 4\}, \{x : x \in R, x \neq 0\}, f^{-1}(x) = 4-\dfrac{1}{x}$ or $\dfrac{4x-1}{x}$

f $\{x : x \in R, x \neq 0\}, \{x : x \in R, x \neq 1\}, f^{-1}(x) = \dfrac{1}{\sqrt[3]{(x-1)}}$

14a $f^{-1}(x) = x-2$ **b** $g^{-1}(x) = \dfrac{2}{x}$ **c** $\frac{1}{2}$ **d** 4 **e** −1 **f** $\frac{2}{3}$ **15** $g(x) = -\frac{1}{2}x$

Geometry — Answers to Chapter 1

Page 92 Exercise 1

1 a 1 **b** 1 **c** 1 **d** 2 **e** 4 **f** $\frac{3}{4}$ **g** $-\frac{4}{3}$ **h** −1 **i** 2

2 $m_{AB} = 1$, $m_{CD} = \frac{1}{3}$, $m_{EF} = -\frac{1}{2}$, $m_{GH} = -\frac{2}{5}$, $m_{KL} = 0$, m_{TV} undefined

a AB, CD **b** EF, GH **c** KL **d** TV

3 a 2 **b** −1 **c** −2 **d** 3 **e** −1 **f** 1 **g** 1 **h** undefined **i** 0

4 If line slopes up from left to right, gradient is positive; if it slopes down, gradient is negative; if it is parallel to the x-axis, gradient is zero; if it is parallel to the y-axis, gradient is undefined.

5 down, up, up, up **6** $m_{AB} = \frac{3}{7} = m_{DC}$; $m_{AD} = 5 = m_{BC}$ **7** $\frac{1}{2}$, −3

8 $m_{AC} = \frac{2}{5} = m_{DB}$; $m_{AD} = \frac{5}{3} = m_{CB}$ **9** $m_{PQ} = -1 = m_{QR}$ $(= m_{PR})$

10 $m_{AC} = \frac{3}{4} = m_{CB}$; $m_{AD} = \frac{2}{5}$, $m_{DB} = \frac{5}{2}$ **11** (0, 7), (1, 5), (3, 1), etc.

12 a 45° **b** 45° **c** 60° **d** 30° **e** 45° **f** 135°

Page 94 Exercise 2

1 a $y = 6x$ **b** $y = \frac{1}{3}x$ **c** $y = -\frac{3}{4}x$ **d** $y = 0$

2 a $y = x$ *b* $y = 2x$ *c* $y = -\frac{3}{5}x$ *d* $y = -\frac{1}{2}x$

3 a F *b* T *c* F *d* T *e* T *4 a* $y = 2x+4$ *b* $y = 2x$ *c* $y = 2x+1$

d $y = 2x-4$ *5 a* $y = -x+4$ *b* $y = -x$ *c* $y = -x+1$ *d* $y = -x-4$

6 $\{y = -x+c, c \in R\}$ *7 a* *b*

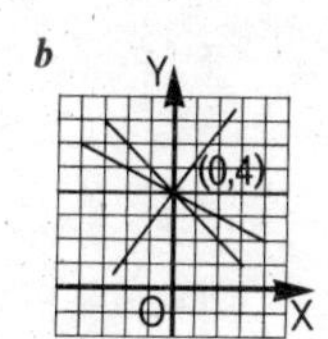

8 a $2, (0, -3)$ *b* $\frac{1}{2}, (0, 1)$ *c* $2, (0, 2\frac{1}{2})$

d $4, (0, -\frac{4}{3})$ *e* $\frac{1}{2}, (0, 3)$ *f* $-1, (0, 4)$

g $-\frac{2}{3}, (0, 3)$ *h* $-1, (0, -1)$ *i* $0, (0, -3)$

10 a $y = 10x+6$ *b* $y = 10x-2$ *11* 5 units parallel to the y-axis

12 a F *b* F *c* T

Page 96 Exercise 3

1 a $2x-3y-3 = 0$ *b* $3x+y-6 = 0$ *c* $3x+4y = 0$ *2* ***a, c, d, g, h***

3 a $5x-y+4 = 0$ *b* $5x-2y+16 = 0$ *4 a* $y = 3$ *b* $x = 2$

5 $x = 1, x = 5, y = 1, y = -2$ *6* $a = 1$ *7* $b = -3\frac{1}{4}$ *8* $3p+4q+5 = 0$

Page 98 Exercise 4

1 a $y = 4x-5$ *b* $y = 2x-3$ *c* $y = -2x+4$ *d* $y = -2x+9$ *e* $y = x$ *f* $2y = x+8$

2 a $y = 6x-4$ *b* $y = -x-3$ *c* $y = x-1$ *d* $3y = -2x$

3 a $3y = -4x+22$ *b* $4y = -x-1$ *c* $y = 3x+3$

4 a $y = x+1$ *b* $5y = 3x+5$ *c* $5y = 3x-12$ *d* $3x+2y = 0$

e $y = -x-3$ *f* $3y = -x-11$

5 $4y = 3x+11, 5y = -2x+8$ *7* $7y = 9x+2, 8y = 3x+12, y = -6x+10$

8 $(0, 4)$; $4y = 3x-9$, $4y = 3x+16$, $3y = -4x+12$, $3y = -4x+37$; $y = 7x-21$, $7y = -x+28$

9 a L(3, 7), M(−1, 2), P(4, 0), Q(8, 5) *b* $4y = 5x+13, 4y = 5x+2, 4y = 5x-20$; they are equal. *10* $2x+3y-9 = 0$

Page 99 Exercise 5

1 a A′(−2, 1), B′(−2, 3), C′(−5, 4) *c* $\frac{2}{3}, -\frac{3}{2}, -1; \frac{5}{4}, -\frac{4}{5}, -1$

2 a D′(−1, −3), E′(1, −2), F′(2, 5) *c* Products of gradients $= -1$.

Page 101 Exercise 6

1 *b, e* *2 a* -3 *b* $\frac{1}{4}$ *c* $-\frac{1}{6}$ *d* $\frac{2}{5}$ *e* -1 *3 a* $y=-3x$ *b* $y=\frac{1}{4}x$ *c* $y=-\frac{1}{6}x$ *d* $y=\frac{2}{5}x$ *e* $y=-x$ *4 a* $3y=-x+5$ *b* $3y=-x$ *c* $3y=-x-16$

5 Products of gradients $=-1$. *6 a* $x=4$ *b* $y=1$

7 a $y=-2x$ *b* $y=-\frac{3}{5}x$ *c* $y=\frac{5}{6}x$ *8 a* $y=-\frac{1}{3}x$ *b* $y=-\frac{8}{5}x$ *c* $y=\frac{1}{3}x$

9 a $y=2x-1$ *b* $y=2x+5$ *c* $y=2x+3$ *10* $y=-1$ *11* $-2, \frac{1}{2}$

12 $y=-2x-3, y=-2x+7, 2y=x+9$ *13* $b=-\frac{1}{3}$

14 $3y=-x+4, y=-3x+4, y=x$ *15* Product of gradients $=-1$.

16 Use products of gradients to show four right angles. *17* Triangle is also isosceles.

Page 103 Exercise 7

1 a $(\frac{1}{2}, 1\frac{1}{2})$ *b* $(-2,1)$ *2* $(3,1),(3,-11),(-1,-3)$ *3 a* $x+2y=5$ *b* $(5,0)$

4 a $2x+3y=12$ *b* $(0,4),(6,0)$ *5 a* $2x+y=8$ *b* $3x-y=-3$ *c* $(1,6)$

6 a $3x-5y=-2, x+3y=4$ *b* $(1,1)$ *7 a* $2x+y=4$ *b* $x-y=2$ *c* $(2,0)$

8 $(4,2),(1\frac{3}{5}, 2\frac{4}{5}),(-4,0)$ *9* The lines are parallel, so do not intersect.

10 $(1\frac{2}{3}, -3)$

Page 105 Exercise 8

1 $x+y=3, x-y=-1$ *2 a* $3x-y=10$ *b* $x-3y=0$

3 $y=3, 5x-2y=-16, 5x+2y=16$ *4 a* $y=\frac{1}{4}x$ *b* $(8,2)$, 8·25

5 a $4x-3y=-19, 3x+4y=17$ *b* $(3,2)$, 5 *6 a* $(2,1)$

7 a $(-1\frac{1}{5}, -1\frac{1}{5})$ *8 a* $(2,1)$ *9* $(6,6)$ *10* $(6,4)$

11a a parallelogram *b* a rectangle *c* a square

12a -1; $y=-x$ *b* -2; $y=-2x$ *c* $\frac{5}{2}$; $5x-2y=0$

13a $y=x$ *b* $y=x$ *c* $y=-x$ *d* $x+y=2$ *e* (*1*) $y=x+3$ (*2*) $y=x-3$

14a $A_1(6,9)$, $B_1(9,-3)$ *b* for AB, gradient $=-4$, equation is $4x+y=11$; for A_1B_1, gradient $=-4$, equation is $4x+y=33$

15a $L_1(-4,4)$, $M_1(-6,5)$, $N_1(-4,0)$

Page 107 Exercise 8B

1 $(6,2)$ *3* T(1, 4). *l* is the image of *m*, and *m* is the image of *l*; $x+2y=9$

4 B$(-4,2)$, C$(-2,-4)$, E$(2,4)$, F$(6,2)$

5 $x-3y=-35,\ x-3y=6,\ 12x+5y=-10,\ 12x+5y=195$; (10, 15), (0, −2)

6 (4, 4), (3, 6), (5, 2) 7 *a* (*1*)

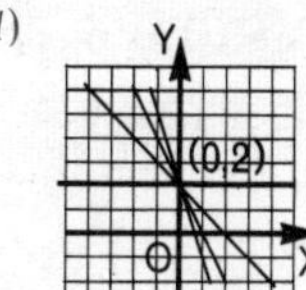

(*2*)

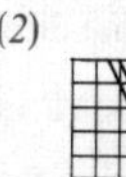

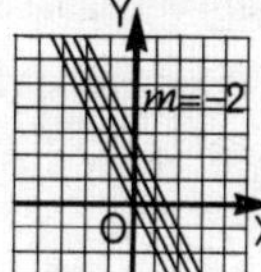

b $\left(\dfrac{2-c}{m-2},\ \dfrac{mc-4}{m-2}\right)$

(*1*) lines meet on y-axis
(*2*) lines are parallel

8 *a* $y=\frac{1}{2}x$ *b* $3x+y=27$; $(7\frac{5}{7}, 3\frac{6}{7})$

9 D(−4, 9); $x-3y=-31,\ x-3y=-53,\ 7x+y=-41,\ 7x+y=-19$;

Q$(-5\frac{1}{2}, 8\frac{1}{2})$; translation $\begin{pmatrix}-3\\-1\end{pmatrix}$, represented by $\overrightarrow{BA}$.

10 $y=x,\ x+y=1,\ 2x-y=3,\ 2x+y=7$; $(2\frac{1}{2}, 2)$, (3, 2)

Geometry — Answers to Chapter 2

Page 111 Exercise 1

1 *a* $\begin{pmatrix}4\\1\end{pmatrix}$ *b* $\begin{pmatrix}4\\-1\end{pmatrix}$ *c* $\begin{pmatrix}8\\0\end{pmatrix}$ 2 *a* $\begin{pmatrix}3\\2\end{pmatrix}$ *b* $\begin{pmatrix}-6\\-9\end{pmatrix}$ *c* $\begin{pmatrix}6\\5\end{pmatrix}$ *d* $\begin{pmatrix}3\\-2\end{pmatrix}$

3 *a* (4, 2) *b* (6, 5) *c* (−1, 6) *d* (3, 0) 4 *a* 10 *b* 1 *c* $4\sqrt{2}$

5 *a* $p=6, q=1$ *b* $p=6, q=2$ 6 *a* $\begin{pmatrix}-2\\3\end{pmatrix}$ *b* $\begin{pmatrix}5\\7\end{pmatrix}$

7 *a* $2\mathbf{b}-3\mathbf{a}$ *b* $\frac{1}{4}(\mathbf{a}-7\mathbf{b})$ 8 *a* $\overrightarrow{BC}$ *b* $\overrightarrow{CD}$ *c* $\overrightarrow{BD}$ *d* $\overrightarrow{ED}$ *e* $\overrightarrow{ED}$ *f* $2\overrightarrow{AD}$

9 $\begin{pmatrix}1\\2\end{pmatrix}, \begin{pmatrix}2\\1\end{pmatrix}, \begin{pmatrix}-2\\0\end{pmatrix}, \begin{pmatrix}0\\-1\end{pmatrix}, \begin{pmatrix}-2\\1\end{pmatrix}, \begin{pmatrix}0\\2\end{pmatrix}$ 10 $\mathbf{u}+2\mathbf{v},\ 2\mathbf{u}+\mathbf{v},\ -2\mathbf{u},\ -\mathbf{v},\ -2\mathbf{u}+\mathbf{v},\ 2\mathbf{v}$

11 They represent vectors in opposite direction to **a**, **b** and **c**; $\mathbf{a}+\mathbf{b}$, $k\mathbf{a}+l\mathbf{b}$, $k=l$

12 *a* $\begin{pmatrix}4\\0\end{pmatrix}, \begin{pmatrix}0\\3\end{pmatrix}, \begin{pmatrix}-4\\-3\end{pmatrix}$ *c* 4, 3, 5; right-angled at Q

13 *a* Sum $=\begin{pmatrix}0\\0\end{pmatrix}$ *b* $\begin{pmatrix}1\\9\end{pmatrix}$ *c* $\begin{pmatrix}\frac{2}{3}\\6\end{pmatrix}$ *d* $\begin{pmatrix}-6\frac{1}{2}\\3\end{pmatrix}, \begin{pmatrix}-4\frac{1}{3}\\2\end{pmatrix}$

Page 116 Exercise 2

1 *a* $\overrightarrow{AN}+\overrightarrow{NP}$, $\overrightarrow{AC}+\overrightarrow{CP}$, $\overrightarrow{AL}+\overrightarrow{LP}$, etc. *b* $\overrightarrow{AB}+\overrightarrow{BN}+\overrightarrow{NP}$, $\overrightarrow{AC}+\overrightarrow{CN}+\overrightarrow{NP}$, $\overrightarrow{AC}+\overrightarrow{CL}+\overrightarrow{LP}$, etc. *c* $\overrightarrow{AB}+\overrightarrow{BC}+\overrightarrow{CL}+\overrightarrow{LP}$, etc.; $\overrightarrow{AB}+\overrightarrow{BC}+\overrightarrow{CD}+\overrightarrow{DM}+\overrightarrow{MP}$, etc. *d* no limit

2 *a* $\overrightarrow{NB}$ *b* $\overrightarrow{BN}$ 3 *a* (*1*) 8 cm, same direction (*2*) 4 cm, opposite direction

(*3*) 1 cm, same direction *b* (*1*) AP produced its own length to X

(*2*) X divides AP in ratio 1:2 (*3*) at P

4 *a* $\overrightarrow{CP}$ *b* $\overrightarrow{PC}$ *c* $\overrightarrow{DP}$

5 *a* $\overrightarrow{AL}-\overrightarrow{AP}$, etc. *b* $\overrightarrow{AD}-\overrightarrow{AP}$, etc. *c* $\overrightarrow{PN}-\overrightarrow{PA}$, etc. *d* $\overrightarrow{NP}-\overrightarrow{NA}$, etc.

e $\overrightarrow{CB}-\overrightarrow{CA}$, etc.

6 *a* $\boldsymbol{u}$, $\overrightarrow{OA}$, $\overrightarrow{PQ}$, $\overrightarrow{CB}$, $\overrightarrow{SR}$; $\boldsymbol{v}$, $\overrightarrow{OC}$, $\overrightarrow{AB}$, $\overrightarrow{PS}$, $\overrightarrow{QR}$; $\boldsymbol{w}$, $\overrightarrow{OP}$, $\overrightarrow{AQ}$, $\overrightarrow{BR}$, $\overrightarrow{CS}$

b $\overrightarrow{OR}=\overrightarrow{OA}+\overrightarrow{AB}+\overrightarrow{BR}$, $\overrightarrow{OA}+\overrightarrow{AQ}+\overrightarrow{QR}$, $\overrightarrow{OC}+\overrightarrow{CB}+\overrightarrow{BR}$, $\overrightarrow{OC}+\overrightarrow{CS}+\overrightarrow{SR}$,

$\overrightarrow{OP}+\overrightarrow{PQ}+\overrightarrow{QR}$, $\overrightarrow{OP}+\overrightarrow{PS}+\overrightarrow{SR}$

7 *a* $-4\boldsymbol{p}+20\boldsymbol{q}-10\boldsymbol{r}$ *b* $-3\boldsymbol{p}-15\boldsymbol{q}$ 8 *b* $\boldsymbol{a}+\boldsymbol{b}$, $\overrightarrow{LM}=\frac{1}{2}\overrightarrow{BA}$

c parallel to the third side and equal to half of it

9 A straight line, bisecting one side of triangle and parallel to a second side, bisects the third side.

Page 120 Exercise 3

1 a (*1*) $\begin{pmatrix}4\\3\\5\end{pmatrix}$ (*2*) $\begin{pmatrix}-1\\2\\1\end{pmatrix}$ 2 *a* $\begin{pmatrix}-6\\12\\0\end{pmatrix}$ *b* $\begin{pmatrix}-1\\2\\0\end{pmatrix}$ *c* $\begin{pmatrix}2\\-4\\0\end{pmatrix}$

3 *a* $\begin{pmatrix}0\\-1\\4\end{pmatrix}$ *b* $\begin{pmatrix}-11\\0\\25\end{pmatrix}$ *c* $\begin{pmatrix}-4\\13\\8\end{pmatrix}$ 4 *a* $\boldsymbol{x}=\begin{pmatrix}0\\-4\\0\end{pmatrix}$ *b* $\boldsymbol{x}=\begin{pmatrix}1\\1\\1\end{pmatrix}$ *c* $\boldsymbol{x}=\begin{pmatrix}1\\-1\\-1\frac{2}{3}\end{pmatrix}$

5 $\begin{pmatrix}1\\2\\3\end{pmatrix}, \begin{pmatrix}2\\4\\6\end{pmatrix}, \begin{pmatrix}0\\5\\-1\end{pmatrix}, \begin{pmatrix}1\\2\\3\end{pmatrix}, \begin{pmatrix}-2\\1\\-7\end{pmatrix}, \begin{pmatrix}-1\\3\\-4\end{pmatrix}, \begin{pmatrix}1\\-3\\4\end{pmatrix}$; $\overrightarrow{OA}=\frac{1}{2}\overrightarrow{OB}$; 1:2

6 $\overrightarrow{OA}=\begin{pmatrix}6\\4\\2\end{pmatrix}=\overrightarrow{CB}$ 7 *a* sum $=\boldsymbol{0}$ *b* $\begin{pmatrix}1\\9\\-3\end{pmatrix}$ 8 $2\overrightarrow{AB}=\overrightarrow{BC}$, 1:2

9 $-3\overrightarrow{AB}=\overrightarrow{BC}$, $-1:3$ *10* $\begin{pmatrix}-3\\-6\\-9\end{pmatrix}, \begin{pmatrix}-1\\-2\\-3\end{pmatrix}$; (1, 5, 5)

11a $(\frac{1}{2}x_1, \frac{1}{2}y_1, \frac{1}{2}z_1), (\frac{1}{2}x_2, \frac{1}{2}y_2, \frac{1}{2}z_2)$ *b* $\overrightarrow{MN}=\frac{1}{2}\overrightarrow{AB}$

Page 122 Exercise 4

1 a $\begin{pmatrix}0\\1\\0\end{pmatrix}$ *b* $\begin{pmatrix}1\\1\\0\end{pmatrix}$ *c* $\begin{pmatrix}0\\1\\1\end{pmatrix}$ *d* $\begin{pmatrix}1\\1\\1\end{pmatrix}$ 2 $l=3, m=2, n=-1$

3 *a* $\boldsymbol{a}=2\boldsymbol{i}+3\boldsymbol{j}+\boldsymbol{k}$, $\boldsymbol{b}=-\boldsymbol{i}+2\boldsymbol{j}$, $\boldsymbol{c}=3\boldsymbol{i}-\boldsymbol{j}+4\boldsymbol{k}$ *b* $9\boldsymbol{j}-2\boldsymbol{k}$, $-16\boldsymbol{i}+5\boldsymbol{j}-17\boldsymbol{k}$

4 a $\begin{pmatrix} 5 \\ -1 \\ -2 \end{pmatrix}$ *b* $\begin{pmatrix} 1 \\ -3 \\ 4 \end{pmatrix}$ *c* $\begin{pmatrix} 12 \\ -1 \\ -7 \end{pmatrix}$ *5* $-4\boldsymbol{i}+2\boldsymbol{j}-6\boldsymbol{k} = -2(2\boldsymbol{i}-\boldsymbol{j}+3\boldsymbol{k})$

6 a $2\boldsymbol{i}+3\boldsymbol{j}$ *b* $2\boldsymbol{i}+3\boldsymbol{j}+4\boldsymbol{k}$ *c* $3\boldsymbol{j}+4\boldsymbol{k}$ *d* $2\boldsymbol{i}+4\boldsymbol{k}$ *e* $-2\boldsymbol{i}+4\boldsymbol{k}$ *f* $\boldsymbol{i}+3\boldsymbol{j}+4\boldsymbol{k}$

Page 124 Exercise 5

1 a 6 *b* 6 *c* 3 *d* 9 *e* 10 *f* 13*a*

2 a 3 *b* 7 *c* $\sqrt{32}$ *d* $\sqrt{84}$ *3* $4\boldsymbol{i}+4\boldsymbol{j}+7\boldsymbol{k}$, 9

4 PQ = 14 = PR *5* PQ = $\sqrt{5}$ = QR, $PQ^2+QR^2 = 10 = PR^2$

6 AB = 3 = AC, $AB^2+AC^2 = 18 = BC^2$; (1, 6, 2)

7 a parallelogram *b* rhombus *8* $\overrightarrow{PQ}+\overrightarrow{QR}+\overrightarrow{RS} = \boldsymbol{0}$,

$|\overrightarrow{PQ}|^2+|\overrightarrow{QR}|^2 = 35 = |\overrightarrow{RP}|^2$

9 Distance of each from (2, 3, 4) = $\sqrt{72}$ = radius.

10 5; equidistant from (1, 2, 3); surface of sphere, centre (1, 2, 3), radius 5

11 $\{(x, y, z): (x-3)^2+(y-4)^2+(z-5)^2 = 16\}$ *12a* 36, 49, 9 *b* 96·4°

Page 126 Exercise 6

1 A S P R Q B *2* C D X Y

3 a 1:1 *b* 1:2 *c* 3:2 *d* 3:5 *e* 5:−2

4 a 1:2 *b* 1:1 *c* 1:4 *d* 1:2 *e* 2:−1 *f* 3:−1 *g* −1:3 *h* −3:2

5 a P K Q L *b* (*1*) 2:3 (*2*) 1:1

6 a 2:1 *b* 3:−1 *c* 3:1 *d* −1:2 *e* 2:−3

Page 128 Exercise 7

1 $\overrightarrow{PQ} = \overrightarrow{PO}+\overrightarrow{OQ}$, etc. *2 a* $2\boldsymbol{c} = \boldsymbol{a}+\boldsymbol{b}$, etc. *b* $\boldsymbol{c} = \boldsymbol{a}+\frac{1}{2}(\boldsymbol{b}-\boldsymbol{a})$, etc.

4 $\frac{1}{3}(2\boldsymbol{b}+\boldsymbol{a})$ *5 a* $\frac{1}{5}(3\boldsymbol{b}+2\boldsymbol{a})$ *b* $3\boldsymbol{b}-2\boldsymbol{a}$ *c* $\frac{1}{3}(\boldsymbol{b}+2\boldsymbol{a})$, $2\boldsymbol{a}-\boldsymbol{b}$

d $\frac{1}{8}(5\boldsymbol{b}+3\boldsymbol{a})$, $\frac{1}{2}(5\boldsymbol{b}-3\boldsymbol{a})$ *6* $\frac{1}{3}(\boldsymbol{u}+\boldsymbol{v})$, 1:2

7 a $\boldsymbol{p} = \frac{1}{2}(\boldsymbol{b}+\boldsymbol{c})$, $\boldsymbol{q} = \frac{1}{2}(\boldsymbol{c}+\boldsymbol{a})$, $\boldsymbol{r} = \frac{1}{2}(\boldsymbol{a}+\boldsymbol{b})$

Page 129 Exercise 7B

1 Each represents $2(\boldsymbol{c}+\boldsymbol{a}-\boldsymbol{b})$

2 $\boldsymbol{a}+\boldsymbol{b}+\boldsymbol{c}$, $\boldsymbol{b}+\boldsymbol{c}$, $\boldsymbol{a}+\boldsymbol{c}$, $\boldsymbol{a}+\boldsymbol{b}$; each $= \frac{1}{2}(\boldsymbol{a}+\boldsymbol{b}+\boldsymbol{c})$; the space diagonals are concurrent and bisect one other.

3 a *(1)* $\frac{1}{2}(\boldsymbol{a}+\boldsymbol{b}), \frac{1}{2}(\boldsymbol{b}+\boldsymbol{c}), \frac{1}{2}(\boldsymbol{c}+\boldsymbol{d}), \frac{1}{2}(\boldsymbol{d}+\boldsymbol{a})$ *(2)* Each $= \frac{1}{4}(\boldsymbol{a}+\boldsymbol{b}+\boldsymbol{c}+\boldsymbol{d})$

b S and T coincide, NR and PM bisect each other. *c* parallelogram

4 a Each $= \frac{1}{4}(\boldsymbol{a}+\boldsymbol{b}+\boldsymbol{c}+\boldsymbol{d})$ *b* S, T and V coincide.
The joins of midpoints of opposite edges are concurrent and bisect one another.

5 a $\frac{1}{2}(\boldsymbol{b}+\boldsymbol{c}), \frac{1}{2}(\boldsymbol{c}+\boldsymbol{a}), \frac{1}{2}(\boldsymbol{a}+\boldsymbol{b})$ *b* each $= \frac{1}{3}(\boldsymbol{a}+\boldsymbol{b}+\boldsymbol{c})$ *c* G, H and K coincide. The medians of a triangle are concurrent and divide one another in the ratio 2:1.

6 b Each $= \frac{1}{4}(\boldsymbol{a}+\boldsymbol{b}+\boldsymbol{c}+\boldsymbol{d})$. *c* G, H, K and L coincide. The joins of the vertices of a tetrahedron to the centroids of the opposite faces are concurrent and divide one another in the ratio 3:1.

Page 132 Exercise 8

1 a (7, 4) *b* (3, 1) *c* (1, 3) *2 a* (8, 19) *b* (3, −6) *c* (9, 2)

3 $(\frac{1}{2}, 1\frac{1}{2})$, (8, −6) *4 a* (4, 3, 8) *b* $(2\frac{1}{5}, -\frac{4}{5}, 3\frac{2}{5})$ *c* (9, −14, 11)

5 (6, 4, 4), (3, −2, −5) *6 a* (2, 5), (5, 2), (8, −1) *b* $\overrightarrow{PQ} = \overrightarrow{QR}$, 1:1

7 a (1, 0), (−4, 4), (−14, 12) *b* $3\overrightarrow{PQ} = \overrightarrow{PR}$, 1:3

8 $\overrightarrow{QR} = -2\overrightarrow{PQ}$; (0, −5, −2) *9 a* (0, −2, 0) *b* $QS^2+SR^2 = 8 = QR^2$

10a (2, 7), (1, 5), $(-1\frac{1}{2}, 6)$ *b* $K(\frac{3}{4}, 6)$ *c* collinear; CK = KN

11b 3:1 *c* $p = -3, q = -4$

Page 136 Exercise 9

1 (i) 6 (ii) 0 (iii) −1 (iv) $28\sqrt{2}$ *2 a* 3 *b* 2 *3 a* −9 *b* 9

4 7 *5* 15 *6 a* (7, −1, 3) *b* $\begin{pmatrix}6\\-3\\6\end{pmatrix}, \begin{pmatrix}4\\-2\\4\end{pmatrix}$ *c* 54

Page 138 Exercise 10

1 a $\frac{1}{2}$ *b* $\frac{5}{14}$ *c* $-\frac{4}{21}$ *2 a* 90° *b* 120° *c* 50·2° *d* 180° *3* −7

4 a 0 *b* 90° *5* scalar product is zero *6* 24·1°, 24·1°, 131·8°

7 $\begin{pmatrix}3\\6\\9\end{pmatrix}, \begin{pmatrix}4\\8\\-2\end{pmatrix}, \begin{pmatrix}4\\-2\\0\end{pmatrix}$; scalar products zero *8 a* (2, 1, 3)

9 a P(1, 2, 1), Q(−1, 2, 3), R(1, 0, 2), S(3, 0, 0) *b* 45°, 45°, 135°, 135°

10a $\boldsymbol{a} = \boldsymbol{0}$ or $\boldsymbol{b} = \boldsymbol{0}$ or $\theta = \frac{1}{2}\pi$ *b* Neither $\boldsymbol{a}$ nor $\boldsymbol{b}$ need be zero.

c $\boldsymbol{p}.\boldsymbol{q} = 0$ *11* $\cos\theta = \frac{1}{2}\sqrt{3}$ *12* $-\frac{34}{11}$

Page 141 Exercise 11B

1 36, $\cos 0° = 1$ *2* 7, 12 *3* 8 *4* $x = 6$ *5* 1, 1, 19; $\sqrt{19}$

6 ***a, d*** *7* $\begin{matrix} 1 & 0 & 0 \\ 0 & 1 & 0 \\ 0 & 0 & 1 \end{matrix}$ *9* B *10* $\frac{2}{7}, \frac{3}{7}, -\frac{6}{7}$

11 2, −2

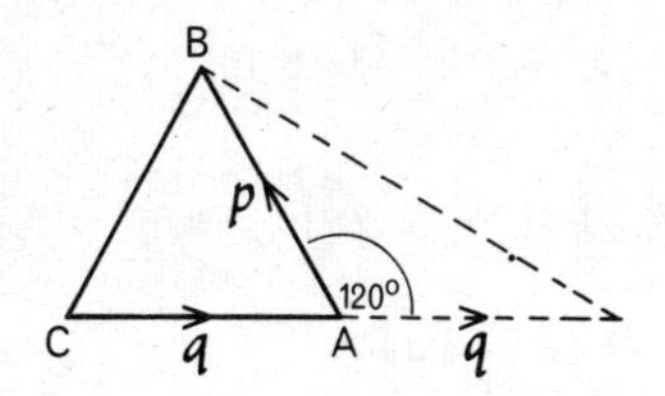

12 $|\boldsymbol{a}|^2 - |\boldsymbol{b}|^2$; 0; they are perpendicular; PQSR is a rhombus.

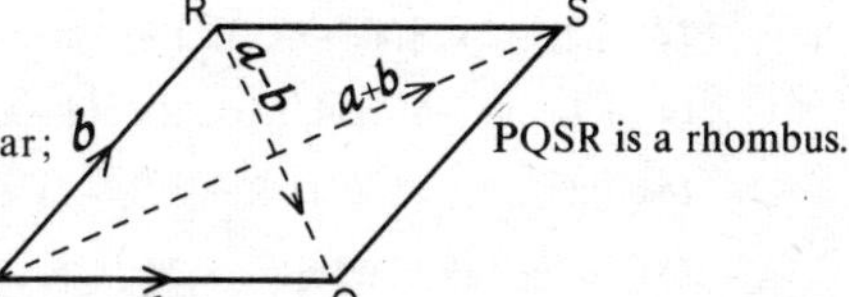

13 $\boldsymbol{u}.\boldsymbol{v} = 0$ *14* Sum of two sides of a triangle is greater than the third; when $\boldsymbol{a}$ and $\boldsymbol{b}$ have the same direction. *15* $\cos\theta \leqslant 1$; $\boldsymbol{a}$ and $\boldsymbol{b}$ have the same direction. *16* 0; AO and BC are perpendicular; the altitudes of a triangle are concurrent.

17 $-\boldsymbol{a}, \boldsymbol{a}-\boldsymbol{b}, -\boldsymbol{a}-\boldsymbol{b}$; an angle in a semicircle is a right angle.

18a $|\boldsymbol{d}-\boldsymbol{a}| = |\boldsymbol{c}-\boldsymbol{b}|$ ***b*** 0 ***c*** If a tetrahedron has three pairs of opposite edges equal, the joins of their midpoints in pairs are mutually perpendicular.

Geometry — Answers to Revision Exercises

Page 146 Revision Exercise 1

1 AC up, DB down *2* ***a*** −1 ***b*** $\frac{3}{2}$ ***c*** $-\frac{1}{2}$ ***d*** $\frac{3}{4}$

3 ***a*** T ***b*** T ***c*** F ***d*** F ***e*** T *4* ***a*** $y = -2x+1$ ***b*** $y = -2x-1$

5 A(−3, 0), B(0, 2); $\sqrt{13}$ *6* C(0, 5), E(−2, −1)

7 $x+5y = 11$, $y = 2$; (1, 2) *8* $4y = 3x+15$; $3y = -4x+5$

9 K(1, 6), L(5, 1), M(7, 9); KL $\doteqdot$ 6·4, LM $\doteqdot$ 8·2, MK $\doteqdot$ 6·7

10 gradients of diagonals 2, $-\frac{1}{2}$; midpoint of PR and QS (−2, 4); rhombus

11 It is a linear equation (of first degree in x and y) ***a*** $-\frac{5}{7}$ ***b*** $-\frac{1}{7}$ ***c*** $\frac{1}{9}$

12 Product of gradients is -1; 20 *13* $\frac{1}{6}\pi$

14 $(3, 8)$, $3x-2y+7=0$, $S(-2\frac{1}{3}, 0)$; $86\frac{2}{3}$ square units

15 $(9, 1)$, $4x-9y=84$ or $16x+3y=24$

Page 148 Revision Exercise 2

1 $\begin{pmatrix}3\\2\end{pmatrix}, \begin{pmatrix}-1\\-6\end{pmatrix}, \begin{pmatrix}-2\\4\end{pmatrix}$ *2* $\begin{pmatrix}-6\\8\end{pmatrix}$; 10 *3* $-3:1$ *5* 0, 1, 1

7 a 13 *b* 3 *c* $\sqrt{126}$ *8 a* $\begin{pmatrix}24\\2\\-4\end{pmatrix}$ *b* $\sqrt{11}, \sqrt{59}$ *9* $\begin{pmatrix}u_1+v_1\\u_2+v_2\\u_3+v_3\end{pmatrix}$

10 $\frac{1}{2}(\boldsymbol{a}+\boldsymbol{b}), \frac{1}{2}(\boldsymbol{b}+\boldsymbol{c}), \frac{1}{2}(\boldsymbol{c}+\boldsymbol{a})$; $\boldsymbol{a}-\boldsymbol{b}+\boldsymbol{c}$ *11* $\boldsymbol{a}+\boldsymbol{c}, \frac{1}{2}(\boldsymbol{a}+\boldsymbol{c})$

12 $\frac{1}{2}(3\boldsymbol{a}+2\boldsymbol{b}), \frac{1}{2}(4\boldsymbol{a}+\boldsymbol{b}), \frac{1}{2}(\boldsymbol{a}+\boldsymbol{b})$; each is $\frac{1}{3}(4\boldsymbol{a}+2\boldsymbol{b})$ *13b* $\frac{1}{2}(\boldsymbol{b}+\boldsymbol{c}), \frac{1}{3}(\boldsymbol{b}+\boldsymbol{c})$

14 $\overrightarrow{KL}$ and $\overrightarrow{HM}$ both represent $\frac{1}{2}(\boldsymbol{d}-\boldsymbol{b})$ *15* $(0, -2, 5), (0, -14, 5)$

16a P(2, 1, 3), Q(2, 0, 1), R(2, −3, −5) *b* 1:3 *17a* C(1, −2, 0), D(4, 2, −1) *b* 1:3

18 90° *19* 70·5(6)° *20* $\sqrt{6}, \sqrt{6}$, 60° *21* −1 *22* D(3, 0, 1)

23 S(1, 4, 3) *24* $p=0, q=r$ *25* $x^2+y^2+z^2=1$; $y=\pm 1/\sqrt{2}$

26 $\boldsymbol{u}+\boldsymbol{v}$; $\boldsymbol{u}-\boldsymbol{v}$

Trigonometry — Answers to Chapter 1

Page 154 Exercise 1

1 a + *b* − *c* − *d* − *e* − *f* + *g* − *h* +

2 a −cos 79° *b* −sin 31° *c* −tan 45° *d* cos 40°

3 a −0·176 *b* 0·500 *c* 0·951 *d* 0·985

4 cos 57° = cos 303°; cos 123° = cos 237°; cos 0° = cos 360°

5 a {30, 150} *b* {60, 300} *c* {45, 225} *d* {131·8, 228·2}

e {48·6, 131·4} *f* {123·7, 303·7} *6* −0·5

7 a 90, 450 *b* 90, 270, 450, 630 *c* −180, 0, 180 *d* 180

8 $\frac{3}{5}; \frac{4}{3}$ *9* $\frac{21}{29}; \frac{21}{20}$

10a *11a* *12*

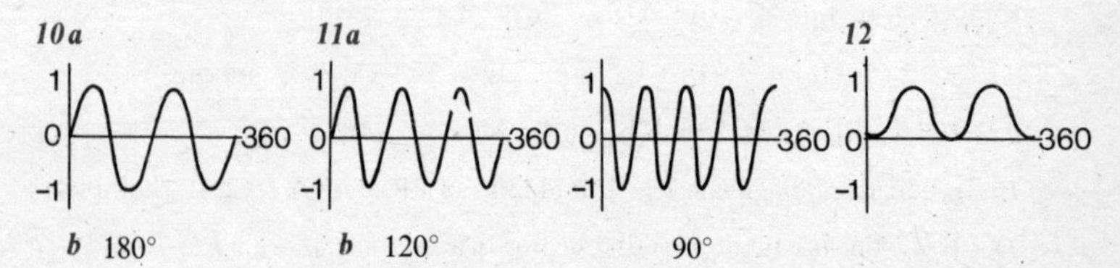

b 180° *b* 120° 90°

13 (3·76, 1·37) *14* (−4·33, −2·5) *15* 6·98, 6·84

Page 157 Exercise 2

1 a $\sin p°$ *b* $-\cos q°$ *c* $-\tan r°$ *d* $\sin s°$ *e* $\cos t°$ *f* $\cos u°$

g $\sin b°$ *h* $\sin c°$ *i* $-\tan d°$ *j* $\sin v°$ *k* $-\sin k°$ *l* $-\cos z°$

2 *a* $\sin 60°$ *b* $-\cos 80°$ *c* $\sin 85°$ *d* $-\cos 1°$

3 *a* 1 *b* $-\tan a°$ *c* $\tan a°$ *d* 1 *e* 1 6 $-\sin a°, \cos a°$

Page 159 Exercise 3

1 $0, \frac{1}{6}\pi, \frac{1}{4}\pi, \frac{1}{3}\pi, \frac{1}{2}\pi, \pi, 2\pi$ 2 45°, 90°, 36°, 240°

3 *a* 0·5 *b* 1 *c* 0 *d* 1 *e* 0·5

4 *a* 115° *b* 0·349 5

4°	66°	308°	37·24°	7·05°
0·070	1·152	5·375	0·65	0·123

6 *a* $\sin\theta$ *b* $-\cos\theta$ *c* $-\tan\theta$ *d* $\cos\theta$ *e* $\sin\theta$ *f* $-\sin\theta$ *g* $\tan\theta$ *h* $-\sin\theta$

7 *a* $\{\frac{1}{4}\pi, \frac{5}{4}\pi\}$ *b* $\{\frac{1}{3}\pi, \frac{5}{3}\pi\}$ *c* $\{\frac{1}{4}\pi, \frac{3}{4}\pi\}$ *d* $\emptyset$ *e* $\{\frac{3}{2}\pi\}$ *f* $\{\frac{1}{6}\pi, \frac{7}{6}\pi\}$

8 *a* 180 *b* π *10* $\frac{1}{3}\pi$

Page 162 Exercise 4

1 $\cos x\cos y+\sin x\sin y$, $\cos A\cos B+\sin A\sin B$, $\cos p\cos q-\sin p\sin q$

$\cos X\cos Y-\sin X\sin Y$ 2 *a* each side = 1 *b* each side = 1

5 *a* $\cos(M-N)$ *b* $\cos 3\alpha$ *c* $\cos 90° = 0$ *d* $\cos 45° = \frac{1}{\sqrt{2}}$ 7 $\frac{63}{65}, -\frac{33}{65}$

9 $\dfrac{\sqrt{3}-1}{2\sqrt{2}}$

Page 163 Exercise 5

1 $\sin x\cos y+\cos x\sin y$, $\sin A\cos B+\cos A\sin B$, $\sin p\cos q-\cos p\sin q$,

$\sin X\cos Y-\cos X\sin Y$ 2 *a* each side = $1/\sqrt{2}$ *b* each side = 0·5

5 *a* $\sin(M+N)$ *b* $\sin\alpha$ *c* $\sin 90° = 1$ *d* $\sin 180° = 0$ 7 $\frac{4}{5}; \frac{44}{125}$

11a (1) $x = r\cos\theta,\ y = r\sin\theta$ (2) $x_1 = r\cos(\theta+\alpha),\ y_1 = r\sin(\theta+\alpha)$

b $y_1 = x\sin\alpha + y\cos\alpha$ *c* $\begin{pmatrix}\cos\alpha & -\sin\alpha\\ \sin\alpha & \cos\alpha\end{pmatrix}$ *d* (1) $(-b, a)$ (2) $(-a, -b)$

(3) $[\frac{1}{\sqrt{2}}(a-b), \frac{1}{\sqrt{2}}(a+b)]$ (4) $[\frac{1}{2}(\sqrt{3}a-b), \frac{1}{2}(a+\sqrt{3}b)]$

12a (1) $\begin{pmatrix}\cos\beta & -\sin\beta\\ \sin\beta & \cos\beta\end{pmatrix}$ (2) $\begin{pmatrix}\cos(\alpha+\beta) & -\sin(\alpha+\beta)\\ \sin(\alpha+\beta) & \cos(\alpha+\beta)\end{pmatrix}$

b The product of matrices corresponds to the composition of the two rotations about O.

c $\cos(\alpha+\beta) = \cos\alpha\cos\beta - \sin\alpha\sin\beta$; $\sin(\alpha+\beta) = \sin\alpha\cos\beta + \cos\alpha\sin\beta$

Page 165 Exercise 6B

1 a $\cos 3a\cos 2b - \sin 3a\sin 2b$ *b* $\sin x\cos 2y + \cos x\sin 2y$

c $\sin 3a\cos 2b - \cos 3a\sin 2b$ *d* $\cos 2x\cos y + \sin 2x\sin y$

2 a $\cos 315^\circ = 1/\sqrt{2}$ *b* $\cos\frac{1}{2}\pi = 0$ *6* $1/\sin a$ *9* $\{30, 210\}$

10 $\{71{\cdot}6, 251{\cdot}6\}$ *13* $\frac{1}{2}(\sqrt{3}-1)$; $1-\frac{1}{2}\sqrt{3}$; $56{\cdot}2$, $26{\cdot}1$

Page 166 Exercise 7

1 $\dfrac{\tan x+\tan y}{1-\tan x\tan y}$, $\dfrac{\tan A+\tan B}{1-\tan A\tan B}$, $\dfrac{\tan p-\tan q}{1+\tan p\tan q}$, $\dfrac{\tan M-\tan N}{1+\tan M\tan N}$

2 a each side $= -\sqrt{3}$ *b* each side $= 0$

3 $\dfrac{\tan x+\tan 2y}{1-\tan x\tan 2y}$, $\dfrac{\tan 3a+\tan 2b}{1-\tan 3a\tan 2b}$, $\dfrac{\tan 2x-\tan y}{1+\tan 2x\tan y}$, $\dfrac{\tan 3p-\tan 2q}{1+\tan 3p\tan 2q}$, $\dfrac{1-\tan x}{1+\tan x}$

4 $\frac{1}{7}$ *6* $-\frac{56}{33}$

Page 168 Exercise 8

1 $2\sin A\cos A$, $\cos^2 A-\sin^2 A = 2\cos^2 A-1 = 1-2\sin^2 A$, $\dfrac{2\tan A}{1-\tan^2 A}$

2 $2\sin\frac{1}{2}\theta\cos\frac{1}{2}\theta$, $\cos^2\frac{1}{2}\theta-\sin^2\frac{1}{2}\theta = 2\cos^2\frac{1}{2}\theta-1 = 1-2\sin^2\frac{1}{2}\theta$, $\dfrac{2\tan\frac{1}{2}\theta}{1-\tan^2\frac{1}{2}\theta}$

3 $2\sin 2\alpha\cos 2\alpha$ *4 a* $2\cos^2 2\alpha-1$ *b* $1-2\sin^2 2\alpha$

5 $2\sin 4A\cos 4A$, $\cos^2 3B-\sin^2 3B = 2\cos^2 3B-1 = 1-2\sin^2 3B$, $\dfrac{2\tan\frac{1}{2}C}{1-\tan^2\frac{1}{2}C}$

6 $\frac{24}{25}, \frac{7}{25}, \frac{24}{7}$ *7* $\frac{4}{3}, \frac{3}{5}, \frac{4}{5}$ *8 a* $\sin 2p$ *b* $\cos 2n$ *c* $-\cos 2x$

d $\sin 70^\circ$ *e* $\cos 2y$ *f* $\cos 10^\circ$ *g* $\tan 2k$ *h* $\tan 100^\circ$

9 a 0·5 *b* 0·5 *c* 0·5 *d* 0·5 *10a* F *b* T *c* T *d* F *e* F *f* F

11a $\frac{4}{3}$ *b* $\frac{3}{4}$ *c* 3 *d* 2

Page 170 Exercise 9

1 $\{0, 120, 180, 240, 360\}$ *2* $\{30, 90, 150, 270\}$ *3* $\{0, 120, 240, 360\}$

4 $\{30, 150, 270\}$ *5* $\{0, 60, 300, 360\}$ *6* $\{0, 180, 360\}$

7 $\{90\}$ *8* $\{0, 180, 210, 330, 360\}$ *9* $\{60, 300\}$

10 $\{120, 180, 240\}$ *11* $\{60, 180, 300\}$ *12* $\{53{\cdot}1, 120, 240, 306{\cdot}9\}$

13 $\{90, 236{\cdot}4, 270, 303{\cdot}6\}$ *14* $\{48{\cdot}2, 104{\cdot}5, 255{\cdot}5, 311{\cdot}8\}$ *15* $\{30, 150, 228{\cdot}6, 311{\cdot}4\}$

16 $\{233{\cdot}1, 306{\cdot}9\}$ *17* $\{0, \frac{1}{3}\pi, \pi, \frac{5}{3}\pi, 2\pi\}$ *18* $\{\frac{1}{2}\pi, \frac{7}{6}\pi, \frac{3}{2}\pi, \frac{11}{6}\pi\}$

19 $\{\frac{1}{3}\pi, \pi, \frac{5}{3}\pi\}$ *20* $\{\frac{1}{2}\pi, \frac{7}{6}\pi, \frac{11}{6}\pi\}$

Page 171 Exercise 10B

12 $\cos 4\theta = 1 - 8\sin^2\theta + 8\sin^4\theta$

14a $\sin^4\theta = \frac{1}{4} - \frac{1}{2}\cos 2\theta + \frac{1}{4}\cos^2 2\theta$ *b* $\sin^4\theta = \frac{3}{8} - \frac{1}{2}\cos 2\theta + \frac{1}{8}\cos 4\theta$

15 $\sin^2\theta\cos^2\theta = \frac{1}{8} - \frac{1}{8}\cos 4\theta$

Trigonometry — Answers to Revision Exercises

Page 173 Revision Exercise 1A

1 a 0·970 *b* −0·242 *c* 0·364 *d* 0·891 *2* $\frac{5}{6}$ *3* $-\frac{3}{4}$

4 a 360° *b* 90° *c* 1080° *5 a* $-v$ *b* $-v$ *c* v *d* v

6 a n *b* n *c* $-n$ *d* n *7 a* $\{4, 184\}$ *b* $\{116, 244\}$

c $\emptyset$ *d* $\{52{\cdot}9, 232{\cdot}9\}$ *e* $\{44{\cdot}4, 315{\cdot}6\}$ *f* $\{221{\cdot}8, 318{\cdot}2\}$

8

180°	45°	225°	360°	36°	270°
π	$\frac{1}{4}\pi$	$\frac{5}{4}\pi$	2π	$\frac{1}{5}\pi$	$\frac{3}{2}\pi$

9 0·5 cm *10* 0·24:1

11a $\{\frac{3}{4}\pi, \frac{7}{4}\pi\}$ *b* $\{\frac{1}{6}\pi, \frac{5}{6}\pi\}$ *c* $\{\frac{2}{3}\pi, \frac{4}{3}\pi\}$ *12a* 1 *b* $-1/\sqrt{2}$

14a $\frac{33}{65}$ *b* $\frac{16}{65}$ *15* $-\frac{3}{4}, -1$ *16a* $\{60, 109{\cdot}5, 250{\cdot}5, 300\}$ *b* $\{53{\cdot}1, 126{\cdot}9\}$

Page 174 Revision Exercise 1B

1 a $\frac{36}{85}$ *b* $\frac{16}{65}$ *5* $\{150, 330\}$ *8* 1 *15a* $5, -3$ *b* $3, -1\frac{1}{2}$ *c* $2, -1\frac{1}{8}$

Calculus — Answers to Chapter 1

Page 182 Exercise 1

1 a 4·5 m/s *b* 4·1 m/s *c* $(4+h)$ m/s, 4 m/s

2 a 8·5 m/s *b* 8·1 m/s *c* $(8+h)$ m/s, 8 m/s

3 a 4·4 m/s *b* $(4+2h)$ m/s, 4 m/s *c* 12 m/s

4 a 40 m/s *b* 80 m/s

Page 185 Exercise 2

1 2 *2* 2 *3* 2 *4* 6 *5* 16

6 15 *7* 8 *8* 4 *9* $10\,cm^2/cm$ *10* $2\pi\,cm/cm$

Page 188 Exercise 4

1 $6x^5$ *2* $8x^7$ *3* $12x^2$ *4* x *5* x^3

6 $-10x^4$ *7* $3ax^2$ *8* 5 *9* 0 *10* 0

11 $2x+2$ *12* $3x^2-14x$ *13* $16x^3-2x$ *14* x^2+x

15 $-20x^4+18x^8$ *16* $2ax+b$ *17* $2x+2$ *18* $2x-4$

19 $18x+24$ *20* $4x+5$ *21* $50x$ *22* $-5+2x$

23 $4x^3+8x$ *24* $6x^5-6x^2$ *25* $8x^7+24x^3$ *26* $3x^2-6x$

27 $6x^2+10x-18$ *28* $3x^2+24x+48$ *29* $1, 0, -1, 21$

30a $f': x \to 6x^5-12x^2$ ***b*** $-18, 144$ ***31a*** 2 ***b*** 3 ***c*** $x>2$

32a $-3, 2$ ***b*** $-2, 1$ ***c*** $-3<x<2$

33a $f': x \to 1$ ***b*** $f': x \to 2x$ ***c*** $f': x \to 3x^2$ ***d*** $f': x \to 4x^3$

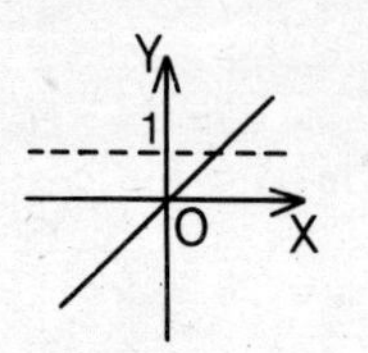

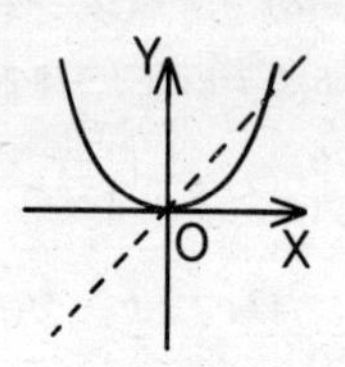

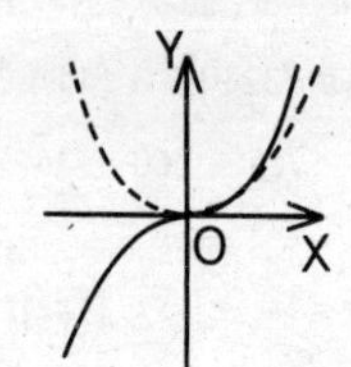

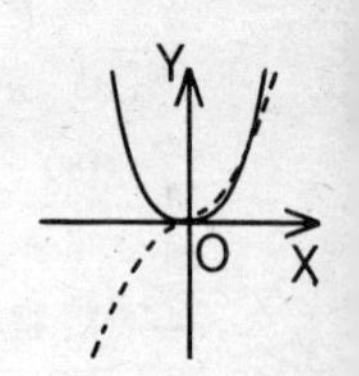

34a 1 cm/s ***b*** 2·5 s ***35a*** 15 m/s ***b*** 3 s

Page 190 Exercise 5

1 $\frac{3}{2}x^{1/2}$ *2* $\frac{4}{3}x^{1/3}$ *3* $\frac{5}{2}x^{3/2}$ *4* $\frac{1}{2x^{1/2}}$ *5* $\frac{1}{3x^{2/3}}$

6 $-\frac{1}{x^2}$ 7 $-\frac{2}{x^3}$ *8* $-\frac{6}{x^7}$ *9* $-\frac{6}{x^4}$ *10* $-\frac{2}{x^5}$

11 $\frac{1}{2x^{1/2}}$ *12* $\frac{1}{3x^{2/3}}$ *13* $\frac{2}{3x^{1/3}}$ *14* $-\frac{2}{x^3}$ *15* $-\frac{4}{x^5}$

16 $-\frac{1}{x^2}$ *17* $-\frac{2}{x^2}$ *18* $-\frac{1}{2x^{3/2}}$ *19* $-\frac{1}{3x^{4/3}}$ *20* $-\frac{6}{x^4}$

21 $-\frac{1}{4x^{3/2}}$ 22 $-\frac{4}{3x^3}$ 23 $-\frac{4}{x^4}$ *24* $-\frac{4}{5x^5}$ 25 $-\frac{1}{4x^{3/2}}$

26 $1-\frac{1}{x^2}$ *27* $\frac{1}{2x^{1/2}}-\frac{1}{2x^{3/2}}$ *28* $2+\frac{2}{x^2}$

29 $4x+\frac{1}{2x^3}$ *30* $2x+\frac{2}{x^3}$ *31* $\frac{1}{5}-\frac{5}{x^2}$

32 $8x^3-\frac{3}{2x^4}$ *33* $\frac{6}{x^{1/4}}+\frac{4}{x^{5/3}}$ *34* $2x+\frac{5}{2}x^{3/2}$

35 $\frac{1}{2x^{1/2}}-1$ *36* $2x-\frac{2}{x^3}$ *37* $4x^3-\frac{4}{x^5}$

38 $50x-\frac{50}{x^3}$ *39* $1-\frac{1}{x^2}$ *40* $2x+\frac{2}{x^3}$

41 $\frac{4}{x^2}$ *42* $1+\frac{1}{x^2}$ *43* $\frac{3}{x^2}-\frac{12}{x^4}$

44 $\frac{1}{2x^{1/2}}+\frac{1}{2x^{3/2}}$ *45* $-\frac{4}{x^3}-\frac{4}{x^2}$ *46* $\frac{1}{3x^{1/3}}-\frac{1}{6x^{4/3}}$

47 $-\frac{1}{2x^2}+\frac{1}{2}$ *48* $-\frac{2}{x^2}+1$ *49* $\frac{1}{2x^{1/2}}+\frac{2}{x^{3/2}}$

50a 0 *b* 6 *c* 12 *d* 2 *51a* -2 *b* 2 *c* $-\frac{1}{4}$ *d* 16

52 $2x+2+\frac{2}{x^3}$ *53* $24x^5+8x-\frac{2}{x^3}$ *54* $\frac{1}{16}$

Page 194 Exercise 6

1 6; $y=6x-9$ *2* 5; $y=5x$ *3* $\frac{1}{4}$; $4y=x+4$

4 -1; $y=-x+2$ *5* $y=12x+16$ *6* $y=-x+1$

7 $y=1$ *8* $y=15x-21$ *9* $y=-8x-6$ *10* $y=7x-15$

11 $y=x+25$ *12* $32y=x+48$ *13* $y=-8x+12$ *14* $y=-3x+4$

15 $y=x-1$ *16* $y=4x-2$, $y=-4x-2$; $(0, -2)$

17 $y=4x+1$ *18* Gradient 3. $(\frac{2}{3}, 0)$, $(0, -2)$; $(-\frac{2}{3}, 0)$, $(0, 2)$

19 $y=-2x+1$, $y=4x-2$; $(\frac{1}{2}, 0)$ *20* $3\sqrt{17}$

21 $(4, 38)$ *22* $(-1, -4)$, $(3, 0)$; $y=9x+5$, $y=9x-27$ *23* $(4, 4)$

24 $y=3x-3$ *25* $-\frac{1}{x^2}$ always negative; $y=-4x+4$, $y=-4x-4$

26 $3(x-2)^2$ never negative; $(2, 9)$

27 $y=2x+8$ *28* $y+2=0$ *29* $(1, 0)$ *30* $(2, 6)$

31 $2y=23x-28$ *32* $a=-2$, $b=4$ *33* $a=6$, $b=-8$

Page 197 Exercise 7

1 Inc. $x > 0$; dec. $x < 0$ *2* Inc. $x > 1$; dec. $x < 1$

3 Inc. $x > 4$; dec. $x < 4$ *4* Inc. $x < \frac{1}{2}$; dec. $x > \frac{1}{2}$

5 Inc. $x > -3$; dec. $x < -3$ *6* Inc. $x > 1$; dec. $x < 1$

7 Inc. $x < 0$ or $x > 0$ *8* Inc. $x < 0$ or $x > 4$; dec. $0 < x < 4$

9 Inc. $-1 < x < 1$; dec. $x < -1$ or $x > 1$

10 Inc. $x < 1$ or $x > 2$; dec. $1 < x < 2$

11 Inc. $x < -1$ or $x > 3$; dec. $-1 < x < 3$

12 Inc. $x < \frac{2}{3}$ or $x > 2$; dec. $\frac{2}{3} < x < 2$

13 Inc. $-1 < x < \frac{1}{3}$; dec. $x < -1$ or $x > \frac{1}{3}$

14 $3(x-1)^2$ is never negative; 1

15 $-\dfrac{c^2}{x^2}$ is always negative; $y = -x+2c$

Page 200 Exercise 8

1 0, min *2* -1, min *3* 4, max *4* 4, max

5 0, min *6* 0, point of inflexion (0, 0) *7* 2, max; -2, min

8 5, max; 4, min *9* $\frac{32}{27}$, max; 0, min *10* -2, max; 2, min

11 $-20\frac{1}{4}$, min; 0, max; $-20\frac{1}{4}$, min *12* 0, point of inflexion (0, 0); 27, max

13 19, max; -13, min *14* $-\frac{1}{2}$, min; 0, max; $-\frac{1}{2}$, min *15* $11\frac{3}{4}$, max; 5, min

Page 202 Exercise 9

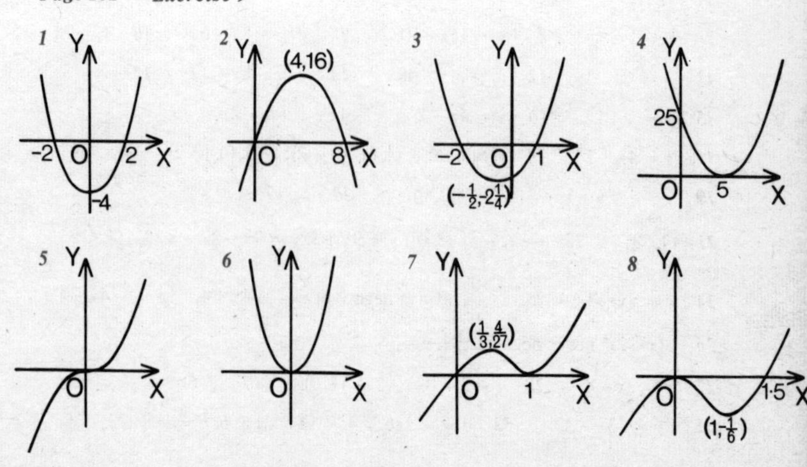

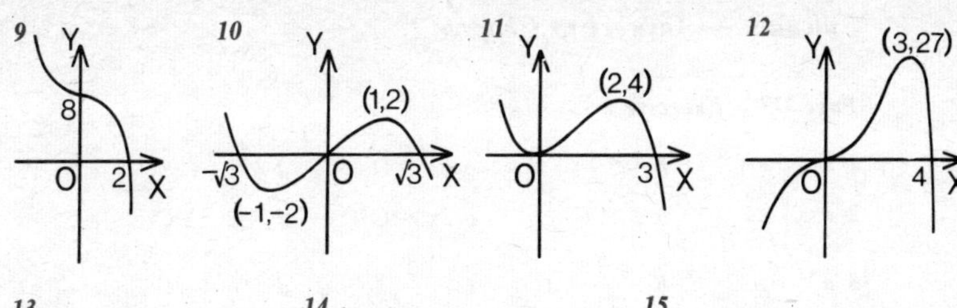

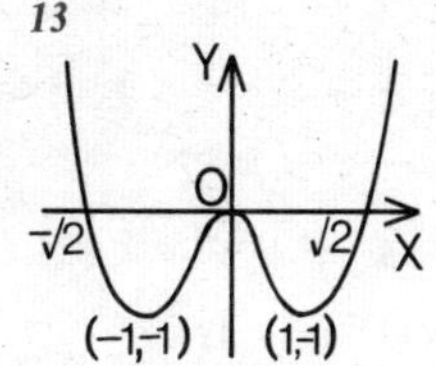

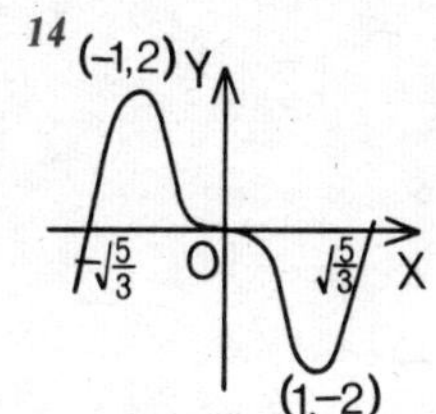

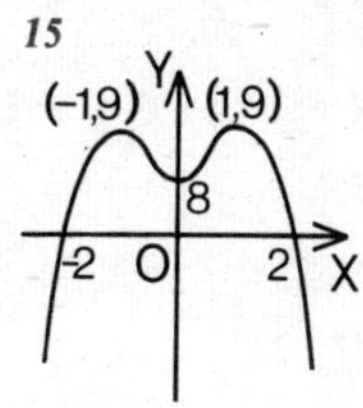

Page 203 Exercise 10

1 $0 \leqslant f(x) \leqslant 16$ *2* $-9 \leqslant f(x) \leqslant 27$ *3* $-3 \leqslant f(x) \leqslant 1$ *4* $0 \leqslant f(x) \leqslant 9$

5 $-54 \leqslant f(x) \leqslant 54$ *6* $-12 \leqslant f(x) \leqslant 4$ *7* $-27 \leqslant f(x) \leqslant 0$ *8* $0 \leqslant f(x) \leqslant \frac{7}{16}$

Page 205 Exercise 11

1 Maximum area = 45 000 m^2, when breadth = 150 m and length = 300 m

2 18 km *3* $x+y = 40$; $P = x(40-x)$; 400 *4* $x+y = 28$; $P = x(28-x)$; 196

5 a $x+y = 50$ *b* $A = xy = x(50-x)$; $x = y = 25$ *6* 28 cm^2

7 5 *8 b* 2, 128 cm^3 *9 b* 10 cm, 20 cm *10* 1 cm by 3 cm by 6 cm

11 $A = x(8-2x)$; (2, 4) *12b* 16 *13a* $x^2h = 32$; $A = x^2+4xh$

b $x = 4$, $h = 2$ *14* gradient = $(t^2+4)/t$; 2 *16* 3 units

Page 209 Exercise 12B

1 a $s = 16$ and 21 *b* $v = 12-3t^2$, $a = -6t$ *c* $t = \pm 2$ *d* $v = 12$

2 a $s = -4$ and 4 *b* $v = 6t^2-6$, $a = 12t$ *c* $t = \pm 1$ *d* $v = -6$

3 $a = 12$ and -12 *4* $v = 150-10t$ *5* 25 m^3/m *6* 31·4 mm^2/mm

7 a $\frac{d\theta}{dt} = 15+14t-t^2$; 48 radians per second *b* 15 s

8 a $F = \frac{k}{x^2}$ *b* $\frac{dF}{dx} = -\frac{2k}{x^3}$ *9* $p = \frac{k}{v}$ *10* 8·54

Calculus — Answers to Chapter 2

Page 217 Exercise 1

1 *a* $\frac{1}{2}x^2+C$ *b* $\frac{1}{3}x^3+C$ *c* $\frac{1}{4}x^4+C$ *d* $\frac{1}{6}x^6+C$ *e* $\frac{1}{9}x^9+C$

2 *a* x^2+C *b* x^3+C *c* $2x^4+C$ *d* $-2x^3+C$ *e* $5x+C$

3 *a* $-\frac{1}{x}+C$ *b* $-\frac{1}{2x^2}+C$ *c* $-\frac{1}{3x^3}+C$ *d* $-\frac{1}{4x^4}+C$ *e* $\frac{2}{x^3}+C$

4 *a* $\frac{2}{3}x^{3/2}+C$ *b* $\frac{2}{5}x^{5/2}+C$ *c* $2x^{1/2}+C$ *d* $\frac{3}{5}x^{5/3}+C$ *e* $4x^{1/2}+C$

5 $\frac{1}{2}x^2+C$ **6** x^2-3x+C **7** $x-\frac{1}{2}x^2+C$

8 $\frac{1}{3}x^3-x+C$ **9** $4x-x^4+C$ **10** x^3+2x^2+5x+C

11 $2x^3-x+C$ **12** $2x^5+x^3+C$ **13** $\frac{1}{3}x^3-4x+C$

14 $\frac{1}{3}x^3-3x^2+9x+C$ **15** $x-3x^2+3x^3+C$ **16** $\frac{1}{4}x^4-\frac{1}{3}x^3-x^2+C$

17 *a* $F(x)=x^2-6$ *b* $F(x)=x-x^2+10$ *c* $F(x)=2x^3$

d $F(x)=\frac{1}{2}x^2+\frac{2}{x}+6$ *e* $F(x)=x-2x^{1/2}+1$ *f* $F(x)=x^3-9x+20$

18 $2x^3-2x+C$ **19** $2x+\frac{1}{2}x^2+C$ **20** $\frac{1}{3}x^3-2x-\frac{1}{x}+C$

21 $\frac{1}{2}x^2-\frac{2}{3}x^3+\frac{1}{4}x^4+C$ **22** $\frac{2}{3}x^{3/2}-4x^{1/2}+C$ **23** $2x^{5/2}+2x^{3/2}+C$

24 $\frac{1}{3}x^3+x+C$ **25** $\frac{1}{3}x^3-\frac{1}{x}+C$ **26** $\frac{1}{3}x^3+2x-\frac{1}{x}+C$

27 $\frac{2}{5}x^{5/2}+4x^{1/2}+C$ **28** $\frac{2}{3}x^{3/2}+\frac{1}{2}x^2+C$ **29** $2x^{1/2}+2x+\frac{2}{3}x^{3/2}+C$

30 $\frac{4}{3}x^3+\frac{4}{x}-\frac{1}{5x^5}+C$ **31** $2x^{5/2}-2x^{3/2}+C$

32 $\frac{2}{5}x^{5/2}+2x^{3/2}+6x^{1/2}-\frac{2}{x^{1/2}}+C$

Page 218 Exercise 2

1 $y=2x+C$ **2** *a* $y=2x^2+1$ *b* $y=x^2-x+6$ *c* $y=x^3-5x^2+6$

d $y=2x^3-3x^2+3x$ **3** $y=3x^2-x^3+6$ **4** $y=x+\frac{4}{x}+1$

5 $s=5t-t^2$ **6** *a* $v=-10t+15$ *b* $s=-5t^2+15t$

7 $v=\frac{1}{3}t^3+t^2$; 18 m/s **8** 15·75 m **9** $M=\frac{1}{2}wlx-\frac{1}{2}wx^2$; $\frac{1}{8}wl^2$

10 Equation of curve is $y=x^3-4x^2+5x-2$ **11** $y=x-x^2+6$ **12** $\frac{1}{3}kl^3$

Page 222 *Exercise 3*

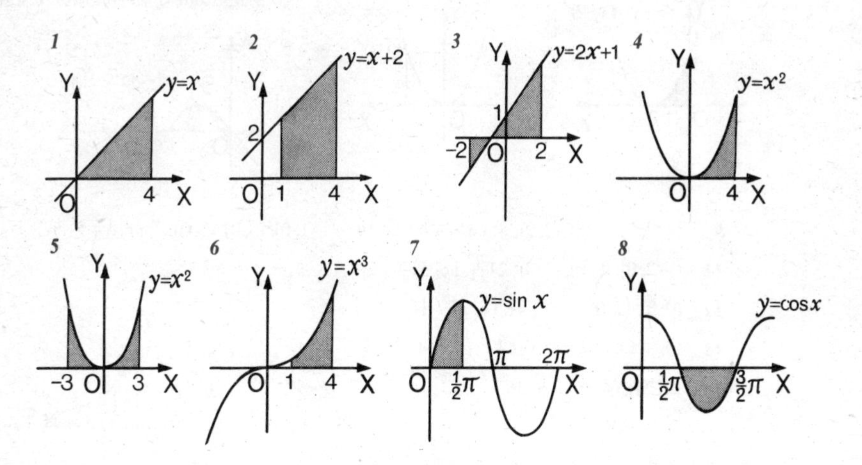

9 (i) $\int_0^5 (5-x)dx$ (ii) $\int_{-2}^2 (4-x^2)dx$ (iii) $\int_2^8 \frac{dx}{x}$ (iv) $\int_0^3 \sqrt{(9-x^2)}dx$

10 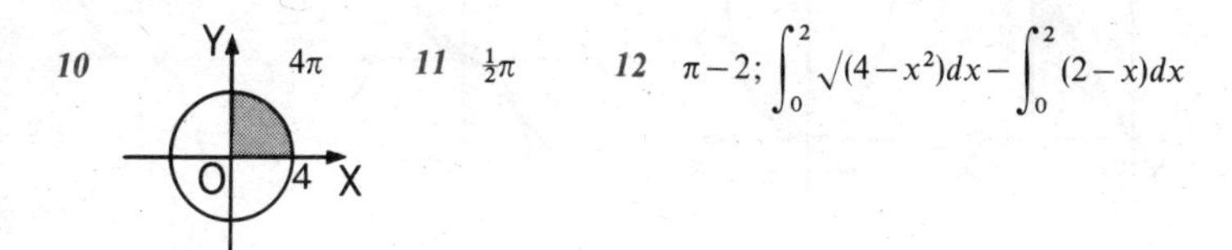 4π **11** $\frac{1}{2}\pi$ **12** $\pi-2$; $\int_0^2 \sqrt{(4-x^2)}dx - \int_0^2 (2-x)dx$

Page 225 *Exercise 4*

1	2	**2**	14	**3**	$5\frac{1}{3}$	**4**	$\frac{2}{3}$	**5**	$3\frac{3}{4}$
6	4	**7**	20	**8**	18	**9**	18	**10**	2
11	$2\frac{2}{3}$	**12**	21	**13**	-3	**14**	8	**15**	$-3\frac{1}{3}$
16	13	**17**	$\frac{1}{4}a^4-a^2$	**18**	$\frac{1}{2}p^2+\frac{4}{3}p^{3/2}+p$	**19**	$1\frac{3}{4}$	**20**	0

21 $-\frac{1}{6}$ **22a** 9 ***b*** 0 or $1\frac{1}{2}$ **23** 128

Page 228 *Exercise 5*

1 (i) 12 (ii) 20 (iii) 16 (iv) 8 **2** (i) 8 (ii) 9 (iii) 26 (iv) 4
(v) 18 (vi) $12\frac{4}{5}$ (vii) $1\frac{1}{3}$ (viii) 36

3 $1\frac{1}{3}$ **4** $4\frac{1}{2}$ **5** $6\frac{3}{4}$

6 $2\frac{2}{3}$

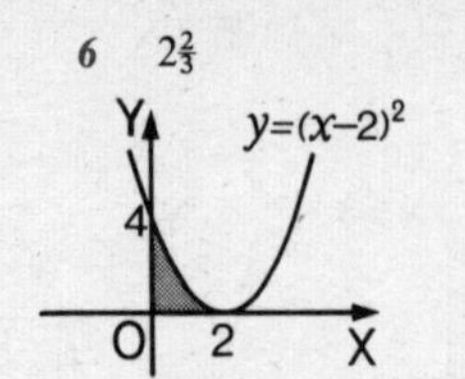

7 9 and 18

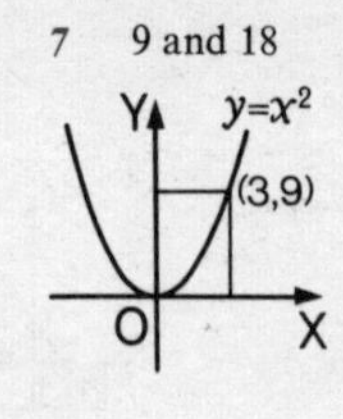

8 0: the two areas are equal in magnitude, but opposite in sign

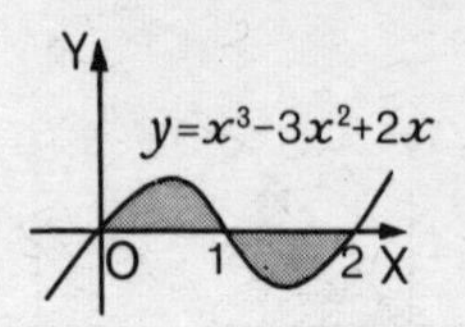

9 $(-2, 0), (0, 0), (2, 0)$; 8, i.e. $4+4$

10 $(-1, 0), (0, 0), (1, 0)$; $\frac{1}{2}$, i.e. $\frac{1}{4}+\frac{1}{4}$

11 $(-2, 0), (0, 0), (3, 0)$; $21\frac{1}{12}$, i.e. $5\frac{1}{3}+15\frac{3}{4}$

12 $(0, 3), (1, 0), (3, 0)$; $2\frac{2}{3}$, i.e. $1\frac{1}{3}+1\frac{1}{3}$

13 $(0, -5), (1, 0), (5, 0)$; 13

14 $9\frac{1}{3}:26\frac{2}{3} = 7:20$

15a $\frac{3}{4}$

b $1\frac{3}{5}$

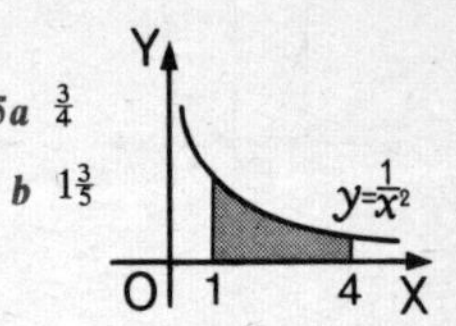

16a $(2, 20), (4, 16)$ *b* 36

Page 230 Exercise 6

1 25 *2* $\frac{1}{6}$ *3* $\frac{1}{2}$ *4* $\frac{1}{12}$

5 $\frac{1}{3}$ *6* $1\frac{1}{3}$ *7* $10\frac{2}{3}$ *8* $21\frac{1}{3}$

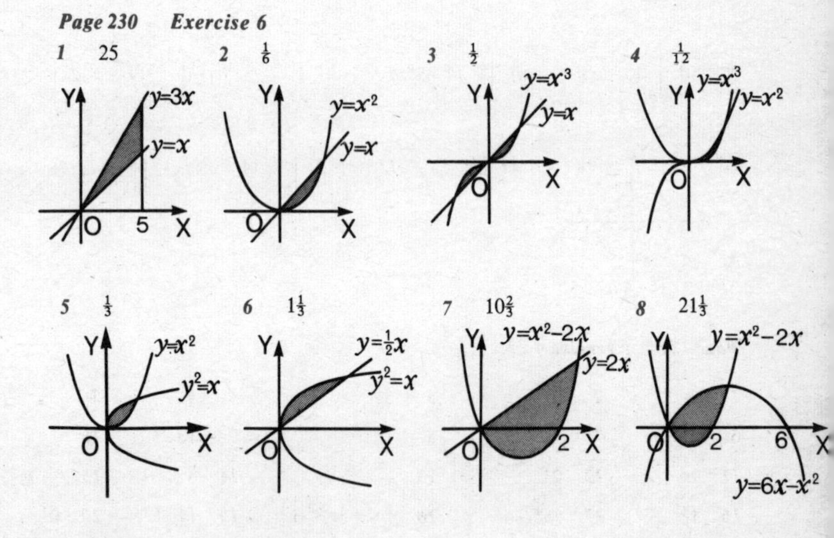

9 $21\frac{1}{3}$ *10* $4\frac{1}{2}$ *11* $1\frac{1}{3}$ *12* $19\frac{1}{5}$

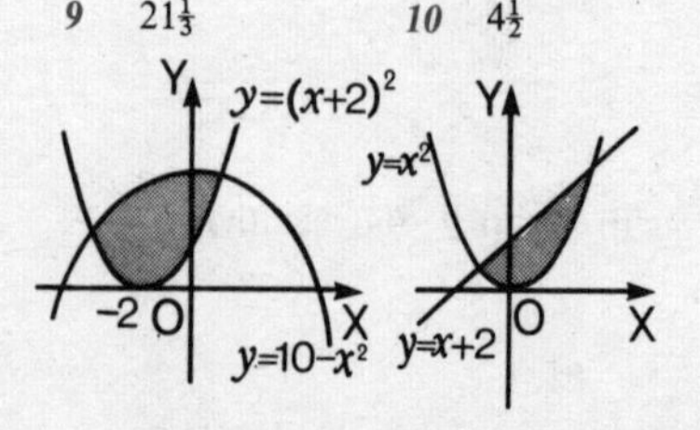

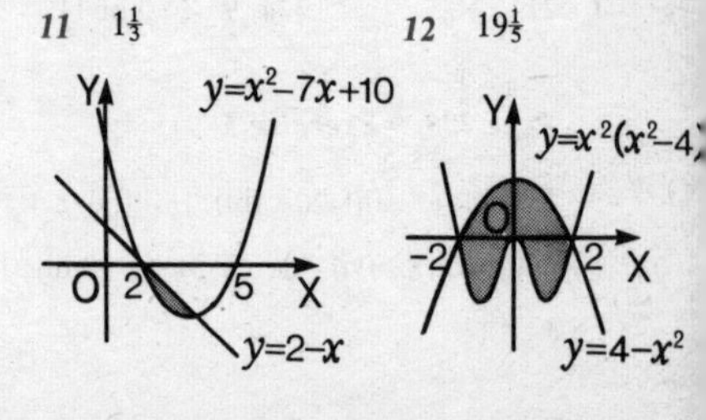

13a 36 ***b*** (0, 0), (3, 9) ***14a*** $y = -2x+3$ ***b*** 1 ***15a*** $(-\frac{1}{2}, 0)$ ***b*** $1\frac{1}{8}+1\frac{1}{8} = 2\frac{1}{4}$

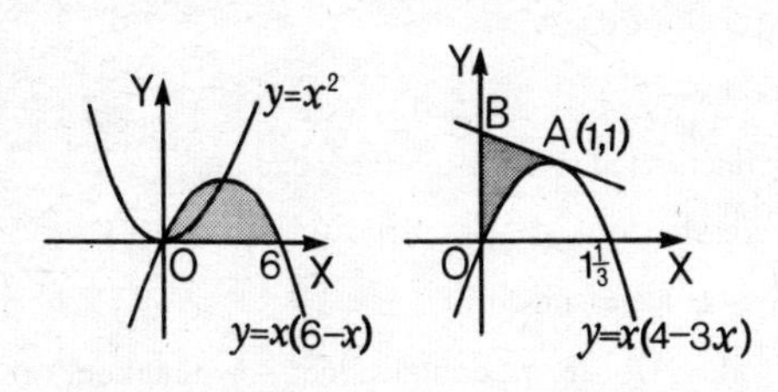

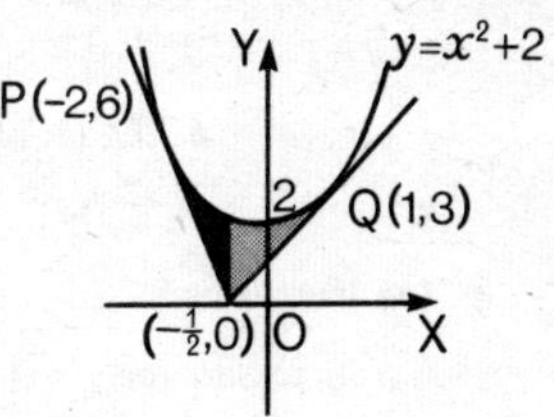

Page 233 ***Exercise 7***

1 (i) 8π (ii) 18π (iii) $\frac{26}{3}\pi$ (iv) 625π (v) $\frac{31}{5}\pi$ (vi) $\frac{8}{3}\pi$ (vii) 18π (viii) 64π

2 12π ***3*** 2π ***4*** 56π ***5*** $\frac{2}{3}\pi$ ***6*** $\frac{19}{3}\pi$

7 $\frac{28}{15}\pi$ ***8*** $\frac{1}{30}\pi$ ***9*** $\frac{16}{15}\pi$ ***12*** 12π ***13*** $179\,\text{cm}^3$

Page 235 ***Exercise 8***

1 8π ***2*** $\frac{1}{5}\pi$ ***3*** $\frac{1}{6}\pi$ ***4*** $\frac{96}{5}\pi$ ***5*** $\frac{3}{2}\pi$

6 32π ***7*** $\frac{7}{48}\pi$ ***8*** 9π ***9*** $\frac{56}{15}\pi$ ***10*** $\frac{1}{12}\pi$

Page 236 ***Exercise 9B***

1 $\frac{2}{15}\pi$ ***2*** $\frac{8}{21}\pi$ ***3*** $\frac{8}{15}\pi$ ***4*** $\frac{3}{10}\pi$ ***5*** $\frac{8}{3}\pi$

6 $\frac{8}{45}\pi$ ***7*** $\frac{4}{3}\pi$ ***8*** $\frac{48}{5}\pi$ ***9*** $\frac{640}{3}\pi$ ***10*** 81π

Calculus — Answers to Revision Exercises

Page 239 ***Revision Exercise 1***

1 12 m/s ***2 a*** $f'(x) = 5$ ***b*** $f'(x) = 6x$ ***c*** $f'(x) = 2x+2$

d $f'(x) = -\dfrac{2}{x^2}$ ***e*** $f'(x) = -\dfrac{2}{3x^3}$ ***f*** $f'(x) = 4-\dfrac{1}{x^2}$

3 a $12x^5+12x^2$ ***b*** $2x^4+7x$ ***c*** $2-\dfrac{1}{2t^2}$ ***d*** $-1+\dfrac{4}{u^3}$ ***e*** $-\dfrac{4}{r^3}-\dfrac{9}{r^4}$

f $1+\dfrac{1}{3x^{4/3}}$ ***g*** $\dfrac{1}{2x^{1/2}}+\dfrac{2}{x^{3/2}}$ ***h*** $2p+1-\dfrac{2}{5p^2}$ ***i*** $2x+\frac{3}{2}x^{1/2}$

4 a $6x+4$ ***b*** $4x^3+4x$ ***c*** $-3+6x-3x^2$ ***d*** $-\dfrac{1}{u^2}$ ***e*** $\frac{1}{2}-\dfrac{1}{2r^2}$

f $4-\dfrac{1}{x^2}$ ***g*** $\dfrac{1}{t^2}+\dfrac{4}{t^3}$ ***h*** $\frac{5}{3}x^{2/3}-\dfrac{2}{x^{1/3}}-\dfrac{4}{3x^{4/3}}$ ***i*** $3x^2+1-\dfrac{1}{2x^{3/2}}$

5 8, −8, 12; −1, −2 *6 a* 2 *b* −14 *c* −6.

$y = 2x-2$, $y = -14x-2$, $y = -6x+2$

7 a 6 *b* 12. (4, 17), $y = 6x-7$

8 $y = -2x+2$, $y = 2x-2$; intersect at (1, 0), i.e. on the x-axis

10a increasing for $x > 2\frac{1}{2}$, decreasing for $x < 2\frac{1}{2}$

b increasing for $x < 0$ or $x > 2$, decreasing for $0 < x < 2$

11 0 *12a* $4\frac{1}{12}$, maximum value *b* 26, maximum value; −6, minimum value

c 4, maximum value; 0, minimum value *d* 3, minimum value

13a (3, −1), minimum turning point *b* (0, 9), point of inflexion

c (−1, 0) and (1, 0), minimum turning points; (0, 1), maximum turning point

d (0, 0), point of inflexion; (6,432), maximum turning point

e (1, 0), point of inflexion *f* (1, −1), minimum turning point

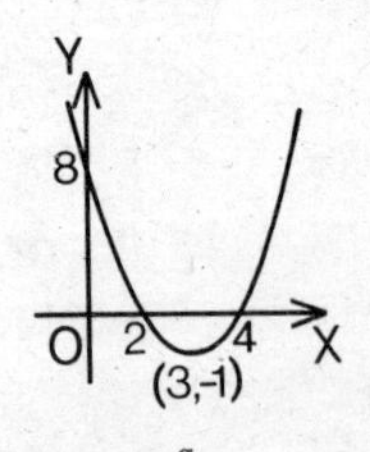

a

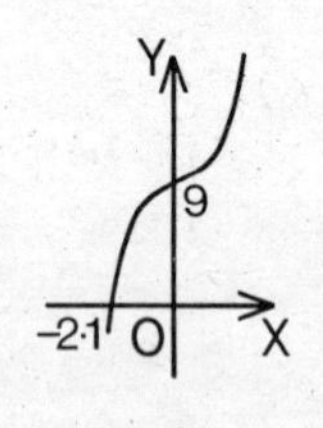

b

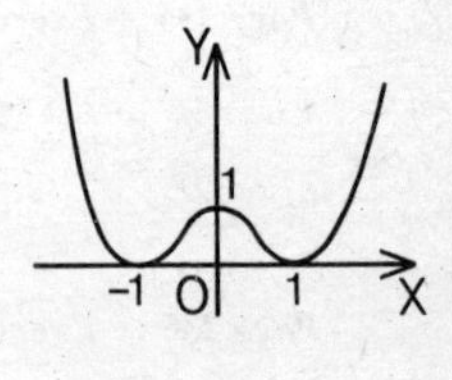

c

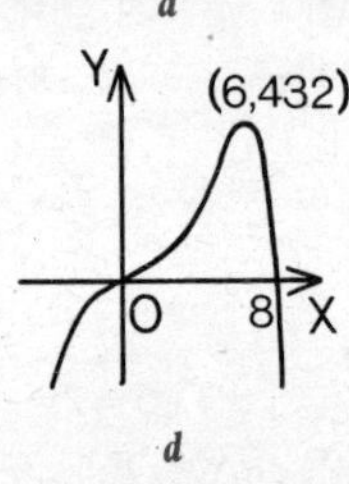

d

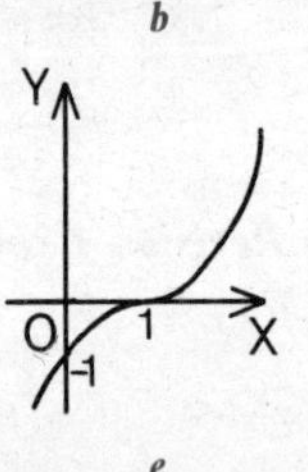

e

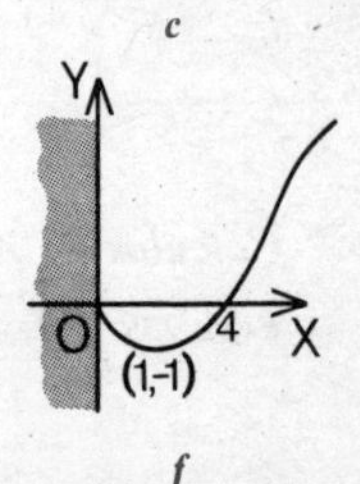

f

14 18 cm^2 *15* Length = breadth = 4 cm; perimeter = 16 cm

16a 0, 20, 0 *b* 10 m/s, 0 m/s, −10 m/s *c* The ball is above its starting point for 4 seconds, reaching its greatest height of 20 m after 2 s; the velocity is zero at the maximum height, negative as the ball descends.

17a 4 m/s *b* 7 m/s *18* 1600π mm^3/mm

Page 242 Revision Exercise 2

1 a $\frac{1}{3}x^3+\frac{1}{2}x^2+x+C$ *b* $\frac{1}{6}x^6-\frac{1}{4}x^4+7x+C$ *c* $3x-x^2-2x^3+C$

d x^3-4x^2+5x+C *e* $3x^3+12x^2+16x+C$ *f* $\frac{1}{4}x^4-2x^2+C$

2 ***a*** $\frac{2}{3}x^{3/2}-2x^{1/2}+C$ ***b*** $\frac{1}{2}x^2-x+C$ ***c*** $\frac{1}{2}x^2+\frac{4}{3}x^{3/2}+x+C$

d $-\frac{1}{x}+\frac{1}{2x^2}+C$ ***e*** $x+\frac{3}{x}+C$ ***f*** $2x^{1/2}-\frac{2}{x^{1/2}}+C$

3 ***a*** $f(x)=4x^2-3x+3$ ***b*** $f(x)=\frac{1}{3}x^3+\frac{1}{x}-1$

4 ***a*** $x^3-\frac{1}{x^2}+C$ ***b*** $2x^{3/2}+\frac{1}{2}x^{2/3}+C$ ***c*** $-\frac{1}{4x}-\frac{2}{x^2}+C$

d $3x^3-\frac{6}{x}-\frac{1}{5x^5}+C$ ***e*** $2x^2-\frac{8}{3}x^{3/2}+x+C$ ***f*** $\frac{6}{5}x^{5/2}+\frac{8}{3}x^{3/2}-10x^{1/2}+C$

5 $y=2x^2+\frac{4}{x}+1$ **6** ***a*** 4 ***b*** 8 ***c*** 3 ***d*** $3\frac{3}{4}$ ***e*** $2\frac{2}{3}$ ***f*** $14\frac{2}{3}$

7 ***a*** -3 or 2 ***b*** $-\frac{1}{3}$ ***c*** 4

8 ***a*** 15 ***b*** 144 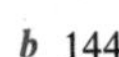 ***c*** 8 

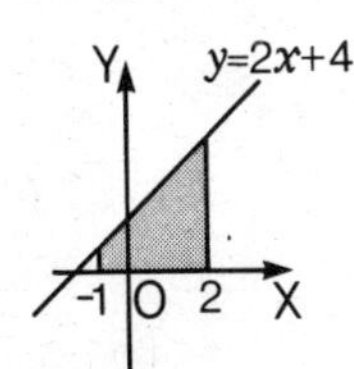

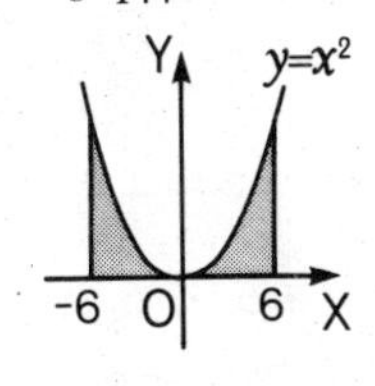

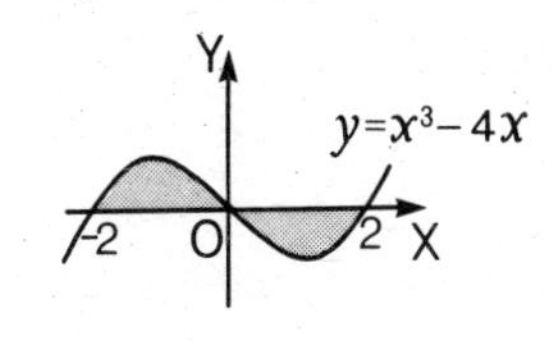

9 ***a*** $4\frac{1}{2}$ ***b*** 36 ***c*** $1\frac{23}{48}$ **10*a*** $21\frac{1}{3}$ ***b*** 1:7

11*a*

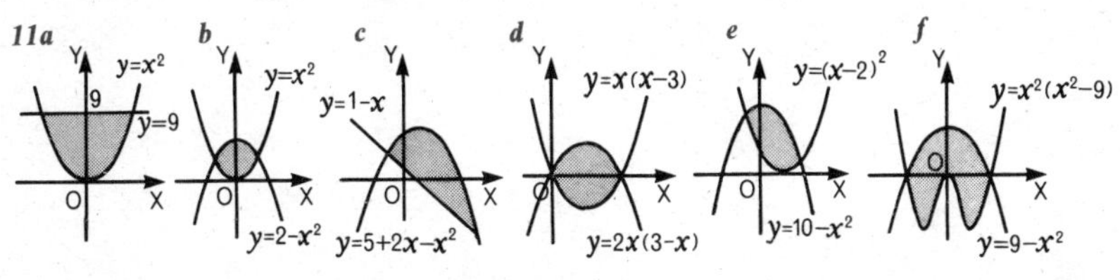

a $(-3,9),(3,9)$; 36 ***b*** $(-1,1),(1,1)$; $2\frac{2}{3}$ ***c*** $(-1,2),(4,-3)$; $20\frac{5}{6}$

d $(0,0),(3,0)$; $13\frac{1}{2}$ ***e*** $(-1,9),(3,1)$; $21\frac{1}{3}$ ***f*** $(-3,0),(3,0)$; $100\frac{4}{5}$

12 $\frac{1}{3}+\frac{1}{3}=\frac{2}{3}$ **13*a*** $\frac{1}{5}\pi$ ***b*** 18π ***c*** $37\frac{11}{15}\pi$

14*a* 18π ***b*** 8π **15*a*** 12π ***b*** 8π

16*a* $\frac{1}{6}\pi$ ***b*** $\frac{3}{10}\pi$ **17*a*** $\frac{62}{3}\pi$ ***b*** $4\pi+5\frac{1}{3}\pi=\frac{28}{3}\pi$